D1423064

Jean-Denis Godet
Plants and Flowers of Great Britain and Europe

God said, "Let th'Earth
Put forth the grass, herb yielding seed,
And fruit-tree yielding fruit after her kind,
Whose seed is in herself upon the Earth!"
He scarce had said when the bare Earth, till then
Desert and bare, unsightly, unadorned,
Brought forth the tender grass, whose verdure clad
Her universal face with pleasant green;
Then herbs of every leaf, that sudden flowered,
Opening their various colours, and made gay
Her bosom, smelling sweet; and, these scarce blown,
Forth flourished thick the clust'ring vine, forth crept
The smelling gourd, up stood the corny reed
Embattled in her field: add the humble shrub,
And bush with frizzled hair implicit: last
Rose, as in a dance, the stately trees, and spread
Their branches, hung with copious fruit, or gemmed
Their blossoms: with high woods the hills were crowned
With tufts the valleys and each fountain-side,
With borders long the rivers; that Earth now
Seemed like to Heaven, a seat where gods might dwell.

John Milton: Paradise Lost

Plants and Flowers of Great Britain and Europe

1st edition ISBN: 3-905039-04-1

ISBN: 3-576-80003-4

Translation: UPS Translations, London

Ektachrome development:	Colorlabor Zumstein, Bern
Photolithography:	Schwitter AG, Basel
Text layout:	Gerard Lennox, Cheltenham
Lighting:	Bund Lighting Service, Berne
Printing and binding:	Brepols Fabrieken N.V., Turnhout

Printed in Belgium

MOSAIK

For Flavia and Pascal

Contents

Foreword

In my first three books I tried to present the woods as they appear in winter, spring and summer in a way that made it easy for amateurs to identify the plants.

Still, trees are not the only plants that help to shape our environment; there are also many species of herbs and shrubs. Their variety of form and their beauty are often overwhelming, but people sometimes find it too difficult to use scientifically defined characteristics to identify a species correctly. Once again this fact has prompted me to demonstrate the most important features of the plants, using as many colour photographs as possible.

The most obvious feature of most herbs and shrubs is the colour of the flower. This leads to five groups of plants with white, yellow, red, blue or green flowers. The ferns, as flowerless plants, come first. There are also borderline cases, in which unambiguous classification by a single colour is impossible without more information. In doubtful cases the species concerned are arranged by their various shades of colour.

Within the groups the flowers can be classified by their shape and their leaves. In arranging the plants in various tables care has been taken as far as possible to bring together those plants which are similar in the shape of flowers and foliage. The leaves in various examples of a given species or plant can be described, even if only two specimens are shown.

Following a precise description of foliage, inflorescence and flowers, there is as full a description as possible of the typical habitat of each individual species.

I should like to give sincere thanks to colleagues at the botanic gardens at Basel, Bern, Geneva, Lausanne, Neuenburg, St Gallen and Zurich for much valuable advice, prompting and support in the last three years. I should like to express particular thanks to Dr Klaus Ammann of the Geobotanical Institute of Bern University and to Robert Goldi of Saas.

May this book be a trusty companion and help to preserve the beauty and variety of plant forms for us and for our children in the future.

Hinterkappelen, March 1991 — Jean-Denis Godet

Part 1: Introduction

1.1. Preface:

The characteristic features of a herbaceous plant or a shrub appear in the way its stem grows, in the shape, structure, and arrangement of its leaves, in the structure and colour of its flowers, and in the form of its fruit.
All these can vary in appearance because of the environmental conditions of specific habitats. For example, the height of a plant grows less with increasing height above sea level. Flowers are also deeper in colour at greater heights.
A plant's flowering time also depends on seasonal weather conditions and therefore on where it grows. Because of this, plants with a wide distribution show a wide range in the time of flowering.
To enable these plants to be identified correctly both the flowers and the foliage are described. Their precise description is meant to ensure that no mistakes are made during fieldwork. This book deals only with those features which can be seen with the naked eye or a good magnifying glass.

1.2. The Growth Axis:

The shoot consists of the growth axis (i.e. the stem) – a cylindrical rod-shaped structure specific to each species of plant; the leaves and the lateral limbs around the growth axis – as a rule their growth is limited. Apart from leaves, flowers, and fruit, the way the plant transmits nutrients and stores them is also important.

1.3. The Leaves:

The leaves are generally green. They are important because the plant breathes through them and photosynthesis takes place in them. They consist of the **leaf proper** (generally thin and spread flat), often with a **stalk**, and the **leaf base**. This can serve as a sheath for the leaf and carry other leaves. In most leaves it has no particular shape and and merges gradually into the leaf stem. In some botanical textbooks the leaf proper and the leaf stem are referred to together as the upper leaf and the leaf base as the lower leaf. In leaves without a stalk the leaf base grows directly from the growth axis.

1 Diagram
2 Arum maculátum, Cuckoo-pint
3 Sedum album, White stonecrop
4 Cirsium oleráceum, Cabbage thistle
5 Lonicera caprifólium, Perfoliate honeysuckle
6 Symphytum officinále, Comfrey

1.3.1. Four kinds of leaf are found on the main stem:

1. Cotyledons	The original shoot of embryo plants carries at the bottom 1, 2, or several simply shaped leaves, generally short-lived.
2. Lower leaves	These are simply formed leaves which lie beneath the foliage leaves and are often like scales.
3. Foliage	These form the bulk of the leaf cover. Very often the first leaves to emerge (early or primary leaves) differ in shape from those appearing later.
4. Bracts	These are situated higher up than the foliage leaves and are usually on the stems of flowers or flower heads. They can also be coloured

1.3.2. Arrangement of foliage leaves on the growth axis:

Leaves grow from the stem in a number of ways. The most important instances are as follows:

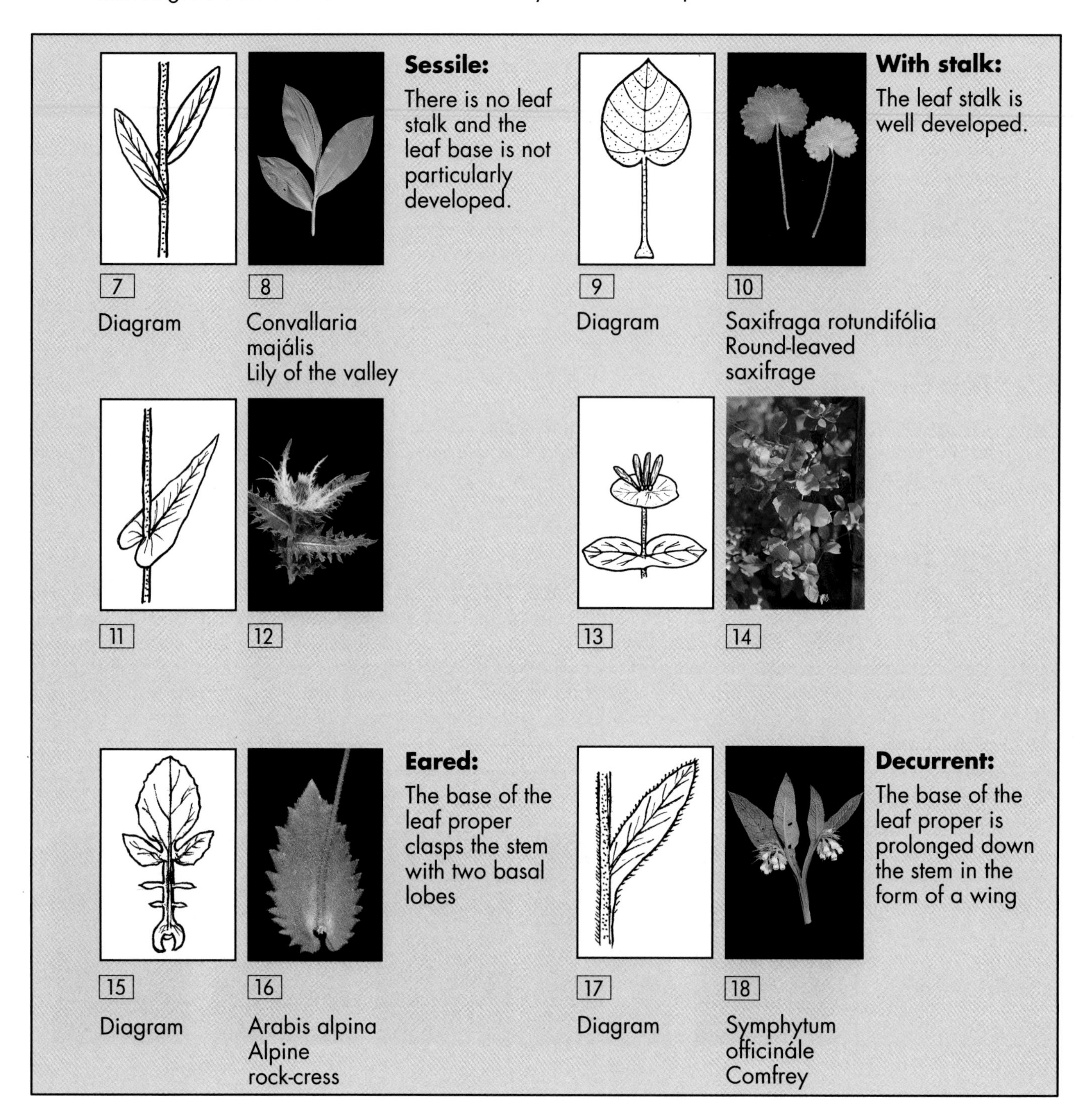

Sessile:
There is no leaf stalk and the leaf base is not particularly developed.

7 Diagram

8 Convallaria majális
Lily of the valley

With stalk:
The leaf stalk is well developed.

9 Diagram

10 Saxifraga rotundifólia
Round-leaved saxifrage

11

12

13

14

Eared:
The base of the leaf proper clasps the stem with two basal lobes

15 Diagram

16 Arabis alpina
Alpine rock-cress

Decurrent:
The base of the leaf proper is prolonged down the stem in the form of a wing

17 Diagram

18 Symphytum officinále
Comfrey

1.3.3. Arrangement of the leaves on the stem:

Leaves are arranged in a number of distinct ways on the stem, as follows:

Alternate or spiral:	At each node there is only one leaf, standing at an angle to the one below of 180 degrees (with e.g. 72 degrees).
Diametrically opposed:	The leaves are in one plane or distichous.
Basal:	The leaves spring from the base of the stem.
Opposite:	Two leaves at the same node on opposite sides of the stem.
Decussate:	Opposite, but successive pairs at right angles.
Whorled:	Three or more leaves in a whorl.

1.4. Life Forms

Plants can be classified into five life forms by both the length of the growing time and where the renewal buds are situated during winter or summer drought. Their relative proportions within a zone change according to the habitat. (In the diagrams the parts of the plant shown in black stay throughout the winter; the rest die off in the autumn.)

1. Phanerophytes:

These carry their buds more than 50 cm above the ground (Fig. 19: diagram). They comprise all evergreen trees and those which are green in summer (Fig. 20: ash – Fráxinus excélsior), together with shrubs, many climbing plants and, in the humid tropics, the epiphytes. These grow on the side or top branches of trees and in this way win a place in the sun. Trees serve them only as a support. Epiphytes in our latitudes are predominantly algae, lichens, and mosses, which can stand drying out for short periods.

2. Chamaeophytes:

These carry their buds 10–15 cm above soil level (Fig. 21: diagram) and in snowy regions are thus largely protected against frost by the covering of snow. They include many low-lying and creeping woody plants on the northern tundra and in high mountains (Fig 22: bilberry – Vaccinium myrtillus; Fig. 23: heather – Calluna vulgaris), many of the Ericaceae and the cushion plants.

23

3. Hemicryptophytes:

The buds of these plants lie close to soil level. They comprise the graminiferous plants (e.g. many grasses); annual and perennial plants forming a rosette like the dandelion (Fig. 25: Taráxacum); plantain species, which winter without a rosette of leaves; stemmed plants (e.g. stinging nettle and loosestrife) whose buds lie at soil level at the end of a stem which has died off; and herbs with runners which are above ground, e.g. strawberries (Fig.26: Fragária).

26

4. Geophytes:

The buds of these plants are very well protected under the soil surface, being situated on the underground growth of the plant. There are differences between rhizomatous geophytes, e.g. wood anemone (Fig. 28: Anémone nemorósa) and bulbous geophytes like the crocus (Fig. 29: Crocus) or autumn crocus (Fig. 30: Cólchicum autumnále)

29

30

5. Therophytes (annual plants):

These plants spend periods hostile to vegetation by abandoning their plant form completely and becoming robust seeds. To this category belong many other plants such as summer cereals and their weeds. (Winter cereals are hemicryptophytes.)

33

34

1.5. Underground storage organs:

For plants to sprout in spring, they need organic woody and vegetable material, which was formed in the last period of growth and stored away, usually in underground storage organs. Together with plants and seeds, these form valuable sources of nourishment for man and beast.
Several kinds are distinguished:

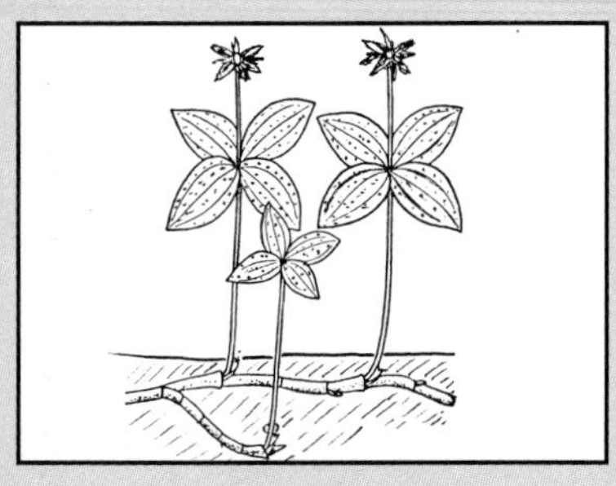
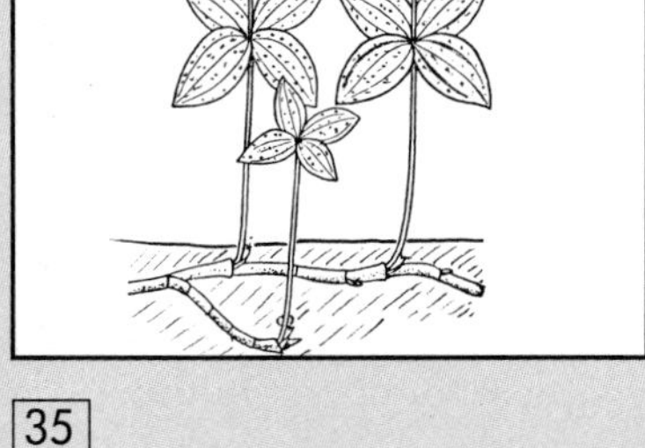

35

1. Rootstocks or rhizomes:

These develop underground, are unlimited in size and may be either branched or unbranched. Over the years they can cover a large area and attain a great age. They may bear on the underside only or all round roots which give rise to the shoots, together with colourless, fleshy leaves. Rhizomes may be distinguished from true roots by these leaves and by the scars left by them, by the development of buds, and by the lack of root covering. (Fig. 35, 36: herb Paris – Páris quadrifólia)

36

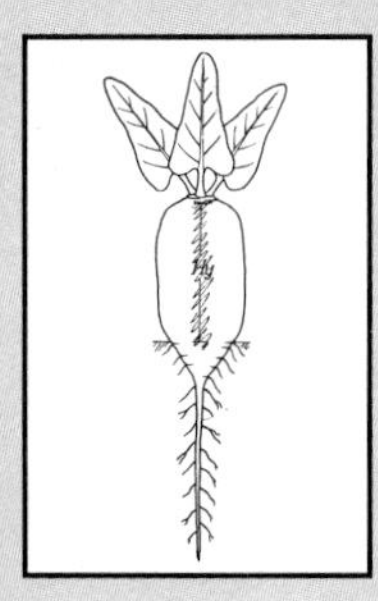

37

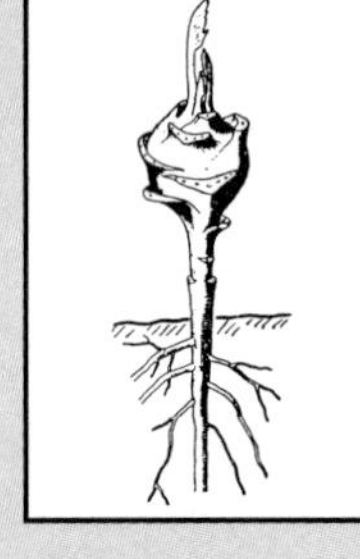

38

2. Tubers:

These develop either by marked primary or secondary thickening of the hypocotyl (Fig. 37: turnip) or from one or several intermediate nodes on the shoot (Fig. 38: kohlrabi). The underground tubers of the potato appear at the end of side shoots through primary thickening of several internodes (Fig. 39: potato).

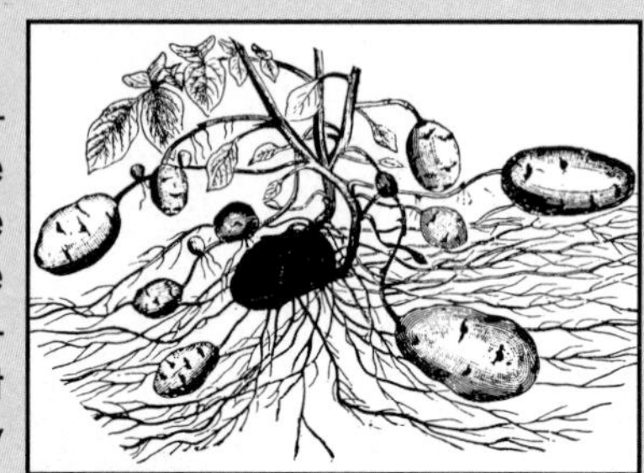

39

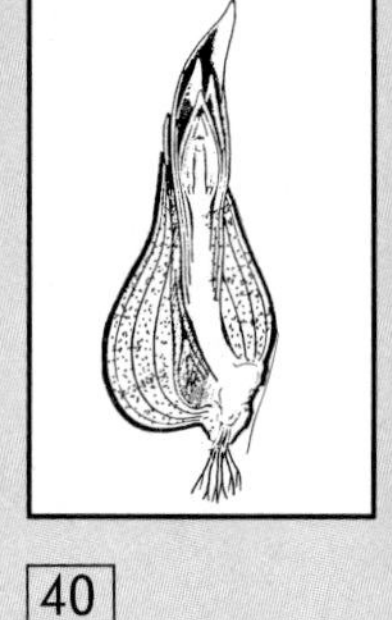

40

41

3. Bulbs:

These are generally mostly underground, very short shoots, with thick, fleshy, scaly leaves. These arise from a short root which is disc-shaped or conical – the bulb root. The shoot appearing above ground arises from its tip (Fig. 42: tulip; Fig. 43: ramsons).

42

43

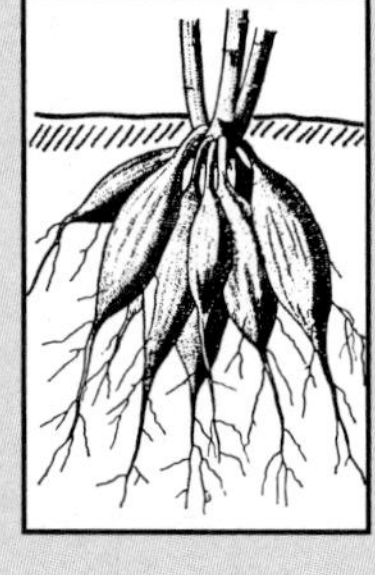

44

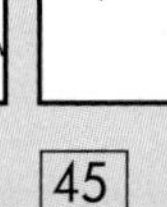

45

4. Root-tubers:

These resemble tubers, but can be distinguished by their root covering, lack of leaf scars, and their structure. All root-tubers store their material in tissue which is very much thickened. (Fig. 44: dahlia; Fig. 45: orchid; Fig. 46: roots of lesser celandine; Fig. 47: lesser celandine – Ranúnculus ficária).

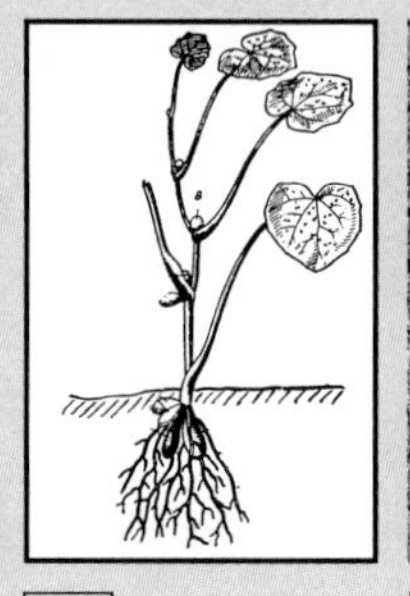

46

47

48

Turnip-like root crops:

In these plants the upper parts of the root are wholly or partly thickened, often being part of the hypocotyl. In the wild carrot (Daúcus caróta) and sugar-beet (Fig. 49: Béta vulgáris) the upper part of the root represents the major part of the storage organ. In celery (Fig. 48: Apium graveólens) the sprouting segment with foliage comes above the hypocotyl as well.

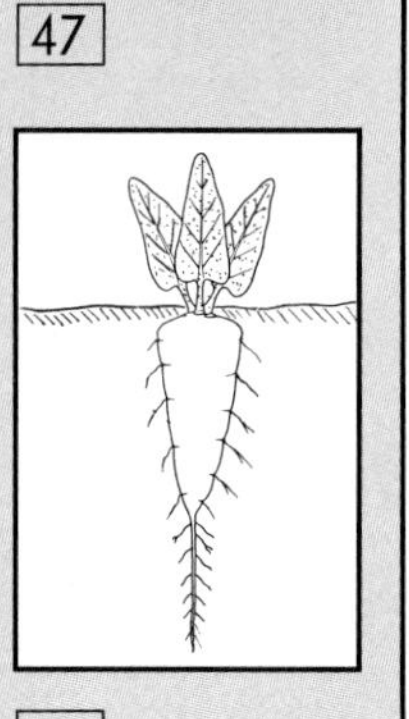

49

1.6. Flowers

A flower is a specialised part of the shoot carrying leaves serving the plant's sexual reproduction and designed accordingly. In primitive flowers (e.g. Fig. 50: magnolia) the axis, an elongated cone, comprises a spirally arranged perianth and numerous stamens and ovaries. In more developed forms (e.g. Fig. 51: round-leaved saxifrage – Saxifraga rotundifólia) we find a circular arrangement.
The flowers of angiosperms are characterised mainly by their very conspicuous perianth. Its development into a vividly coloured means of display was clearly an adaptation to the visits of the insects which fertilise them. The process of interaction with the evolution of flower-visiting insects took place mainly in the Tertiary era, leading to the development of the multifarious shapes and colours of today's flowers.
In **homochlamydeous** perianths all the segments (petals) are similar. There can be 1, 2, or several rings. If there is only one, it is called a simple perianth. If there are several the perianth is compound (e.g. Fig. 52: tulip – Túlipa; Fig. 53: martagon lily – Lilium mártagon).
Heterochlamydeous flowers have a double perianth consisting of greenish sepals with petals inside them, which are soft and usually brightly coloured (Fig. 54: yellow mountain saxifrage – Saxifrága aizoides).
Apochlamydeous flowers have no perianth.

1. The sepals:

These develop as a rule from bracts in the region of the flower. In many plants (e.g. Fig. 55:, stinking hellebore – Helleborus foétidus) the transition from bracts to sepals can be seen clearly.
While flowers are in bud, the sepals act as protection for the inner organs. After the flowers have unfolded they often fall away. Sometimes the sepals can harden after fruition and later surround the ripe fruit. Sepals can be free- standing (Fig. 56: alpine honesty – Lunária redivíva) or fused together (Fig. 57: cowslip – Primula véris). If they are surrounded by a ring of small bracts the latter are referred to as outer sepals (e.g. in the mallows and many of the Rosaceae).

2. The petals:

These are softer and usually larger than the sepals. Though they can be of various colours, they are the most conspicuous part of the flower – its means of display. This visual means of attraction is often accompanied by chemical stimuli – the flower's scent.
Petals are often metamorphosed stamens. The transition between the two organs of the flower is well seen in the white waterlily (Fig. 58: Nymphaéa álba). The "doubling" of garden flowers shows the same transformation.
Petals can be free-standing (Choripetalae) or fused into a tube (Sympetalae). The number of petals is determined by the number of independent petal ends: whether the petals are fused or not, their number can be found by pulling the petals in their entirety away from the sepals.

3. The stamens:

All the sepals together form the androecium. They can be arranged spirally or in circles (Fig. 60:

59 60 61 62 63 64

yellow mountain saxifrage – Saxifrága aizoides). Often there are 2 circles, standing in the gaps between the sepals and the petals respectively. As Fig. 59 shows, the stamen consists of a stalk-like supporting element, the **filament** (=F), the **anther** (=A), and a sterile intervening **space** (=K). Each anther consists of two **thecae** (=T).

There are often horn-shaped appendages to the stamens, as in the Ericaceae. **Staminodes** are degenerate stamens playing a part other than their original one. They vary widely in form. A special type is represented by the nectaries, which lie between the perianth and the stamens. In some plants they are small and inconspicuous (e.g. globeflower); in others they resemble petals and are bigger than the petals in the perianth (e.g. columbine).

4. The ovaries:

The ovaries form the female part of plants: the whole assemblage is known as the gynoecium. In angiosperms they are arranged in one organ holding the seeds.

In **apocarpous** gynoecia the numerous carpels are free-standing (Fig. 61: lesser celandine – Ranúnculus ficária); each has an ovary, together with a style and stigmata forming the pistil.

In **coenocarpous** gynoecia the carpels are fused together into one organ (Figs. 62, 63: yellow mountain saxifrage – Saxifraga aizoides). Here the swollen basal part holding the seeds is known as the ovary (Fig. 64, diagram). It often extends upwards in a column terminating in a stigma which is sticky so as to receive the pollen grains and generally has a lobe at the top.

The arrangement of the ovary can be of great importance in the identification of plants. Three forms are shown here:

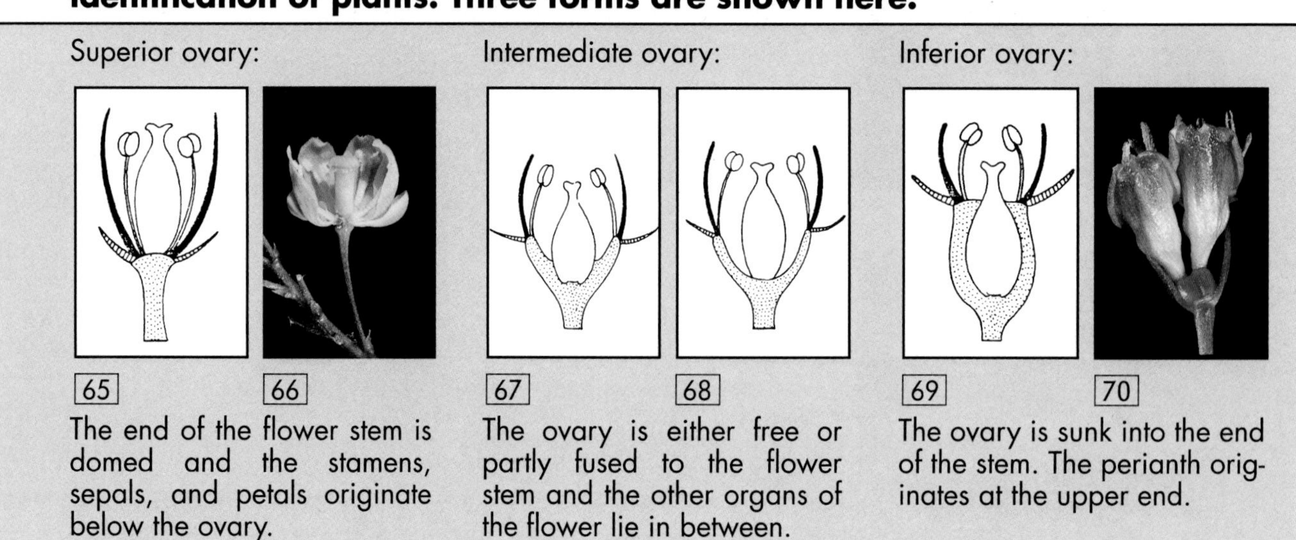

Superior ovary:

65 66

The end of the flower stem is domed and the stamens, sepals, and petals originate below the ovary.

Intermediate ovary:

67 68

The ovary is either free or partly fused to the flower stem and the other organs of the flower lie in between.

Inferior ovary:

69 70

The ovary is sunk into the end of the stem. The perianth originates at the upper end.

5. Differentiation by sex:

Hermaphrodite flowers contain both stamens and ovaries. Male flowers have stamens only, female ones ovaries only. There are also flowers which display both sexes (i.e. are morphologically hermaphrodite), but in which one has stopped working (i.e. they are functionally of one sex). If male and female flowers are found on the same plant this is referred to as monoecious. In dioecious plants the two kinds of flower are on different plants.

1.7. Flower heads (inflorescences):

Many plants carry only one flower on each stem (e.g. dandelion, tulip). In all other species there are several or many flowers united in flower heads (inflorescences) of specific types. These are the result of major changes in the stem, which is strongly differentiated from the purely vegetative parts of the plant. The development from single flower to flower head must have originated millions of years ago, starting from lateral flowers on the stem.

The way the main stem of the flower head is divided is important for the morphological classification of types of inflorescence. In cymose inflorescences the stem ends in a terminal bud which characteristically comes into bloom before all the neighbouring lateral flowers. In racemose inflorescences a terminal flower never develops.

Monopodial or racemose inflorescences:			**Sympodial or cymose inflorescences:**	
During development of the inflorescence the main stem grows over a long period, being thicker than the side stems and dominant over them. It either ends in a terminal flower or grows without forming one. A distinction can be made between simple (1–7) and compound racemose (8 –11) inflorescences.			The main stem completes its growth very early and can end in a flower. Lateral stems predominate in further growth of the inflorescence. A distinction can be made between pleiochasia (several lateral stems), dichasia (two lateral stems), and monochasia (one lateral stem)	
1. Racemes 2. Corymbs 3. Spikes 4. Catkins	5. Cones 6a. spadices 6b. capitula 7. Umbels	8. Panicles 9. Corymbous cymes 10. Combined spikes and corymbs	12. False umbels – Dichasium – Reel – Double reel	– Spiral – Double spiral – Thyrse

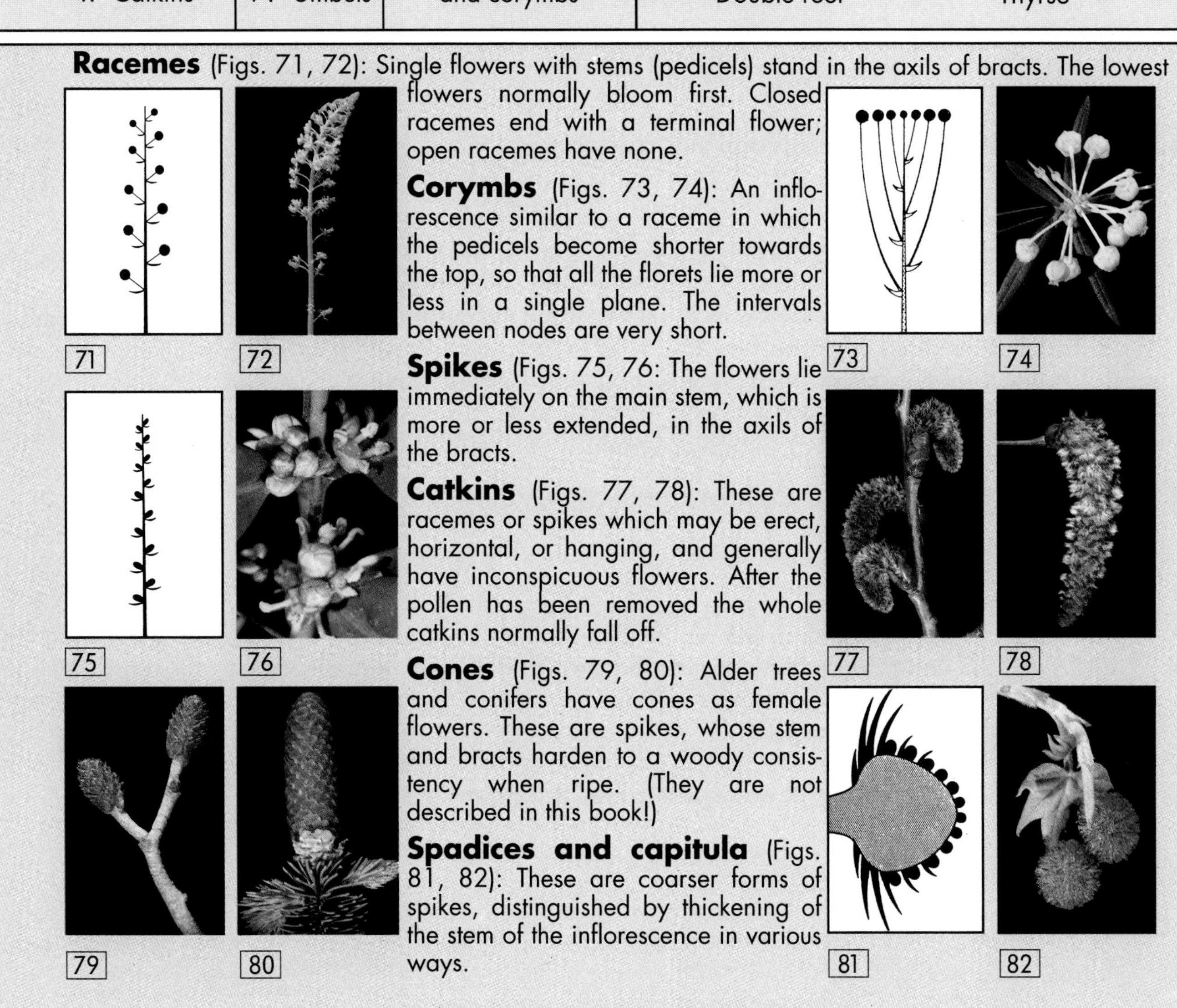

Racemes (Figs. 71, 72): Single flowers with stems (pedicels) stand in the axils of bracts. The lowest flowers normally bloom first. Closed racemes end with a terminal flower; open racemes have none.

Corymbs (Figs. 73, 74): An inflorescence similar to a raceme in which the pedicels become shorter towards the top, so that all the florets lie more or less in a single plane. The intervals between nodes are very short.

Spikes (Figs. 75, 76: The flowers lie immediately on the main stem, which is more or less extended, in the axils of the bracts.

Catkins (Figs. 77, 78): These are racemes or spikes which may be erect, horizontal, or hanging, and generally have inconspicuous flowers. After the pollen has been removed the whole catkins normally fall off.

Cones (Figs. 79, 80): Alder trees and conifers have cones as female flowers. These are spikes, whose stem and bracts harden to a woody consistency when ripe. (They are not described in this book!)

Spadices and capitula (Figs. 81, 82): These are coarser forms of spikes, distinguished by thickening of the stem of the inflorescence in various ways.

71 72 73 74 75 76 77 78 79 80 81 82

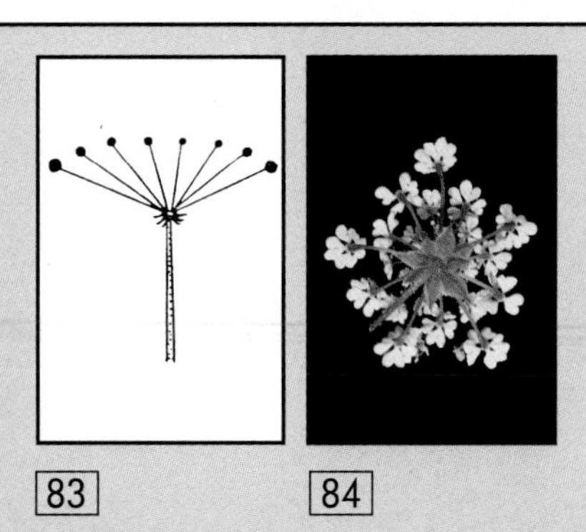

[83] [84]

Umbels (Figs. 83, 84): As opposed to corymbs, there are no intervals between the nodes, so the stemmed florets all rise to an equal height on the stem from bracts set in the form of a rosette. Their number is often less than that of the flowers. If the stems of the florets are all the same length, the inflorescence forms a hemisphere or sphere.

[85] [86]

Panicle (Fig. 85, 86): In closed panicles all stems of the many-branched raceme end in a terminal flower. This is probably the oldest type of inflorescence. Open panicles lack terminal flowers.
In **corymbous cymes** the branches are of different length, so that the flowers lie flat in one plane or form a slightly vaulted surface.
In **compound spikes** subordinate spikes with several flowers replace the individual flowers.
In **compound umbels** the first-order umbel stalks end in miniature umbels, called second-order umbels. As opposed to umbels, false umbels do not have lateral branches rising to the same height.

1.8. Pollination of flowers:

In **gymnosperms**, the pollen grains on the fertilising organs remain stuck to the micropyle of the seed plants on pollination. In **angiosperms** the pollen is held fast by a sticky papillose stigma. This makes sexual reproduction largely independent of moisture content.
Pollen can be transported in the following ways:

1.8.1. Fertilisation by animals (zoophily, zoogamy):

Requirements:

- The animals carrying out the fertilisation must visit the flowers regularly and stay long enough.
- The plants must have evolved to meet the mechanical needs involved.
- The pollen grains must be able to stick firmly to the carriers (pollen sacs in the case of bees).
- Plants fertilised by animals must have, in order to attract them, a lure (e.g. pollen, nectar)and a means of stimulation (colour, scent), and they must also have sticky pollen.
- Fertilisation is achieved by insects, birds, or bats.

1.8.2. Wind fertilisation (anemophily, anemogamy):

Requirements:

- The pistil and stigmata must be much enlarged so that they can easily capture the pollen.
- So that they can drift in the air as long as possible, the pollen grains should be light, dry, smooth, and provided with devices that help them to float (e.g. air sacs).
- Lures and attractive devices are no longer necessary, but to guarantee pollination an enormous output of pollen is necessary.
- If the flowers ripen in early spring pollination can take place before leaves emerge to get in the way (e.g. hazel bushes, poplars, willows, and alders).

1.8.3. Self-fertilisation (autogamy):

In unfavourable circumstances many plants can resort to self-fertilisation. This takes place inside one flower or between different flowers of the same plant.

1.9. Vegetation zones, defined by height:

87 View from the hills up to the alpine zone: in the background are the Eiger, the Mönch, and the Jungfrau (Bernese Oberland, Switzerland).

88 89

90 91

92 In the subalpine forest

93 The Aletsch glacier: VS/Switzerland

Hill zone:

- To 600 m in central Germany, to 700 m in the northern Alps, to 800 and 900 m in the central and southern Alps.
- Mean annual temperature between, 8 and 12 degrees C.
- Vegetation period over 250 days.
- Characterised by mixed deciduous woods (Fig.88).
- In low-lying places oak and beech woods, in the warmest regions on calcareous soil evergreen oak woods (Fig. 89, the Rebburg), in dry regions pine woods, in higher regions mixed beech woods.
- On the southern foothills of the Alps mixed oak woods; nowadays these are often planted with sweet chestnut trees.

Montane zone:

- To 1200–1300 m on the northern side of the Alps, to 1300–1500 m in the central Alps, to 1500–1700 m in the southern Alps.
- Mean annual temperature between 4 and 8 degrees C.
- Vegetation period over 200 days.
- Characterised by beech, fir, and mixed beech/fir woods (Figs. 90, 91).
- Pine woods in the central Alps.
- In more continental areas the spruce marks the natural boundary.

Subalpine zone:

- To1700–1900 m in the northern Alps, to 1900–2400 m in the central Alps, to 1800–2000 m in the southern Alps.
- Mean annual temperature between 1 and –2 degrees C.
- Vegetation period 100–200 days.
- Characterised by subalpine spruce woods.
- In the central alpine chains larch woods are found above spruce woods; at the transition to the alpine zone grow thickets of dwarf bushes composed of alpine roses, junipers, crowberries, etc.; alpine alders are found in damp, shady places (mostly sites facing north); in calcareous soil mountain pine woods and stands of Swiss mountian pines.

Alpine zone:

- The treeless topmost parts of the Alps, from the tree line to the natural snow line (= the line at which snow remains on horizontal surfaces in summer).
- In many books the region above the natural snow line is distinguished from the alpine zone as the snow zone.
- The snow line lies between 2400 m and 3200 m.
- Vegetation period under 100 days.
- Characterised at lower levels by thickets of dwarf bushes (composed of alpine roses, junipers, crowberries, etc.), and above them by alpine meadows: the number of species on these continually decreases with greater height.

Part 1: Ferns (pp. 36–157)

A1

94

Sporangia not on the edge or the underside of the pinnae (pp. 38/39)

A1

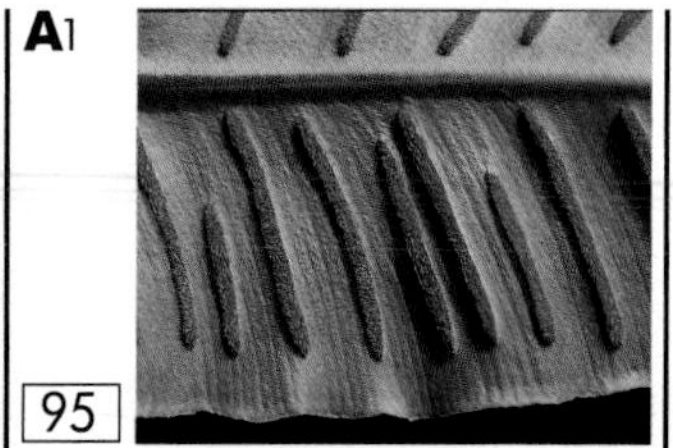

95

Sporangia on the edge or the underside of the pinnae in small groups or rows (pp. 38–53)

A2

96

Marsh ferns with clover-like leaves growing in soil (pp. 52/53)

A3

97

Free-floating water ferns without roots (pp. 52/53)

Part 2: Herbs and shrubs with white flowers (pp. 54–101)

A1

98

Simple perianths formed of 6 petals or bristles; leaves with parallel veins (pp. 54–59)

A3

99

Simple perianths with 6 petals; leaves in three parts with branching veins (pp. 76/77)

A4

100

Double perianths: 3 + 3 petals (different!) or 2–4 sepals/petals (pp. 58–67)

A5

101

Double perianth with 5 radial free-standing sepals and petals (pp. 66–87)

A13

102

Perianths with 5 sepals and 5 partly or wholly fused petals (pp.78–89)

A14

103

Composite flowers; 5 fused petals, sepals metamorphosed (pp. 88–97)

A9
A10
A11

104

Perianth with 5–33 radially arranged and free petals (pp. 70–73 and 86/87)

A12
A15

105

Flowers zygomorphic, with 5 petals and 5 sepals (pp. 86/87 and 98–101)

Part 3: Herbs and shrubs with yellow flowers (pp. 102–157)

A1

106

Simple perianth formed of 6 petals; leaves with parallel veins (pp. 102/103)

A2

107

Plants with flowers lacking perianths; bracts replace perianths (pp 104/105)

A3

108

Simple perianth with 4 sepals and 5 or more petals; leaves with branching veins

A4

109

Double perianth: 4 petals, radial or zygomorphic; leaves with parallel veins (pp. 104–109)

A5
A6
A7
A8

110

Up to 6 unfused sepals and petals or outer + inner rings of sepals and 5 (or 6) petals (pp. 108–125)

A13
A14

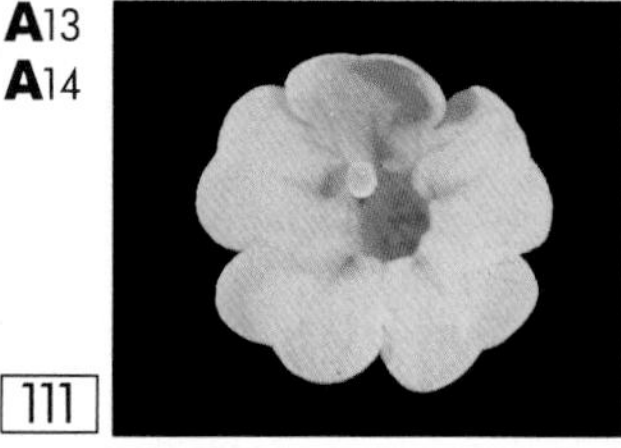

111

5 sepals, absent or metamorphosed; 5 fused petals (pp. 126–147)

A9
A10
A11

112

6–20 radial petals; 3–6 sepals (pp 124/125)

A12
A15

113

Flowers zygomorphic, with 5 petals and 5 sepals (pp. 124–127, 148–157)

Summary of Artificial Key, Parts 4–6 Main Text: pp.158–249

Part 6: Herbs and bushes with green flowers (pp. 236–249)

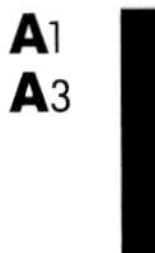

A1
A3

114

3–6 petals and leaves with parallel veins or 2–4 sepals (pp. 236/237, 244–249)

A2

115

Plants without perianths; leaves with branching veins; flowers set one above other in a spadix (pp.236–239)

A4

116

4 + 4 petals, 4 + 4 sepals or 4 sepals + petals (fused) (pp. 238–243)

A5
A13
A14

117

Double perianth with 5 sepals and 5 petals (sepals also metamorphosed into hairs) (pp. 244–249)

Part 5: Herbs and shrubs with blue flowers (pp. 210–235)

A1

118 a

Simple perianths formed of 6 petals; leaves with parallel veins (pp. 210/211)

A3

118 b

Simple perianths with 5–10 petals; leaves with branching veins (pp. 214–215, 220–221)

A4

119

Double perianths with 4 or 5 sepals and 3 or 4 petals (pp. 212–215)

A5

120

Double perianth with 5 radially arranged, free sepals and petals (pp. 214/215)

A13

121 a

Petals fused; 5 sepals; when sepals are represented only by hairs, there are 4 stamens (pp 216 - 229)

A14

121 b

Composite flowers: 5 fused petals; sepals metamorphosed (pp. 230/231)

A12

122 a

Perianth zygomorphic: 5 each of petals and sepals; flowers are single (pp. 220–223)

A15

122 b

Flowers are zygomorphic: over- and under-lips or with 2 wings, a standard, and a keel (pp. 232–235)

Part 4: Herbs and shrubs with red flowers (pp. 158–209)

A1
A1a
A3a
A3b
A17

123

Perianth of 3–6 petals. Leaves with parallel or branching veins (pp. 158, 160, 178, 180, 206, 208)

A3

124

Simple perianth with 4 dark-red sepals; leaves unevenly pinnate (pp 164 - 165)

A4

125

Double perianth with 2 or 4 sepals and 4 petals (pp. 162–167)

A5
A16

126

5 sepals and 5 petals or 5–10 petals; flowers not composite (pp. 166–181, 208/209)

A13

127

4 or 5 sepals (as bristles when in fruit); 4 or 5 petals, fused (pp. 184/185)

A14

128

Composite flowers: ray-florets with 5 fused petals; (pp. 186–195)

A19

129 a

Perianth with 12–16 radially arranged sepals and petals (pp. 166/167)

A12
A15

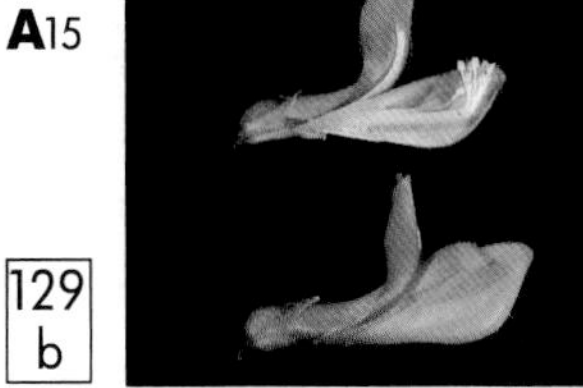

129 b

Flowers zygomorphic, with 5 petals and 5 sepals (pp. 182/183, 194–209)

Key Part 1: Pages 20, 22, 38–53 — **Ferns**

A1 Land plants growing in soil — True, see **B1(1)** or **B1(2)**; False, see **A2** or **A3**

B1(1) Sporangia on the top part of a doubly pinnate leaf **C1(1)** or a separate branch of a singly pinnate leaf **C1(2)** — True, see **C1(1)** or **C(1)2**; False, see **B1(2)**

> **C1(1)** **Royal fern – Osmúnda** (pp. 38/39)
> **C1(2)** **Moonwort – Botrychium** (pp. 38/39)

B1(2) Sporangia on the edge or the underside of the pinnae, arranged in brownish groups or rows. — True, see **C1(3)–C1(7)**; False, back to **B1** or see **A2**

C1(3) Leaf not pinnate, almost entire, mostly stalked, 15–60 cm long — True, see **D1(1)**; False, see **C1(4)–C1(7)**

> **D1(1)** **Hart's tongue fern – Phyllitis** (pp. 38/39)

C1(4) Leaf singly pinnate — True, see **D1(2)** or **D1(3)**; False, see **C1(5)–C1(7)**

D1(2) First-order pinnae entire or more or less entire — True, see **E1(1)**; False, see **D1(3)**

> **E1(1)** **Hard fern – Blechnum** (pp. 38/39)
> **Rusty-back fern – Céterach** (pp. 38/39)
> **Sensitive fern – Onoclea** (pp. 38/39)
> **Polypody – Polypodium** (pp. 40/41)
> **Saw-toothed polypody – Polypodium** (pp. 40/41)

D1(3) First-order pinnae with pointed teeth or wavy — True, see **E1(2)**; False, see **C1(5)**

> **E1(2)** **Shield fern – Polystichum** (pp. 44–47)
> **Spleenwort – Asplénium** (pp. 46/47)

C1(5) Leaf doubly or triply pinnate — True, see **D1(4)** or **D1(5)**; False, see **C1(6)**

D1(4) First-order pinnae with no or slight stems — True, see **E1(3)** or **E1(4)**; False, see **D1(5)**

E1(3) Second-order pinnae entire or serrated only in the top region — True, see **F1(1)**; False, see **E1(4)**

> **F1(1)** **Buckler fern – Dryópteris** (pp. 40/41)
> **Oak fern – Lastrea** (pp. 46/47)
> **Ostrich fern – Matteúccia** (pp. 40/41)
> **Marsh fern – Thelypteris** (pp. 40/41)

E1(4) Second-order pinnae all distinctly toothed — True, see **F1(2)**; False, back to **E1(3)** or see **D1(5)**

> **F1(2)** **Buckler fern – Dryópteris** (pp. 42–45)
> **Lady fern – Athyrium** (pp. 44/45)
> **Spleenwort – Asplénium** (pp. 50/51)

D1(5) First-order pinnae with distinct stems — True, see **E1(5)**; False, back to **D1(2)** or see **C1(6)**

> **E1(5)** **Oak fern – Lastréa** (pp. 48/49)
> **Spleenwort – Asplénium** (pp. 48/49)
> **Bladder fern – Cystópteris** (pp. 50/51)
> **Parsley fern – Cryptográmma** (pp. 50/51)

C1(6) Leaves regularly divided into 3, 5 or more sections — True, see **D1(6)**; False, see **C1(7)**

> **D1(6)** **Forked spleenwort - Asplénium** (pp. 52/53)

C1(7) Leaf with more than 3 orders of pinnae — True, see **D1(7)**; False, back to **C1(3)**

> **D1(7)** **Bracken – Pterídium** (pp. 52/53)

A2 Marsh plants with clover-shaped leavesgrowing in soil — True, see **B2**; False, see **A3**

> **B2** **Pepperwort – Marsilea** (pp. 52/53)

A3 Water ferns: free-floating plants — True, see **B3**; False, back to **A1**

> **B3** **Salvinia – Salvínia** (pp. 52/53)
> **Azolla – Azólla** (pp. 52/53)

Key Part 2: Pages 20, 23–26, 54–101 **White flowers**

Key	Characteristics	Result
A1	Simple perianth with 4 or 6 petals or bristles; leaves have parallel veins	True, see **B**1(1) to **B**1(3) False, see **A**3
B1(1)	4 radially arranged petals	True, see **C**1(1) False, see **B**1(2)
C1(1)	**May lily – Maianthemum** (pp. 54/55)	
B1(2)	6 petals radially arranged	True, see **C**1(2) or **C**1(3) False, see **B**1(3)
C1(2)	3 inner petals shorter than 3 outer petals	True, see **D**1(1) False, see **C**1(3)
D1(1)	**Snowdrop – Galánthus** (pp. 54/55)	
C1(3)	All 6 petals of equal length	True, see **D**1(2)–**D**1(4) False, see **B**1(3)
D1(2)	Flowers solitary at end of stem	True, see **E**1(1) False, see **D**1(3)
E1(1)	**Snowflake – Leucójum** (pp. 54/55) **Crocus – Crocus** (pp. 58/59)	
D1(3)	Flowers in flat or domed umbels	True, see **E**1(2) False, see **D**1(4)
E1(2)	**Leek and allium – Allium** (pp. 54/55)	
D1(4)	Several flowers to each stem: singly or in pairs on the stem or in racemes, corymbs or spikes	True, see **E**1(3) False, back to **D**1(2) or see **B**1(3)
E1(3)	**St Bernard's lily – Anthéricum** (pp. 56/57) **Star of Bethlehem – Ornithógalum** (pp. 56/57) **St Bruno's lily – Paradísea** (pp. 56/57) **Lily of the valley – Convallária** (pp. 56/57) **Solomon's seal – Polygonátum** (pp. 56/57) **Helleborine – Cephalánthera** (pp. 58/59)	
B1(3)	Perianths consist of hairs forming woolly heads when ripe	True, see **C**1(4) False, back to **B**1(1) or see **A**3
C1(4)	**Cotton-grass – Erióphorum** (pp. 58/59)	
A3	Flowers with single perianths; leaves with branching veins	True, see **B**3 False, see **A**4
B3	Perianths of 6 petals; leaves in 3 parts with branching veins	True, see **C**3 False, see **A**4
C3	**Anemone – Anemone** (pp. 76/77)	
A4	Perianths double: flowers have 3 + 3 petals or 2–4 sepals and petals	True, see **B**4(1)–**B**4(5) False, see **A**5
B4(1)	Water plants with 3 white and 3 green petals or 4 petals and 4 sepals; leaves have parallel or net-like veins	True, see **C**4(1) False, see **B**4(2)
C4(1)	**Frog-bit – Hydrocháris** (pp 58/59) **Water plantain – Alísma** (pp 60/61)	
B4(2)	Land plants with 4 free-standing sepals and 4 radial petals: leaves have net-like veins	True, see **C**4(2) False, see **B**4(3)
C4(2)	Leaves entire, undivided, spathulate, lanceolate, or narrow oval	True, see **D**4(1) False, see **C**4(4)
D4(1)	**Penny cress – Thlaspi** (pp. 60/61)	
C4(3)	Leaves entire, narrow oval or round, with a toothed or wavy edge	True, see **D**4(2) False, see **C**4(4)
D4(2)	**Gipsywort – Lycopus** (pp. 60/61) **Garlic mustard – Alliária** (pp. 60/61) **Scurvy grass – Cochleária** (pp. 60/61) **Rock cress – Arabis** (pp. 62/63)	
C4(4)	Basal leaves and most leaves on the stem have pinnate divisions to the middle vein	True, see **D**4(3) False, see **C**4(6)
D4(3)	**Meadow cress – Cardaminópsis** (pp. 62/63) **Hedge mustard – Ráphanus** (pp. 62/63)	
C4(5)	Leaves unequally pinnate	True, see **D**4(4) False, see **C**4(6)
D4(4)	**Toothwort – Dentária** (pp. 62/63) **Bittercress – Cardámine** (pp. 62/63) **Watercress – Nasturtium** (pp. 64/65)	
C4(6)	Leaves mostly whorled; flowers arranged in thyrses	True, see **D**4(5) False, see **B**4(3)
D4(5)	**Woodruff – Aspérula** (pp. 64/65) **Bedstraw – Galium** (pp. 64/65)	

B4(3)	Perianths radial, with 2 free-standing sepals (dropping off when flower blooms) and 4 free-standing petals; leaves doubly pinnate	True, see **C**4(7) False, see **B**4(4)

C4(7) **Poppy – Papáver** (pp. 64/65)

B4(4)	Perianths zygomorphic, with 4 sepals and 4 petals; leaves doubly or triply pinnate	True, see **C**4(8) False, see **B**4(5)

C4(8) **Fumitory – Corydalis** (pp. 64/65)

B4(5)	Radial perianth with 4 sepals and 4 petals fused into a tube; leaves entire, with main veins parallel	True, see **C**4(9) False, back to **B**4(1) or see **A**5

C4(9) **Plantain – Plantágo** (pp. 66/67)

A5	Free-standing perianth with 5 sepals and 5 petals, radially arranged	True, see **B**5(1)–**B**5(9) False, see **A**9
B5(1)	Plants with undivided entire leaves	True, see **C**5(1)–**C**5(7) False, see **B**5(2)
C5(1)	Flowers solitary at end of long stem; leaves narrow, oval to heart-shaped, often with somewhat wavy edges	True, see **D**5(1) False, see **C**5(2)

D5(1) **Rock-rose – Cistus** (pp. 66/67)
Grass of Parnassus – Parnássia (pp. 66/67)

C5(2)	Flowers in several spikes attached to the leaf stem; leaves heart to arrow-shaped	True, see **D**4 False, see **C**5(3)

D5(2) **Buckwheat – Fagopyrum** (pp. 66/67)

C5(3)	Flowers in false umbels or balls; leaves oval to cordate or longish	True, see **D**5(3) False, see **C**5(4)

D5(3) **Swallow wort – Vincetóxicum** (pp. 66/67)
Chickweed – Stellária (pp. 66/67)

C5(4)	Long-stemmed single flowers in umbel-like panicles, forming a head or with 2–3 at end of stem; leaves narrow, lanceolate and opposite	True, see **D**5(4) False, see **C**5(5)

D5(4) **Baby's breath – Gypsóphila** (pp. 68/69)
Pink – Diánthus (pp. 68/69)
Soapwort – Saponária (pp. 68/69)
Sandwort – Moehríngia (pp. 68/69)

C5(5)	Flowers in umbel-like panicles; leaves cylindrical, somewhat fleshy and sessile	True, see **D**5(5) False, see **C**5(6)

D5(5) **Stonecrop – Sedum** (pp. 68/69)

C5(6)	Flowers in terminal spikes; leaves lanceolate with edges bent downwards	True, see **D**5(6) False, see **C**5(7)

D5(6) **Knotweed – Polygonum** (pp. 70/71)

C5(7)	Flowers in umbels with 2–8 blooms; leaves in rosettes, lanceolate, with long hairs 2 mm	True, see **D**5(7) False, back to **C**5(1) or see **B**5(2)

D5(7) **Rock jasmine – Andrósace** (pp. 70/71)

B5(2)	Leaves entire but toothed; flowers not in umbels	True, see **D**5(8) False, see **B**5(3)
C5(8)	Flowers in many-flowered panicles or racemes with 3–15 flowers; basal leaves do not form a rosette	True, see **D**5(8) False, see **C**5(9)

D5(8) **Saxifrage – Saxifrága** (pp. 70–73)

C5(9)	Flowers in corymbs standing on the stem or in racemes to one side; leaves not in a rosette	True, see **D**5(9) False, back to **C**5(8) or see **B**5(3)

D5(9) **Stonecrop – Sédum** (pp. 70/71)
Wintergreen – Pyrola (pp. 72/73)

B5(3)	Flowers not in umbels; leaves divided into 3 or more segments with entire lobes	True, see **C**5(10) False, see **B**5(4)

C5(10) **Wood sorrel – Oxalis** (pp. 72/73)
Bogbean – Menyánthes (pp. 72/73)

B5(4)	Flowers not in umbels; leaves divided into 3 or more segments with toothed or deeply divided leaves	True, see **C**5(11) False, see **B**5(5)

C5(11) **Buttercup – Ranúnculus** (pp. 74/75
Windflower – Anemone (pp.74–77)
Cinquefoil – Potentilla (pp. 74/75)
Cranesbill – Geranium (pp. 76/77)

B5(5) Flowers not in umbels; leaves unequally divided — True, see **C**5(12) False, see **B**5(6)

C5(12) **Cinquefoil – Potentilla** (pp. 76/77)
Meadowsweet – Filipendula (pp. 76/77)
Danewort – Sambucus (pp. 76/77)

B5(6) Flowers in racemes with 4–8 blooms; leaves with 3–7 blunt sections — True, see **C**5(13) False, see **B**5(7)

C5(13) **Saxifrage – Saxifrága** (pp. 78/79)

B5(7) Flowers in umbels; leaves close to the stem and divided into 5–7 parts — True, see **C**5(14) False, see **B**5(8)

C5(14) **Sanicle – Sanícula** (pp. 78/79)
Masterwort – Astrántia (pp. 80/81)

B5(8) Flowers in umbels; leaves 3 in number, with stalked, lobate leaflets — True, see **C**5(15) False, see **B**5(9)

C5(15) **Masterwort – Peucédanum** (pp 80/81)

B5(9) Flowers in umbels; leaves unequally pinnate — True, see **C**5(16) False, back to **B**5(1) or see **A**9

C5(16) **Angelica – Angélica** (pp. 80/81)
Cow parsley – Herácleum (pp. 80/81)
Hogweed – Laserpítium (pp. 82–85)
Water parsnip – Síum (pp. 82/83)
Chervil – Chaerophyllum (pp 82/83)
Cow parsley – Anthriscus (pp. 82/83
Moon carrot – Séseli (pp. 84/85)
Caraway – Cárum (pp. 84/85)
Masterwort – Peucédanum (pp. 84/85)
Candy carrot – Athamánta (pp. 86/87)
Carrot – Daucus (pp. 84/85)

A9 Free-standing sepals 7–9; petals usually 8; leaves narrow oval,notched or toothed, with edges rolled downwards and covered in white hairs on the underside — True, see **B**9 False, see **A**10

B9 **Mountain avens – Dryas** (pp. 70/71)

A10 Perianth of 5–10 free-standing petals and 3 sepal-like bracts; most leaves have 3 leaflets — True, see **B**10 False, see **A**11

B10 **Liverwort – Hepática** (pp. 72/73)

A11 Perianth of at most 4 free-standing sepals and 14–33 petals; leaves broad oval to round, and entire; water plants — True, see **B**11 False, see **A**12

B11 **Waterlily – Nymphaéa** (pp. 86/87)

A12 Flowers zygomorphic, 1–2, with 5sepals and 5 petals, on the leafstems; leaves narrow, oval to lanceolate, and toothed — True, see **B**12 False, see **A**13

B12 **Violet – Víola** (pp. 86/87)

A13 Perianth of 5 radial and partly or wholly fused sepals and 5 petals; flowers not fused when in spadices or capitula; in globe thistles and teasels flowers form globes or cylindrical capitula — True, see **B**13(1) False, see **A**14

B13(1) Water plants; sepals fused at base; petals fused in the bottom part; leaves pinnate with linear lobes — True, see **C**13(1) False, see **B**13(2)

C13 (1) **Water violet – Hottónia** (pp. 78/79)

B13(2) Petals deeply divided, with long claw; flowers in panicles or cymes; leaves entire and basal; land plants — True, see **C**13(3) False, se **B**13(4)

C13(2) **Campion – Siléne** (pp. 78/79)

B13(3) Flowers in many-flowered balls, with petals fused nearly to the end; leaves only scales and through lack of chlorophyll whitish to yellowish in colour; land plants without roots — True, see **C**13(3) False, see **B**13(4)

C13(3) **Dodder – Cuscúta** (pp. 86/87)

B13(4) Flowers in cymes with stalks; petals fused at the middle and with pointed ends; leaves entire or toothed and opposite; land plants — True, see **C**13(4) False, see **B**13(5)

C13(4) **Nightshade – Solánum** (pp. 86/87)

B13(5)	Flowers in cymes; petals wholly fused together; leaves unequally pinnat	True, see **C**13(5) False, see **B**13(6)
B13(5)	**Potato – Solánum** (pp. 86/87)	
B13(6)	Petals fully fused together; flowers solitary on long stalks; leaves heart-shaped or arrow-shaped and entire	True, see **C**13(6) False, see **B**13(7)
C13(6)	**Bindweed – Convólvulus** (pp. 88/89) **Bindweed – Calystégia** (pp. 88/89)	
B13(7)	Petals fused at the beginning of blooming period at both the base and the end; flowers in egg-shaped or cylindrical spikes; leaves lanceolate	True, see **C**13(7) False, see **B**13(8)
C13(7)	**Rampion – Phyteúma** (pp. 88/89)	
B13(8)	Flowers in globes or cylindrical capitula; leaves pinnate	True, see **C**13(8) False, back to **B**13(1) or see **A**14
C13(8)	**Globe thistle – Echinops** (pp. 88/89) **Teasel – Dípsacus** (pp. 88/89)	
A14	Flowers composite, capitula or spadices, sepals generally metamorphosed to hairs or bristles; petals fused together	True, see **B**14(1)–**B**14(5) False, see **A**15
B14(1)	Head 5–12 cm wide with tubular florets; bracts coloured silver, imitating ray-florets; leaves deeply pinnate, spiny and arranged in a rosette	True, see **C**14(1) False, see **B**14(2)
C14(1)	**Carline thistle – Carlina** (pp. 88/89)	
B14(2)	Numerous ray- and tubular florets in a flat capitulum	True, see **C**14(2) False, see **B**14(3)
C14(2)	**Michaelmas daisy – Bellidiástrum** (pp. 90/91) **Daisy – Béllis** (pp. 90/91) **Fleabane – Erígeron** (pp. 90–93) **Ox-eye daisy – Chrysánthemum** (pp 90–93) **Chamomile – Anthemis** (pp. 90/91) **Mayweed – Tripleurospérmum** (pp. 90/91) **Wild chamomile – Matricária** (pp. 92/93)	
B14(3)	Tubular florets in capitula and few ray-florets	True, see **C**14(3) False, see **B**14(4)
C14(3)	**Quickweed – Galinsóga** (pp. 92/93) **Yarrow – Achilléa** (pp. 94/95)	
B14(4)	Spadices and capitula with hairlike or inconspicuous tubular ray-florets	True, see **C**14(4) False, see **B**14(5)
C14(4)	**Edelweiss – Leontopódium** (pp. 92/93) **Cat's foot – Antennária** (pp. 96/97) **Cudweed – Gnaphálium** (pp. 96/97) **Fleabane – Erígeron** (pp. 96/97)	
B14(5)	Capitula with tubular florets only	True, see **C**14(5) False, back to **B**14(1) or see **A**15
C14 (5)	**Butterbur – Petasítes** (pp. 96/97) **Cabbage thistle – Cirsium** (pp. 96/97) **Spiniest thistle – Cirsium** (pp. 96/97)	
A15	Zygomorphic flowers	True, see **B**15(1) or **B**15(20) False, back to **A**1
B15(1)	5 fused sepals; 5 petals fused and arranged in an upper and lower lip	True, see **C**15(1) False, see **B**15(2)
C15(1)	**Hyssop – Hyssópus** (pp. 98/99) **Dead-nettle – Lámium** (pp. 98/99) **Hemp-nettle – Galeópsis** (pp. 98/99) **Catnip – Nepéta** (pp. 98/99) **Eyebright – Euphrásia** (pp. 98/99)	
B15(2)	Flowers with standard, 2 wings and keel	True, see **C**15(2) False, back to **A**1
C15(2)	**Canary clover – Dorycnium** (pp.100/101) **Clover – Trifólium** (pp.100/101) **Melilot – Melilótus** (pp.100/101) **Milk vetch – Astrágalus** (pp.100/101) **Vetch – Vícia** (pp.100/101)	

A1 Simple perianth with 6 petals; leaves with parallel veins; 2,3, or 6 stamens — True, see **B**1(1) False, see **A**2

B1(1) One or several flowers per stem — True, see **C**1(1) False, see **B**1(2)

C1(1) **Lady's slipper – Cypripédium** (pp. 102/103)
Tulip – Tulipa (pp. 102/103)
Iris or flag – Iris (pp. 102/103)
Daffodil – Narcissus (pp. 102/103)

B1(2) Cylindrical raceme with thickly bunched flowers at end of stem — True, see **C**1(2) False, back to **C**1(1) or see **A**2

C1(2) **Small white orchid – Leucórchis** (pp.102/103)
Scotch asphodel – Tofiéldia (pp.102/103)

A2 Plants without perianths; plants enclosed by bracts; leaves entire and fleshy; 1 stamen — True, see **B**2 False, see **A**3

B2 **Spurge – Euphórbia** (pp. 104/105)

A3 Simple perianth with 4 sepals and 5 or more petals — True, see **B**3(1)–**B**3(4)

B3(1) Flowers with 4 yellowish sepals and 8 stamens — True, see **C**3(1) False, see **B**3(2)

C3(1) **Golden saxifrage – Chrysosplénium** (pp. 104/105)

B3(2) Perianth of 5 petals; leaves entire; water plants — True, see **C**3(2) False, see **B**3(3)

C3(2) **Waterlily – Nuphar** (pp. 108/109)

B3(3) Perianth of 5–15 petals; leaves with 3–7 leaflets; flowers solitary on stem; land plants — True, see **C**3(3) False, see **B**3(4)

C3(3) **Globeflower – Tróllius** (pp.118/119)
Anemone – Pulsatílla (pp.124/125)
Winter aconite – Eránthis (pp.122/123)

B3(4) Perianth of 5 petals at most; leaves whole and notched or serrated; stem carries several flowers; land or marsh plants — True, see **C**3(4) False, back to **B**3(1) or see **A**4

C3(4) **Kingcup – Cáltha** (pp.114/115)

A4 Double perianth, divided into calyx and corolla; perianth with 4 petals — True, see **B**4(1) or **B**4(2) False, see **A**5

B4(1) Perianth composed of 2 sepals and 4 petals — True, see **C**4(1) or **C**4(2) False, see **B**4(2)

C4(1) Flowers zygomorphic; 4 stamens; leaves triply ternate — True, see **D**4(1) False, see **C**4(2)

D4(1) **Yellow fumitory – Corydalis** (pp.108/109)

C4(2) Flowers radial; numerous stamens; leaves pinnately cut or singly or doubly pinnate — True, see **C**4(3)–**C**4(5) False, back to **B**4(1)

D4(2) **Greater celandine – Chelidónium** (pp.106/107)
Alpine poppy – Papáver (pp.108/109)

B4(2) Radial perianth with 4 sepals and 4 petals — True, see **C**4(3)–**C**4(5) False, back to **B**4(1)

C4(3) 6 stamens; leaves entire, somewhat wavy in outline, toothed or pinnate — True, see **D**4(3) False, see **C**4(4)

D4(3) **Wood – Isatis** (pp.104/105)
Rape – Brássica (pp.104/105)
Bladderwort – Alyssoídes (pp.104/105)
Buckler mustard – Biscutélla (pp.106/107)
Whitlow grass – Drába (pp.104/105)
Bedstraw – Gálium (pp.108/109)
Yellow rocket – Barbárea (pp.106/107)
Charlock – Sinápis (pp.106/107)

C4(4) 8 stamens; leaves lanceolate, entire or slightly toothed — True, see **D**4(4) False, see **C**4(5)

D4(4) **Evening primrose – Oenóthera** (pp. 104/105)

C4(5)	Numerous stamens; leaves divided into 3 or 5 leaflets	True, see **D**4(5) False, back to **C**4(1) or see **A**5
D4(5)	**Cinquefoil – Potentílla** (pp.106/107)	
A5	Double perianth: 5 free-standing, radially arranged sepals and petals	True, see **B**5(1) or **B**5(2) False, see **A**6
B5(1)	Water plants	True, see **C**5(1) False, see **B**5(2)
C5(1)	**Fringed waterlily – Nymphoídes** (pp.108/109)	
B5(2)	Land plants	True, see **C**5(2)–**C**5(6) False, back to **B**5(1)
C5(2)	5 stamens; flowers not in umbels; leaves entire, wavy outline or finely notched, not fleshy and longer than 10 mm	True, see **D**5(1) False, see in **C**5(3)–**C**5(6)
D5(1)	**Yellow loosestrife – Lysimáchia** (pp.108–111) **Mullein – Verbáscum** (pp.110/111)	
C5(3)	5 stamens; flowers in umbels; leaves entire or pinnate	True, see **D**5(2) False, see **C**5(4)–**C**5(6)
D5(2)	**Hare's ear – Bupleúrum** (pp.120–123) **Celery – Apium** (pp.122/123) **Dill – Anéthum** (pp.122/123)	
C5(4)	10 stamens; leaves entire, may be fleshy or ternate	True, see **D**5(3) False, see **C**5(5) or **C**5(6)
D5(3)	**Saxifrage – Saxifraga** (pp.110/111) **Stonecrop – Sédum** (pp.112–115) **Wood sorrel – Oxalis** (pp.114/115)	
C5(5)	Numerous stamens; leaves entire and not fleshy	True, see **D**5(4) False, see **C**5(6)
D5(4)	**St John's wort – Hypéricum** (pp.110/111) **Rock-rose – Heliánthemum** (pp.112/113) **Buttercup – Ranúnculus** (pp.112/113)	
C5(6)	Numerous stamens; leaves lobed, with 3 or 5 leaflets or pinnate	True, see **D**5(5) False, back to **C**5(2) or see **A**6
D5(5)	**Buttercup – Ranúnculus** (pp.114–121) **Goose-grass – Agrimónia** (pp.120/121)	
A6	Perianths radial; outer and inner calyx each with 5 sepals; 5 petals; numerous stamens	True, see **B**6 False, see **A**7
B6	**Cinquefoil – Potentilla** (pp.114–121) **Avens – Géum** (pp.114/115)	
A7	Calyx with 2–6 teeth; 5 or 6 radial petals;stamens generally 6; leaves elliptical with main veins parallel	True, see **B**7 False, see **A**8
B7	**Gentian – Gentiána** (pp.122/123)	
A8	Calyx generally with 5 teeth; 5 or 6 petals set radially; 5 stamens; leaves pinnate; flowers in panicles	True, see **B**8 False, see **A**9
B8	**Tomato – Solánum** (pp.124/125)	
A9	6 sepals; 6 petals; numerous stamens; leaves divided into 3 or more parts	True, see **B**9 False, see **A**10
B9	**Mignonette – Réseda** (pp.124/125)	
A10	Perianth radial; 3–5 sepals; 8–10 petals;numerous stamens	True, see **B**10 False, see **A**11
B10	**Lesser celandine – Ficária** (pp.124/125)	
A11	Perianth radial; 5 sepals; 10–20 petals;numerous stamens	True, see **B**11 False, see **A**12

B11 **Pheasant's eye – Adónis** (pp.124/125)

A12 Perianth zygomorphic; 5 free-standing sepals; 5 free-standing petals; flowers solitary — True, see **B**12; False, see **A**13

B12 **Violet – Viola** (pp.124–127)

A13 Perianth of 5 fused sepals and 5 fused petals; flowers not composite — True, see **B**13(1) or **B**13(2); False, see **A**14

B13(1) Single flowers or flowers in umbels — True, see **C**13(1); False, see **B**13(2)

C13(1) **Primrose – Prímula** (pp,126/127)

B13(2) Flowers are racemes, panicles, or spikes, solitary or at end of stem in groups of 1–3 — True, see **C**13(2); False, back to **B**13(1) or see **A**14

C13(2) **Honeywort – Cerínthe** (pp.128/129)
Gromwell – Lithospérmum (pp.128/129)
Comfrey – Symphytum (pp.128/129)
Foxglove – Digitális (pp.128/129)
Bellflower – Campánula (pp.128/129)

A14 Composite flowers with ray-florets and tubular florets formed of 5 fused petals; sepals metamorphosed; flowers in clusters; 5 stamens — True, see **B**14(1)–**B**14(3); False, see **A**15

B14(1) Florets tubular only; composite flowers in panicles, umbel-shaped panicles or racemes — True, see **C**14(1); False, see **B**14(2)

C14(1) **Ragwort – Senécio** (pp.130/131)
Wormwood – Artemísia (pp.130/131)
Tansy – Chrysánthemum (pp.130/131)
Goldilocks – Aster (pp.130/131)

B14(2) Flowers with both ray- and tubular florets — True, see **C**14(2)–**C**14(5); False, see **B**14(3)

C14(2) Composite flowers solitary or 1–5 at end of stalk; leaves entire, toothed or more or less deeply dentate — True, see **D**14(1); False, see **C**14(3)

D14(1) **Mountain tobacco – Arnica** (pp.130–133)
Spikenard – Inula (pp.132/133)
Yellow ox-eye – Buphthálmum (pp.132/133)
Bur-marigold – Bidens (pp.132/133)
Coltsfoot – Tussilágo (pp.134/135)
Leopard's bane – Dorónicum (pp.134/135)
Ragwort – Senécio (pp.136/137)
Corn marigold – Chrysánthemum (pp.136/137)

C14(3) Composite flowers in loose umbel-shaped panicles or panicles pointing to one side; leaves entire or toothed — True, see **D**14(2); False, see **C**14(4)

D14(2) **Fleabane – Pulicária** (pp.132/133)
Golden rod – Solidágo (pp.132–135)
Ragwort – Senécio (pp.134/135)

C14(4) Solitary composite flowers at end of long stem; pinnate leaves — True, see **D**14(3); False, see **C**14(5)

D14(3) **Yellow chamomile – Anthemis** (pp.136/137)

C14(5) Composite flowers in umbel-shaped panicles; leaves heavily toothed or entire — True, see **D**14(4); False, back to **B**14(2) or see **A**14(3)

D14(4) **Ragwort – Senécio** (pp.136–139)

B14(3) Composite flowers of ray -florets only — True, see **C**14(6)–**C**14(10); False, back to **B**14(1) or see **A**15

C14(6) Composite flowers mostly solitary at end of stem; leaves entire — True, see **D**14(5); False, see **C**14(7)

D14(5) **Goat's beard – Tragópogon** (pp.138/139)
Hawkweed – Hierácium (pp.138/139)

C14(7) Composite flowers mostly solitary at end of stem; leaves entire, crenate, or toothed — True, see **D**14(6); False, see **C**14(8)

D14(6) **Hawk's beard – Crepis** (pp.140/141)
Hawkbit – Leóntodon (pp.140/141)
Cat's ear – Hypochoéris (pp.140/141)
Hawkweed – Hierácium (pp.140/141)

C14(8) Composite flower heads in pairs or larger numbers or in panicles or umbel-shaped panicles; leaves entire or toothed — True, see **D**14(7) False, see **C**14(9)

D14(7) **Ox-tongue – Picris** (pp.140/141)
Hawkweed – Hierácium (pp.142/143)
Hawk's beard – Crepis (pp.142/143)
Cat's ear – Hypochoéris (pp.142/143)

C14(9) Composite flowers solitary; leaves heavily toothed or dentate, divided to middle vein or pinnate — True, see **D**14(8) False, see **C**14(10)

D14(8) **Hawkbit – Leóntodon** (pp.144–147)
Dandelion – Taráxacum (pp.144/145)
Aposeris – Apóseris (pp.144/145)
Hawk's beard – Crepis (pp.146/147)

C14(10) Composite flowers in panicles — True, see **D**14(9) False, back to **C**14(6) or see **A**15

D14(9) **Hawk's beard – Crepis** (pp.144/145)
Sow thistle – Sonchus (pp.146/147)
Prickly lettuce – Lactúca (pp.146/147)

A15 Zygomorphic flowers — True, see **B**15(1)–**B**15(6) False, back to **A**1

B15(1) Top and bottom lip; 4 or 5 sepals and 5 petals; 4 stamens — True, see **C**15(1) False, see **B**15(2)

C15(1) **Toadflax – Linária** (pp.148/149)
Eyebright – Euphrásia (pp.152/153)
Cow-wheat – Melámpyrum (pp.148/149)
Yellow rattle – Rhinánthus (pp.150/151)
Monkey flower – Mímulus (pp.150/151)

B15(2) Perianth of 3 sepals and 5 petals; lateral petals fused in pairs, so that only 3 petals are apparent; 5 stamens — True, see **C**15(2) False, see **B**15(3)

C15(2) **Balsam – Impátiens** (pp.150/151)

B15(3) 3 sepals and 5 petals; 4 stamens; flowers in whorls one above the other — True, see **C**15(3) False, see **B**15(4)

C15(3) **Broom – Genísta** (pp.148/149)
Betony – Betónica (pp.148/149)
Woundwort – Stáchys (pp.150/151)
Iron woundwort – Sideritis (pp.148/149)
Dead nettle – Lámium (pp.150/151)

B15(4) Perianth of 5 petals, the topmost shaped like a helmet; numerous sepals — True, see **C**15(4) False, see **B**15(5)

C15(4) **Wolfsbane – Acónitum** (pp.152/153)

B15(5) Perianth formed of standard, 2 wings, and keel; 10 stamens — True, see **C**15(5)–**C**15(7) False, see **B**15(6)

C15(5) Leaves entire or divided into 2 parts with a tendril; flowers in racemes of 3 or more blooms — True, see **D**15(1) False, see **C**15(6)

D15(1) **Broom – Genísta** (pp.148/149)
Vetchling – Láthyrus (pp.154/155)

C15(6) Leaves divided into 3 or 5; flowers in panicles or umbels of 2 or more — True, see **D**15(2) False, see **C**15(7)

D15(2) **Winged pea – Tetragonóbolus** (pp.154/155)
Restharrow – Onónis (pp.154/155)
Black medick – Medicágo (pp.152/153)
Clover – Trifólium (pp.152/153)
Melilot – Melilótus (pp.152/153)
Birdsfoot trefoil – Lotus (pp.154/155)

C15(7) Leaves multiply and unevenly pinnate; flowers in capitula, umbels or racemes — True, see **D**15(3) False, back to **C**15(5) or see **B**15(6)

D15(3) **Kidney vetch – Anthyllis** (pp.152/154)
Crown vetch – Coronílla (pp.154–157)
Horseshoe vetch – Hippocrépis (pp.154/155)
Crazy weed – Oxytrópis (pp.156/157)
Milk vetch – Astrágalus (pp.156/157)

B15(6) Perianth of 5 sepals and 5 petals, formed into an upper and a lower lip; 4 stamens; flowers in racemes; leaves pinnate — True, see **C**15(8) False, back to **B**15(1)

C15(8) **Lousewort - Pediculáris** (pp.156/157)

Key Part 4: Pages 21, 31–34, 158–209 — **Red flowers**

A1	Single perianth with 6 petals; leaves entire, with parallel veins; land or water plants	True, see **B**1(1) or **B**1(2) False, see **A**1a
B1(1)	Radial, symmetrical petals; flowers solitary or 2–3 at end of stem, in racemes of 3–10 or in umbels	True, see **C**1(1) False, see **B**1(2)
C1(1)	**Fritillary – Fritillária** (pp.158/159) **Meadow saffron – Bulbocódium** (pp.158/159) **Autumn crocus – Cólchicum** (pp.158/159) **Martagon lily – Lílium** (pp.158/159) **Leek – Allium** (pp. 160/161) **Flowering rush – Bútomus** (pp.160/161)	
B1(2)	Flowers zygomorphic, 1 petal with a spur; flower in dense spikes	True, see **C**1(2) False, back to **B**1(1) or see **A**1a
C1(2)	**Spotted orchid – Dactylorhíza** (pp.158/159) **Black orchid – Nigritélla** (pp.158/159)	
A1a	Perianth 3 fused petals; leaves round to kidney-shaped with net-like veins	True, see **B**1a False, see **A**3
B1a	**Asarabacca – Asarum** (pp.160/161)	
A3	Single perianth of 4 dark red sepals; 2 or 3 bracteoles below each flower; flowers in racemes; leaves unequally pinnate	True, see **B**3 False, see **A**4
B3	**Burnet – Sanguisórba** (pp.164/165)	
A4	Perianth divided into calyx and corolla; perianth with 4 petals; leaves with net-like veins	True, see **B**4(1)–**B**4(3) False, see **A**5
B4(1)	Perianth of 2 sepals and 4 petals; sepals drop off when flower opens; leaves singly or doubly pinnate	True, see **C**4(1) False, see **B**4(2)
C4(1)	**Poppy – Papáver** (pp.162/163)	
B4(2)	Radial perianth with 4 sepals and 4 petals; land plants	True, see **C**4(2)–**C**4(4) False, see **B**4(3)
C4(2)	Flowers in umbels or racemes; leaves entire or toothed	True, see **D**4(1) False, see **C**4(3)
D4(1)	**Penny cress – Thlaspi** (pp.162/163) **Honesty – Lunária** (pp.162/163) **Willow-herb – Epilóbium** (pp.164/165)	
C4(3)	Flowers solitary or axillary at upper end of leaf stems; leaves entire and toothed	True, see **D**4(2) False, see **C**4(4)
D4(2)	**Willow-herb – Epilóbium** (pp.162–165)	
C4(4)	Flowers solitary or in racemes; leaves deeply cut, in three parts or unevenly pinnate	True, see **D**4(3) False, back to **C**4(2) or see **B**4(3)
D4(3)	**Rock beauty – Petrocállis** (pp.164/165) **Cuckoo flower – Cardámine** (pp.164/165)	
B4(3)	Flowers zygomorphic; 4 sepals and 4 petals; leaves ternate, each segment toothed and divided	True, see **C**4(5) False, back to **B**4(1) or see **A**5
C4(5)	**Fumitory – Corydális** (pp.166/167)	
A5	Double perianth; 5 sepals and 5 petals, radially arranged; flowers not composite	True, see **B**5(1)–**B**5(10) False, see **A**6
B5(1)	Leaves elliptical or oval, entire, on stalks and arranged in a rosette, 10 to 35 mm long; flowers in false umbels; 10 stamens	True, see **C**5(1) False, see **B**5(2)
C5(1)	**Shy stonecrop – Sedum** (pp.166/167)	
B5(2)	Leaves oval, generally entire, up to 25 mm long and only basal; stem with 1 flower or up to 3 in an umbel; 5 stamens	True, see **C**5(2) False, see **B**5(4)
C5(2)	**Primrose – Prímula** (pp.168/169)	
B5(3)	Leaves oval, entire, opposite, over 35 mm long and sessile; flowers in loose panicles or false umbels; 10 stamens	True, see **C**5(3) False, see **B**5(4)

C5(3) **Soapwort – Saponária** (pp.166/167)
Cow-cockle – Vaccária (pp.166/167)

B5(4) Leaves generally oval, entire, opposite, over 35 mm long, on stalks and hairy; 10 stamens — True, see **C**5(4); False, see **B**5(5)

C5(4) **Red campion – Siléne** (pp.168/169)
Catchfly – Lychnis (pp.168–172)

B5(5) Leaves lanceolate or linear, entire, opposite, smooth and over 35 mm long; 10 stamens — True, see **C**5(5); False, see **B**5(6)

C5(5) **Pink – Diánthus** (pp.168–173)
Petrorhagia – Petrorhágia (pp.168–172)
Catchfly – Viscária (pp.170/171)
Corn cockle – Agrostémma (pp.170/171)

B5(6) Leaves narrow lanceolate, entire, smooth, opposite and over 35 mm long — True, see **C**5(6); False, see **B**5(7)

C5(6) **Thrift – Arméria** (pp.170/171)

B5(7) Leaves narrow oval to oval, entire, smooth or rough-haired, over 35 mm long, alternate, and with net-like veins; 5 stamens — True, see **C**5(7); False, see **B**5(8)

C5(7) **Tobacco flower – Nicotiána** (pp.182/183)
Comfrey – Symphytum (pp.182/183)
Lungwort – Pulmonária (pp.184/185)
Scopolia – Scopólia (pp.184/185)

B5(8) Leaves narrow oval to oval, entire, smooth, opposite, over 35 mm long, with main veins parallel; 5 stamens — True, see **C**5(8); False, see **B**5(9)

C5(8) **Gentian - Gentiána** (pp.182/183)

B5(9) Leaves lanceolate to narrow oval, 2–12 mm long, entire, with bristles at edges, in a rosette; 10 stamens; solitary flowers on long stem — True, see **C**5(9); False, see **B**5(10)

C5(9) **Moss campion – Silene** (pp172/173)
Saxifrage – Saxifraga (pp.174/175)

B5(10) Leaves whole, finely to coarsely toothed — True, see **C**5(10)–**C**5(22); False, back to **B**5(1) or see **A**3a

C5(10) Leaves 1–2 cm long, in a rosette and alternately on the stem, narrow oval to spathulate, with short hairs and toothed edges; 4 stamens — True, see **D**5(1); False, see **C**5(11)

D5(1) **Alpine balsam – Erínus** (pp.174/175)

C5(11) Leaves over 2 cm long, alternate, with deeply, irregularly toothed or lobed; flowers solitary or several together on leaf stems; numerous stamens — True, see **D**5(2); False, see **C**5(12)

D5(2) **Marsh mallow – Althaéa** (pp.174/175)
Mallow – Malva (pp.174/175)

B5(12) Leaves over 2 cm long, toothed (particularly in upper parts), with wide stems and in a rosette; flowers on stalks without leaves, in a rosette; 5 stamens — True, see **D**5(3); False, see **C**5(13)

D5(3) **Primrose – Prímula** (pp.172/173)

C5(13) Leaves over 2 cm long, round or kidney-shaped, with regular fine serrations and white markings on the upper side; solitary flowers; 5 stamens — True, see **D**5(4); False, see **C**5(15)

D5(4) **Cyclamen – Cyclamen** (pp.172/173)

C5(14) Leaves over 2 cm long and with 3,5,7, or 9 coarsely toothed or incised lobes; solitary flowers or 2 to a stem, often forming a bunch; 10 stamens — True, see **D**5(5); False, see **C**5(15)

D5(5) **Cranesbill – Geranium** (pp.174–177)

C5(15) Leaves over 2 cm long and double or trebly pinnate; 5 stamens; flowers in umbels — True, see **D**5(6); False, see **C**5(16)

D5(6) Storksbill – Erodium (pp.176/177)

C5(16)	Stem leaves over 2 cm long, with 5–7 deep, rounded lobes; flowers solitary in upper leaf axils; numerous stamens	True, see **D5(6)** False, see **C5(17)**

D5(6) Mallow – Malva (pp.176/177)

C5(17)	Leaves 3–8 cm long, pinnate, with toothed leaflets; numerous stamens	True, see **D5(7)** False, see **C5(18)**

D5(7) Avens – Geum (pp.178/179)

C5(18)	Leaves 1–4 cm long, narrow oval, entire, opposite, with lanceolate stipules; flowers in bunches; numerous stamens	True, see **D5(8)** False, see **C5(19)**

D5(8) Rock-rose – Heliánthemum (pp.168/169)

C5(19)	Leaves more than 2 cm long, lanceolate, entire, opposite, in threes or alternate; flowers in long spikes; 5–6 petals; 12 stamens	True, see **D5(9)** False, see **C5(20)**

D5(9) Purple loosestrife – Lythrum (pp.168/169)

C5(20)	Leaves near the ground, ternate, each leaflet often having 2–5 lobes; solitary flowers; numerous stamens	True, see **D5(10)** False, see **C5(21)**

D5(10) Glacier crowfoot – Ranúnculus (pp.178/179)

C5(21)	Leaves with 5–7 toothed leaflets; umbel-like flower heads; numerous stamens	True, see **D5(11)** False, see **C5(22)**

D5(11) Marsh cinquefoil – Comárum (pp.178/179)

C5(22)	Pinnate leaves with toothed leaflets; flowers in umbels; 5 stamens	True, see **D5(12)** False, back to **C5(10)** or see **A3a**

D5(12) Burnet saxifrage – Pimpinélla (pp.178/179)
Angelica – Angélica (pp.180/181)
Lovage – Ligústicum (pp.180/181)

D5(12) Turnip-rooted chervil – Chaeróphyllum (pp.180/181)

A3a	Leaves at ground level, in 5, 7, or 9 segments; 5 petals; flowers solitary or in twin-flowered umbels; numerous stamens	True, see **B3a** False, see **A3b**

B3a Hellebore – Helleborus (pp.178/179)

A3b	Leaves lanceolate, 3–20 cm long, entire; 4–5 petals; 4–8 stamens; flowers in cylindrical spikes 3-8 cm long; land or water plants	True, see **B3b** False, see **A12**

B3b Knotweed – Polygonum (pp.180–183)

A12	Flowers zygomorphic; 5 sepals, 3 outer ones brownish-red; 2 sepals form violet wings; corolla has fringed points; 8 stamens; flowers in racemes	True, see **B12** False, see **A13**

B12 Milkwort – Polygala (pp.182/183)

A13	4 or 5 sepals, forming bristles on ripe fruit; 4 or 5 fused petals; flowers in umbrella-shaped panicles; 3 or 4 stamens; leaves entire, toothed, lobed, or unevenly pinnate	True, see **B13** False, see **A14**

B13 Valerian – Valeriána (pp.184/185)

A14	Composite flowers consisting of ray- and/or tubular florets concisting of 5 fused petals; 5 stamens; sepals metamorphosed	True, see **B14(1)–B14(3)** False, see **A15**
B14(1)	All florets in composite flower head tubular or thread-like	True, see **C14(1)** False, see **B14(2)**

C14(1) Alpine coltsfoot – Homógyne (pp.186/187)
Cotton thistle – Onopórdum (pp.188/189)
Giant knapweed – Rhapónticum (pp.188/189)
Scotch thistle – Cárduus (pp.188/189)
Knapweed – Centaúrea (pp.190/191)

C14(1) **Thistle – Cirsium** (pp.188–191)
Butterbur – Petasítes (pp.186/187)
Wild lettuce – Adenostyles (pp.186/187)
Cat's foot – Antennaria (pp.186/187)
Hemp agrimony – Eupatórium (pp.186/187)
Saw-wort – Serrátula (pp.190/191)
Burdock – Arctium (pp.188/189)
False saw-wort – Crupina (pp.192/193)

B14(2) Ray- and tubular florets in composite flower — True, see **C**14(2) False, see **B**14(3)

C14(2) **Aster – Aster** (pp.192/193)
Fleabane – Erígeron (pp.192/193)
Yarrow – Achilléa (pp.192/193)

B14(3) All florets ray-florets — True, see **C**14(3) False, see **B**14(1) or see **A**15

C14(3) **Purple lettuce – Prenánthes** (pp.192/193)
Hawkweed – Hierácium (pp.194/195)
Hawk's beard – Crepis (pp.194/195)

A15 Flowers zygomorphic, sepals and 5 petals — True, see of 5 **B**15(1)–**B**15(6) False, see **A**16

B15(1) Flowers consist of keel, standard, and 2 wings — True, see **C**15(1) False, see **B**15(2)

C15(1) **All papilionaceous flowers** (pp.194–199)

B15(2) Flowers with top and bottom lips; 4 stamens enclosed by upper lip; leaves pinnately cut as far as middle vein — True, see **C**15(2) False, see **B**15(3)

C15(2) **Lousewort – Pediculáris** (pp.198–201)

B15(3) Flowers with top and bottom lips; 4 stamens; leaves entire, toothed or divided into 3, 5, or 7 lobes (pp.200–207) — True, see **C**15(3) False, see **B**15(4)

C15(3) **Plants of the mint and figwort families (Labiatae and Scrophulariaceae)**: thyme, savory, alpine calamint, marjoram, basil, bastard balm, bartsia, water-mint, dead-nettle, betony, self-heal, hemp-nettle, skullcap, germander, motherwort

B15(4) Upper and lower lips; lateral petals fused in pairs; with spur; leaves lanceolate, oval, entire, or toothed; flowers in racemes; 5 stamens — True, see **C**15(4) False, see **B**15(5)

C15(4) **Touch-me-not – Impátiens** (pp.206/207)

B15(5) Top and bottom lip; corolla with long spur; 4 stamens; leaves lanceolate, or with 3–5 leaflets in a whorl, or with 5–7 teeth; flowers in loose racemes or solitary on leaf stems — True, see **C**15(5) False, see **B**15(6)

C15(5) **Toadflax – Linária** (pp.206/207)
Ivy-leaved toadflax – Cymbalária (pp.208/209)

B15(6) Corolla forms tube with long spur; 1 stamen; point shaped panicles; leaves lanceolate, sessile, and opposite — True, see **C**15(6) False, back to **B**15(1) or see **A**16

C15(6) **Red valerian – Centránthus** (pp.208/209)

A16 5 sepals; 5–10 petals; numerous stamens fused at the base into a fleshy nectar ring;leaves doubly ternate — True, see **B**16 False, see **A**17

B16 **Peony – Paeónia** (pp.208/209)

A17 Perianth deep violet, with 5 petaloid sepals and 5 petals, each with a long hollow nectar-secreting spur pointing backwards; numerous stamens; ternate leaves — True, see **B**17 False, see **A**18

B17 **Columbine – Aquilégia** (pp.206/207)

A18 Either 4–5 petals, inconspicuous and falling off at flowering time, or 6 well-formed ones; numerous stamens; leaves pinnate; flowers in thick panicles or solitary — True, see **B**18 False, see **A**19

B18 **Meadow rue – Thalíctrum** (pp.178/179)
Mountain anemone – Pulsatílla (pp.208/209)

A19 Sepals and petals 12–16 each — True, see **B**19 False, see **A**1

B19 **Houseleek – Sempervivum** (pp.166/167)

A1 Simple perianth of 6 petals; leaves with parallel veins; flowers solitary, or several at end of stem, or in racemes — True, see **B**1; False, see **A**3 (there is no **A**2 in the key to this part)

B1 **Squill – Scilla** (pp.210/211)
Spanish hyacinth – Scilla (pp.210/211)
Tassel hyacinth – Muscári (pp.210/211)
Iris – Iris (pp.210/211)
Commelina – Commelína (pp.210/211)

A3 Perianth of 5–10 petals — True, see **B**3; False, see **A**4

B3 **Fennel flower – Nigélla** (pp.214/215)
Liverwort – Hepática (pp.220/211)

A4 Perianth double, with 4 or 5 sepals and 3 or 4 petals — True, see **B**4(1) or **B**4(2); False, see **A**5

B4(1) Flowers of 3 small blue and 2 large, wing-shaped sepals and 3 fused blue petals; leaves entire and lanceolate — True, see **C**4(1); False, see **B**4(2)

C4(1) **Milkwort – Polygala** (pp.212/213)

B4(2) Flowers of 4 or 5 sepals and 4 free-standing petals; 2 stamens — True, see **C**4(2); False, back to **B**4(1) or see **A**5

C4(2) **Speedwell - Verónica** (pp.212–215)

A5 Perianth double; 5 each of free-standing radial sepals and petals; flowers form bunches — True, see **B**5; False, see **A**12

B5 **Flax – Línum** (pp.214/215)

A12 Perianth zygomorphic; 5 free-standing sepals; 5 free-standing petals; flowers solitary — True, see **B**12; False, see **A**13

B12 **Violets – Víola** (pp.220–223)

A13 Radial, fused petals; sepals form bristles, there are only 4 stamens — True, see **B**13(1)–**B**13(10); False, see **A**14 (see p. 36)

B13(1) Perianth of bristles formed from sepals and 4 petals fused into a tube; 4 stamens; flowers in capitula — True, see **C**13(1); False, see **B**13(2)

C13(1) **Wood scabious – Knaútia** (pp.216/217)

B13(2) Perianth of an outer calyx, sepals in form of bristles, and 5 petals fused into a tube; flowers in capitula; 4 stamens — True, see **C**13(2); False, see **B**13(3)

C13(2) **Scabious – Scabiósa** (pp.216/217)

B13(3) Five lanceolate sepals and 5 petals initially fused but later divided into ribbon-shaped points — True, see **C**13(3); False, see **B**13(4)

C13(3) **Sheep's-bit – Jasíone** (pp.216/217)

B13(4) Five sepals; 5 fused petals; ends of stamens much longer than corolla tube and spread flat — True, see **C**13(4); False, see **B**13(5)

C13(4) **Periwinkle – Vinca** (pp.218/219)
Forget-me-not – Myosótis (pp.218/219)
Felwort – Swértia (pp.218/219)
Borage – Borágo (pp.220/221)
Venus's looking-glass – Legoúsia (pp.220/221)

B13(5) Five sepals and 5 petals; flowers in cylindrical or spherical composite heads; 5 stamens — True, see **C**13(5); False, see **B**13(6)

C13(5) **Sea holly – Eryngium** (pp.220/221)

B13(6) Fused sepals and petals, 5 of each; tube of corolla much longer than ends of stamens; leaves entire, toothed, or pinnate — True, see **C**13(6); False, see **B**13(7)

C13(6) **Gentian – Gentiána** (pp.222–225)
Bellflower – Campánula (pp.224–227)
Bugloss – Anchúsa (pp.226/227)
Stoneweed – Buglossoídes (pp.226/227)
Viper's bugloss – Echium (pp.228/229)
Vervain – Verbéna (pp.228/229)

B13(7) Fused sepals and petals, 5 of each; flowers in spherical heads; leaves entire — True, see **C**13(7); False, see **B**13(8)

C13(7) **Globe daisy – Globulária** (pp.232/233)

B13(8) Fused sepals and petals, 5 of each; petals have ragged edges; 2–3 flowers at end of each stem — True, see **C**13(8) False, see **B**13(9)

C13(8) **Snowbell – Soldanélla** (pp.228/229)

B13(9) Fused sepals and petals, 5 of each; flowers in panicles 10–20 cm long; leaves unequally pinnate — True, see **C**13(9) False, see **B**13(10)

C13(9) **Jacob's ladder – Polemónium** (pp.228/229)

B13(10) Fused sepals and petals, 5 of each; flowers form heads or cylindrical spikes with many flowers; leaves entire and finely toothed — True, see **C**13(10) False, back to **B**13(10) or see **A**14

C13(10) **Horned rampion – Phyteúma** (pp.228/229)

A14 5 fused petals; calyx metamorphosed to hairs or bristles; composite flowers or capitula; 5stamens — True, see **B**14(1)–**B**14(3) False, see **A**15

B14(1) Tubular flowers only in composite flower; leaves entire — True, see **C**14(1) False, see **B**14(2)

C14(1) **Cornflower – Centaúrea** (pp.230/231)

B14(2) Tubular and ray-florets present; leaves entire or somewhat toothed; flowers solitary or in loose racemes — True, see **C**14(2) False, see **B**14(3)

C14(2) **Aster – Aster** (pp.230/231)

B14(3) Only radial florets in composite flower; leaves pinnate or cut almost to the middle vein; composite flowers form spikes or racemes — True, see **C**14(3) False, back to **B**14(1) or see **A**15

C14(3) **Chicory – Cichórium** (pp.230/231)
Alpine sow-thistle – Cicérbita (pp.230/231)

A15 Zygomorphic flowers — True, see **B**15(1)–**B**15(6) False, back to **A**1

B15(1) Perianth divided into top and bottom lip, bottom distinctly longer than top and with a white mark; leaves 1–2 cm long, entire and hairy — True, see **C**15(1) False, see **B**15(2)

C15(1) **Winter savory – Saturéja** (pp.232/233)

B15(2) Perianth divided into top and bottom lip, bottom longer than top lip and with no white mark; leaves 2–8 cm long — True, see **C**15(2) False, see **B**15(3)

C15(2) **Ground ivy – Glechóma** (pp.234/235)
Bugle – Ajuga (pp.234/235)

B15(3) Top and bottom lip about equal; leaves entire; flower heads form spikes or racemes — True, see **C**15(3) False, see **B**15(4)

C15(3) **Sage – Salvia** (pp.232/233)
Dragon mouth – Hormínum (pp.232/233)

B15(4) Perianth of standard, 2 wings and keel combined; 10 stamens; leaves ternate — True, see **C**15(4) False, see **B**15(5)

C15(4) **Lucerne – Medicágo** (pp.234/235)

B15(5) Perianth of standard, 2 wings and keel combined; 10 stamens; leaves unevenly pinnate — True, see **C**15(5) False, see **B**15(6)

C15(5) **Alpine milk-vetch – Astrágalus** (pp.234/235)
Vetch – Vícia (pp. 236/237)

B15(6) Perianth of 5 petals; numerous stamens; leaves in 3–7 parts at middle of plant or in 5 parts at ground level — True, see **C**15(6) False, see **B**15(1) or **A**1

C15(6) **Larkspur – Delphínium** (pp.234/235)
Monkshood – Aconítum (pp.234/235)

A1 Single perianth with 3–6 petals and linear leaves with parallel veins; water plants — True, see **B1**; False, see **A2**

B1 **Bur reed – Spargánium** (pp.236/237)

A2 Plants without perianths; male and female flowers lie one above the other on a spadix enclosed by a sheath — True, see **B2**; False, see **A3**

B2 **Cuckoo-pint – Arum** (pp.238/239)
Spurge – Euphórbia (pp.236–239)

A3 Simple perianth with either sepals or petals; leaves with branching veins — True, see **B3(1)–B3(6)**; False, see **A4**

B3(1) Water plants with a perianth in 2–4 parts round rim of ovary — True, see **C3(1)**; False, see **B3(2)**

C3(1) **Mare's tail – Hippúris** (pp.248/249)

B3(2) Flowers with 3 free-standing sepal points, 3 green staminodes and in male plants up to 20 stamens; leaves long, oval, and toothed — True, see **C3(2)**; False, see **B3(3)**

C3(2) **Dog's mercury – Mercuriális** (pp.238/239)

B3(3) Perianth of 4 free-standing petals or petals partly fused at base; flowers in panicles; leaves unequally pinnate — True, see **C3(3)**; False, see **B3(4)**

C3(3) **Burnet – Sanguisórba** (pp.244/245)

B3(4) Perianth of 4 free-standing petals; flowers in panicles; leaves crenate or toothed — True, see **C3(4)**; False, see **B3(5)**

C3(4) **Mountain sorrel – Oxyria** (pp.248/249)
Stinging nettle – Urtica (pp.244/245)

B3(5) Perianth of 3–5 free-standing petals; flowers solitary, in racemes, false panicles or panicles; green and brown plants — True, see **C3(5)**; False, see **B3(6)**

C3(5) **Goose-foot – Chenopódium** (pp.240/241)
Hellebore – Hellebórus (pp.244/245)
Bird's-nest orchid – Neóttia (pp.248/249)

B3(6) Perianth of 6 petals; flowers in panicles; leaves with net-like veins or parallel main veins — True, see **C3(6)**; False, back to **B3(1)** or see **A4**

C3(6) **False hellebore – Verátrum** (pp.236/237)
Sorrel – Rumex (pp.246 - 249)

A4 Perianth double: 4 inner and 4 outer petals, 4 inner and 4 outer sepals, or 4 each of sepals and petals — True, see **B4(1)** or **B4(2)**; False, see **A5**

B4(1) Perianth of 4 inner and 4 outer petals or 4 inner and 4 outer sepals; leaves entire or with 7–11 lobes — True, see **C4(1)**; False, see **B4(2)**

C4(1) **Herb Paris – Paris** (pp. 238/239)
Lady's mantle – Alchemílla (pp.240–243)

B4(2) Perianth of 4 sepals fused at the base and 4 fused petals; flowers in long spikes; leaves have parallel main veins — True, see **C4(2)**; False, back to **B4(1)** or see **A5**

C4(2) **Plantain – Plantágo** (pp..242/243)

A5 Perianth double: 5 sepals and 5 petals; petals free-standing; leaves narrow, lanceolate — True, see **B5**; False, see **A13**

B5 **Hare's ear – Bupleúrum** (pp.246/247)

A13 Perianth of 5 fused sepals and petals — True, see **B13(1)** or **B13(2)**; False, see **A14**

B13(1) Flowers in umbels or panicles; leaves lobed — True, see **C13(1)**; False, see **B13(2)**

C13(1) **Bryony – Bryónia** (pp.244/245)

C13(1) Flowers composite; leaves with prickles and ternately cut — True, see **C13(2)**; False, see **A14**

C13(2) **Sea holly – Eryngium** (pp.246/247)

A14 Flowers composite, in panicles; tubular flowers only present; leaves pinnately cut and without prickles — True, see **B14**; False, back to **B1**

B14 **Pineapple weed – Matricária** (pp.246/247)
Mugwort - Artemísia (pp.246/247)

130 **Moonwort – B. lunária**
Plants 5–30 cm high, often variable in form and with worldwide distribution

Botrychium lunária (L.)Sw.
Moonwort

Ophioglossáceae – Adder's tongue family

Leaves: Carry no spores, and branch off between a third and two-thirds of the height of the stem; 3–12 cm long, up to 4 cm wide, mostly short-stalked, narrow oval in outline, rounded at tip, glossy, smooth, yellow-green and with up to 9 pinnae on each side; these are mostly overlapping, semicircular in outline, entire or with blunt teeth, and rarely having second-order pinnae

Sporangiophore: Generally long-stemmed, with 1–3 branches, mostly narrow oval in outline, standing above the part of the plant not carrying spores and panicular in form, with many branches

Sporangia: Brown to reddish-brown in colour; spores ripe in summer

Habitat: Scattered over hill to alpine zones in rough grass, rough pasture, upland meadows, waste ground, by the roadside and in thickets, on moderately damp to dry, mostly lime-poor or lime-free, generally acid and sandy or loamy soil; plants need light; often only found in isolation or in small groups

131 **Royal fern – O. regális**
Plants 60 cm to 2 m high, with dark brown, short, thick rhizome and worldwide distribution

Osmúnda regalis L.
Royal fern

Osmundáceae – Royal fern family

Leaves: Oval in outline, 30 to 160 cm long, up to 40 cm wide, yellow-green to dark green and doubly pinnate; first-order pinnae are stalked up to the 7th or 9th, pointing forward, upper ones often partly overlapping, up to 30 cm long, up to 15 cm wide and narrow oval to lanceolate; second-order pinnae are only short-stalked, rounded at tip and entire or finely toothed

Sporangiophore: Sporangia in highly branched panicles

Sporangia: Spherical, up to 0.5 mm in diameter; short-stemmed and only half opening when ripe; spores mostly similar; ripe in summer

Habitat: In hill zone in alder woods, meadows, open woods, undrained hollows, in ditches and bogs, by reservoir discharges, on well aired, lime-poor, acid and peaty sand – and clay beds in humid conditions where winters are mild; also south-facing alpine positions in dry scrub; plants found in half-shade; protected species; fern known from ancient times; also used as ornamental plant

132 **Hart's tongue fern – P. scolopéndrium**
Plants 30–50 cm high, the rhizome having hard, narrow scales; many varieties

Phyllitis scolopéndrium (L.) Newman
Hart's tongue fern

Aspleniaceae – Spleenwort family

Leaves: Arranged in tufts, ascending and often bending over, with dark brown stalks and mostly having narrow scales, 15–55 cm long, thin and lanceolate, not divided, entire, heart-shaped at base; foliage remains throughout the winter

Sporangiophore: Leaves (underside)

Sori: Forming multiple strips over the whole width between central vein and edge of leaf, standing out and pointing forwards; covering the sporangia, colourless and entire at first, later brownish and bent back; spores ripe in summer

Habitat: Scattered from hill to alpine zones in woods and ravines, by wells, damp and shady rubbish dumps containing lime, in rock crevices, on shady walls and rocks on wet, mostly calcareous, very acid, rocky and stony soil containing humus, in conditions with a damp atmosphere and a mild winter climate; plants found in shade; easy coloniser of rock crevices; ornamental plants

133 **Hard fern – B. spicant**
Plants 20–70 cm high, with ascending or upright rhizome; foliage spreading out

Bléchnum spícant (L.) Roth
Hard fern

Blechnáceae – Hard fern family

Leaves: Not spore-bearing, lying on the ground or forming a slightly rising rosette; lanceolate, up to 60 cm long, 4–8 cm wide, leathery, dark green, smooth and generally simply pinnate; generally with 30 to 50 pinnae somewhat forward-pointing, entire, round to slightly pointed in the form of a comb; spore-bearing leaves appearing first after the former are up to 70 cm high, standing upright, with a long stem and very narrow pinnae

Sporangiophore: On the underside of rigidly upright leaves rising from the middle of the rosette of non-spore-bearing leaves

Sori: When ripe, covering the whole underside of the narrow pinnae; covering bent back when the spores are ripe in summer

Habitat: In mountain and subalpine (rarely hill) zones in pine woods and in oak woods with few other species, in meadows on the remains of former woods and in sparse woodlands, on moist to damp soil poor in nutrients, base-poor, slightly acidic, and containing humus

134 **Rusty-back fern – C. officinárum** Plant 4–10 cm high, with short upright rhizome, living several years; genus' only European representative.

Céterach officinárum DC.
Asplénium céterach L.
Rusty-back fern.

Aspleniáceae – Spleenwort family

Leaves: Forming thick bundles, narrowly lanceolate in outline, 4–10 cm long, simply pinnate or only pinnately cut, fleshy, grey-green on top and underneath clearly covered with scales arranged like roof tiles; pinnae entire or wavy in outline, arranged alternately and folding upwards in drought; stem up to 3 cm long, brown and thickly covered in scales

Sporangiophore: Pinnae (all the underside)

Sori: Linear in form, at first covered by the scales and only visible when ripe; there is no covering; spores ripe in summer and autumn

Habitat: In hill (rarely mountain) zones on walls and rocks in hot and dry situations with mild winter conditions, on lime and lime-free rock; as southern European plants the main range is the whole Mediterranean area, but it extends northwards to England; on the north side of the Alps it is scattered in regions where the föhn wind blows; commoner in warmer areas on the south side of the Alps

135 **Sensitive fern – O. sensibilis L.**
Plants up to 30 cm in height with a long creeping rhizome; sterile and fertile leaves

Onoclea sensibilis L.
Sensitive fern

Athyriáceae – Spleenwort family

Leaves: Springing individually from rhizome and definitely dimorphous: sterile leaves up to 30 cm long, oval or broad oval in shape, deeply pinnately cut or pinnate at the base of leaf, with spindle widening towards the tip and a pinnately cut point; pinnae lobed, sharply defined and the larger ones more or less lobate; leaf veins form a network (no independent veins present); fertile leaves are more narrowly oval, green in colour when young, later segments brownish and singly or doubly pinnate; segments are lanceolate and rolled up like a pearl round the sori, appearing brown to black when the spores are ripe

Sporangiophore: Fertile leaves (underside of pinnae)

Sori: Laid out asymmetrically and mostly spread over the whole surface of the pinnae; spores ripe in autumn, but released only in following spring

Habitat: Often planted as winter-hardy plants in parks and gardens in hill zone; naturalised in England and Central Europe

136 **Polypody – P. vulgáre**
Perennial plants 10–40 cm high, with rhizome 40 cm long

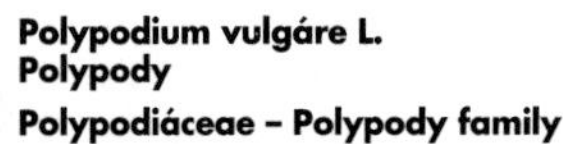

Polypodium vulgáre L.
Polypody

Polypodiáceae – Polypody family

Leaves: 10–40 cm long, up to 8 cm wide, lanceolate to oval, pinnately cut or pinnate, with long stems; pinnae alternate, 2–5 times as long as wide, lanceolate, round at tip, entire or finely toothed, and mid-green in colour; lateral veins of lowest leaf segments/pinnae forked
1–3 times; in sheltered places lasting through the winter and only dying down in spring; growing again in early summer

Sporangiophore: Underneath the leaves

Sori: Unripe and ripe sori round and distributed in great numbers in two rows on the underside; spores ripe in summer and autumn

Habitat: In hill to subalpine (rarely alpine) zones in mixed deciduous woods, by shady rocks and tree-trunks, by the mossy bases of old trees, on moderately dry, lime-poor, soil containing rotted matter mostly in low-lying, acidic soils in humid locations enjoying mild winters; plants found in half-shadow; earlier used as medicinal plants; remnant of Tertiary Era

137 **Saw-toothed polypody**
Perennial plants 10–50 cm high, with rhizome 40 cm long

Polypokium austrále Fée
P. serrátum (Willd.) Sauter
P. cambricum
Saw-toothed polypody

Polypodiáceae – Polypody family

Leaves: 10–50 cm long, triangular to oval in outline and pinnate or pinnately cut; pinnae/segments narrow and lanceolate, gradually drawing to a point, with clearly visible fine teeth; light green, particularly on the underside; lateral veins on pinnae/segments 3–6 forked 3–6 times; dying down in early summer and springing up again in late summer or autumn

Sporangiophore: Underneath the leaves

Sori: Unripe and ripe sori elliptical and long, distributed in great numbers in two rows on the underside; spores ripe in summer and autumn

Habitat: In hill zone, particularly in the Mediterranean region, in crevices in rocks and walls, shady woods and ravines; north of the Alps in districts with mild winters, e.g. South Ticino, the Rhone valley below Geneva, around Lake Geneva and Lake Neuchatel

138 **Ostrich fern – M. struthiópteris**
Plants up to 170 cm high, with rhizome 8 cm thick and sending out long subsoil runners

Matteúcia struthiópteris (L.)
Ostrich fern
Athyriáceae – Spleenwort family

Leaves: Non-spore-bearing leaves up to 170 cm long, 10–30 cm wide, sharply pointed at tip, narrowing gradually towards the ground, with short stalk, with 30–50 pairs of pinnae, dying down in winter and forming a narrow funnel; pinnae towards the end of the leaf pointing forward, in the middle pointing out horizontally, and in the lower part pointing downwards, pinnately cut and sessile; segments 3–10 mm long, slightly pointed or rounded at the end, entire or somewhat wavy and downy grey on the underside

Sporangiophore: Leaves up to 35 cm long and 5–15 cm wide, linear to lanceolate, drawing to a blunt point, with cylindrically rolled-up segments, rigidly upright, simply pinnate; lasting through winter; pinnae up to 4 cm long

Sori: Arranged in dense groups inwards from the edges of the pinnae, which are closed till ripe; spores ripe in late summer and autumn

Habitat: Forest clearings, by brooks and rivers and in hollows by springs in the mountains on moist, generally lime-free, humus-containing, and sandy or clay soils

139 **Marsh fern – T. palústris**
Plants up to 70 cm high, with elongated and spreading rhizomes

Thelypteris palústris Schott
Marsh fern

Thelypteridáceae – Beech fern family

Leaves: 20–70 cm long, up to 25 cm wide, hardly narrowing at base or not at all (but first pair of pinnae is often much smaller!), rigidly upright and simply (rarely doubly) pinnate; first- order pinnae when young are hairy on the underside and glandular, yellow-green, entire, wavy, or rather finely serrated and generally blunt or barely pointed; stalk thin, bare, and brownish

Sporangiophore: Pinnae with the sori clearly rolled up underneath, generally triangular in appearance

Sori: Generally arranged in 2 rows between the central vein and the rolled edge of the pinnae and remaining; covering falling away long before the spores are ripe in summer and autumn.

Habitat: In hill and mountain (rarely subalpine) zones in alder woods, marshes, moors, undrained hollows, in willow scrub, on boggy soil rich or poor in nutrients which is moderately acid, peaty clay, or broken clay with rotted matter; plants found in half shade

140 **Male fern – D. filix-mas**
Plants up to 100/120 cm high, with thick, erect rhizome and leaves dying off in autumn

Dryópteris filix-mas (L.) Schott
Male fern

Aspidiáceae – Spleenwort family

Leaves: 30–110 cm high, a narrow oval in outline, clearly narrowing towards the ground, gradually coming to a point at tip and simply pinnate; pinnae in 20–40 pairs and except for topmost wholly or nearly divided at central vein; segments nearly rectangular, rounded or blunt at the end, finely and sharply toothed or notched

Sporangiophore: Leaves (underside of pinnae)

Sori: Generally formed only on lower two-thirds of leaf; covering grey at first and only later brown, shrinking well before spores are ripe and later falling away; spores ripe between June and September

Habitat: Hill and subalpine zones in deciduous or coniferous forests, among bushes, in mountain pasture, in subalpine scrub and scree on temperate, lime-rich and lime-poor or loose loamy soils rich in nutrients and humus; plants grow in shade and on waste ground

141 **Shield fern – D. affinis**
Plants up to 150 cm high, with thick rhizome and leaves generally lasting through winter.

Dryópteris affinis Fraser-Jenkins
D. borreri Newm.
Shield fern

Aspidiáceae – Spleenwort family

Leaves: 40–150 cm long, narrow to oval in outline, narrowing toward base, generally coming to an obtuse point at tip, coarsely and simply pinnate; pinnae in 20–40 pairs and except for topmost clearly divided at middle vein; segments nearly rectangular, rounded or blunt at the end, at base mostly attached to each other and less sharply toothed than the male fern; base of the spindle dark violet when fresh

Sporangiophore: Leaves (underside of pinnae)

Sori: Generally formed in lower two-thirds of leaf; covering coarse, bent round downwards and inwards, sporangia providing a covering till ripe , not falling away and remaining until early in the year; spores ripe in autumn

Habitat: In hill and subalpine zones in beech and oak woods, on shady banks and slopes on moist, lime-poor, rather acid, humus-containing, sandy or stony soil in humid locations; also found on scree

142 **False male fern – D. pseúdomas**
Plants up to 150 cm high, with rising or vertical rhizome and leaves surviving the winter

Dryópteris pseúdomas Holub and Pouzar
D. affinis Fraser-Jenkins
False male fern

Aspidiáceae – Spleenwort family

Leaves: 40–150 cm long, narrow to oval in outline, narrowing toward base, generally coming to an obtuse point at tip, coarsely and simply pinnate; pinnae in 20–40 pairs and except for topmost clearly divided at middle vein; segments nearly rectangular, rounded or blunt at tip, at base mostly attached to each other and **in this plant more distinctly toothed than those on page 40**; base of the spindle dark violet when fresh

Sporangiophore: Leaves (underside of pinnae)

Sori: Generally formed in lower two-thirds of leaf; covering coarse, bent downwards and inwards, sporangia providing a covering till ripe, not falling away, and remaining until early in year; spores ripe in autumn

Habitat: From hill to subalpine zones in beech and oak woods, on shady banks and slopes on moist, lime-poor, rather acid, humus-containing soils

143 **False male fern – D. pseúdomas**
Plants up to 150 cm high, with ascending or vertical rhizome and leaves surviving the winter

Dryópteris pseúdomas Holub and Pouzar
D. affinis Fraser-Jenkins
False male fern

Aspidiáceae – Spleenwort family

Leaves: 40–150 cm long, narrow to oval in outline, narrowing toward base, generally coming to an obtuse point at tip, coarsely and simply pinnate; pinnae in 20–40 pairs and except for topmost clearly divided at middle vein; segments nearly rectangular, rounded or blunt at tip, at base mostly attached to each other and **in this plant more distinctly toothed than those on page 40**; base of the spindle dark violet when fresh

Sporangiophore: Leaves (underside of pinnae).

Sori: Generally formed in lower two-thirds of leaf; covering coarse, bent downwards and inwards, sporangia providing a covering till ripe, not falling away, and remaining until early in year; spores ripe in autumn

Habitat: From hill to subalpine zones in beech and oak woods, on shady banks and slopes on moist, lime-poor, rather acid, humus-containing soils

144 **Rigid buckler fern – D. villárii**
Plant up to 60 cm high, with horizontal or ascending rhizome and with fragrant leaves.

Dryópteris villárii (Bell.) Woynar
Rigid buckler fern

Aspideáceae – Spleenwort family

Leaves: 15–60 cm long, up to 15 cm wide, narrow oval, bunched in a funnel shape, rigidly upright, with small, yellowish, spherical glands on upper side and doubly pinnate; stalk black at bottom and with light reddish-brown scales; first-order pinnae generally up to 25 in number, touching each other – particularly in upper part of leaf – or overlapping slightly and often perpendicular to the surface of the leaf because of rotation of the stalk; second-order pinnae with blunt but distinctly toothed point and with teeth singly or in pairs

Sporangiophore: Leaves (underside of pinnae)

Sori: Fairly round and generally touching till ripe; covering thick with yellow to light brown glands; spores ripe in late summer or autumn

Habitat: In subalpine zone on limestone screes (covered from eight to nine months of the year in snow) in regions with high air humidity; often on fine subsoil; not found in central areas

145 **Short buckler fern – D. abbreviáta**
Plant up to 100 cm high, with thick rhizome and leaves dying off in spring

Dryópteris abbreviáta (DC.) Newm.
D. oreádes
Short buckler fern

Aspideáceae – Spleenwort family

Leaves: 30–100 cm high, mostly narrow oval in outline, distinctly tapering towards the ground, pointed at the end and simply pinnate; pinnae 20–40 pairs, clearly divided from the middle vein (or slightly separated) in most cases pointing forward on the living plant and almost forming a tube; segments rectangular, rounded at the end, distinctly toothed and with small glands on the underside

Sporangiophore: Leaves (underside of pinnae)

Sori: Generally formed only on lower two-thirds of leaf, spherical, with a coarse covering and glands; spores ripen in autumn

Habitat: In mountain and subalpine zone, generally on north-facing, exposed screes of lime-free stone, forming continuous thickets; probably safely identified with D. abbreviata Newm.; reaches northward to England, Ireland, Iceland, and Greenland, and southward to the southern Alps, the Pyrenees, the Apennines and the island of Elba

146 **Crested wood fern – D.cristáta**
Plant up to 100 cm high, with horizontal or ascending rhizome and leaves in loose bundles

Dryópteris cristáta (L.) Gray
Crested wood fern

Aspidiáceae – Spleenwort family

Leaves: 30–100 cm long, up to 20 cm wide, very narrow oval or lanceolate, pointed, stalked and simply pinnate; non-spore-bearing leaves on the outside bending down, with thin, finely scaled stalk and 10–20 pairs of pinnae, triangular or egg-shaped and often grooved; the lowest pinnae are clearly separated; spore-bearing leaves stand upright, pinnae are perpendicular to the surface of leaf through rotation of the stalk

Sporangiophore: Fertile leaves (underside of pinnae)

Sori: Very often cover whole surface of the pinnae, very large and touching when ripe; spores ripen in summer

Habitat: In hill and mountain zones in hollows, peat moors, wooded marshes, alder and birch woods, willow copses, and by old rootstocks on soils which are wet, more or less rich in nutrients, base-rich, moderately acidic and containing humus, on peaty or broken clay; plants grow in half-shade; little variability in form

147 **Narrow buckler fern – D. carthusiána**
Plant up to 90 cm high with horizontal or upwardly sloping rhizome; leaves occasionally surviving the winter

Dryópteris carthusiána (Vill.) H. P. Fuchs
D. spinulósa (Mueller) Watt
Narrow buckler fern

Aspidiáceae – Spleenwort family

Leaves: 50–90 cm long, doubly pinnate and generally rigidly upright; stalk relatively thin, yellow-green in colour, with a few light brown fine scales and often as long as leaves proper; these are up to 45 cm long, 2–4 times as long as wide, narrow oval in outline and narrowing towards the base only slightly or not at all; lowest first-order pinnae 5–15 cm long, generally triangular in outline and up to 10 cm apart; second-order pinnae are pinnately cut, blunt and have pronounced teeth at tip

Sporangiophore: Leaves (underside of pinnae)

Sori: Do not touch when ripe; coverings very often toothed at edge and without glands; spores ripen in summer

Habitat: From hill and subalpine zones in oak and pine forests poor in other species, in alder copses, wooded marshes, high moors, and heathland on damp to wet, acidic soil poor in nutrients and chemicals, fairly peaty, humus-rich, and moorland soils; plants live in half-shade

148 **Broad buckler fern – D. dilatáta**
Plants up to 150 cm high, with erect or ascending rhizome and remaining green into winter

Dryópteris dilatáta A. Gray
Broad buckler fern

Aspidiáceae – Spleenwort family

Leaves: 10–150 cm long, soft to the touch and bending over; leaf stalk well covered in fine scales, having yellowish glands, shorter than leaf proper; this is up to 70 cm long, forming a broad triangle or wide oval, often with glands on both sides, and doubly or trebly pinnate; second-order pinnae often touching, especially in upper region, and rounded or awned at tip; third-order pinnae pinnately cut, with toothed segments having spiny points

Sporangiophore: Leaves (underside of pinnae)

Sori: Not touching when ripe; indusium often toothed at the edge; spores ripe in summer and autumn

Habitat: In mountain and subalpine (rarely alpine) zones in beech, oak, and spruce woods rich in vegetation and grass, on boulders in shady places on wet to moist ground, often trickling with water; soil often more or less rich in nutrients, moderately acidic, loose, filled with rotted matter or humus, and generally sandy loam or stony clay or scree; plants found in shade; often forms large thickets

149 **Remote fern – D. remóta**
Plant up to 90 cm high, generally with ascending rhizome; leaves die off in winter

Dryópteris remóta Druce
Remote fern

Aspidiáceae – Spleenwort family

Leaves: 15–90 cm long, with only sparse glands, yellow-green when young, later dark green in colour; leaf proper narrow oval or narrow triangle or lanceolate and 10–70 cm long; lowest pair of first-order pinnae generally clearly distanced from the others and lanceolate in form; first-order pinnae with deeply furrowed stalks; second-order pinnae generally with parallel sides, blunt or pointed and slightly overlapping; segments are cut short and the largest ones toothed

Sporangiophore: Leaves (underside of pinnae)

Sori: Distributed over the whole of the underside; indusium is thick, does not fall away, and is free from glands

Habitat: Often found in large concentrations in hill and mountain zones in shady beech, oak, or spruce woods and mixed deciduous woods on steep slopes on wet to moist soil which is often stony, also on moors with springs; soil lime-poor, rich in nutrients and rotted matter or humus; often found with ash trees and mountain maples

150 **Lady fern – A. filix-fémina**
Perennial plant up to 120 (150) cm high, with horizontal or upright rhizome

Athyrium filix-fémina (L.) Roth
Lady fern

Athyriáceae – Lady fern family

Leaves: 30–120 cm long, with short stalk and doubly or trebly pinnate; stalk up to 2 cm wide at ground, coloured dark brown or black and covered with dark brown fine scales; leaf proper up to 100 cm long, up to 40 cm wide, oval and without hairs; up to 30 first-order pinnae; these are narrowly lanceolate, coming to a fine point, pointing somewhat forwards and often touching or overlapping; second-order pinnae up to 3 cm long, coming to a fine point, hardly touching except at base and deeply toothed or pinnately cut; segments are distinctly toothed

Sporangiophore: Leaves (underside of pinnae)

Sori: Longish, horseshoe-shaped or oval; indusium toothed at the edge or lidded, generally still present when spores are ripe in summer to autumn

Habitat: From hill to subalpine (rarely alpine) zones in deciduous and coniferous woods with plenty of undergrowth, in scree, in shady places on well drained to wet soil which is lime-poor, rich in humus, more or less base-rich, sandy, stony, or pure loam or clay

151 **Alpine lady fern – A. distentifólium**
Plants up to 100 cm high, hardy and with horizontal to erect, generally thick, rhizome

Athyrium distentifólium Opiz Tausch
Alpine lady fern

Athyriáceae – Lady fern family

Leaves: 30–100 cm long, arranged in a bundle, with a stalk and doubly or trebly pinnate; stalk widens at ground level, is dark in colour and covered in fine scales; leaf proper up to 80 cm long, oval and hairless; up to 30 pairs of first-order pinnae; these are narrowly lanceolate, coming to a fine point, pointing forward (especially in the upper part of the leaf), and very often touching or overlapping; second-order pinnae very often touching and generally with a rounded point.

Sporangiophore: Leaves (underside of pinnae)

Sori: Round in shape; indusium very small and already disappearing berfore spores ripen; spores ripe in summer

Habitat: In subalpine (rarely mountain or alpine) zones in mixed mountain woodland with plenty of undergrowth, thick scrub, scrubby heathland and on scree, in shaded locations on loose, wet, more or less base-rich, lime-poor, moderately acidic and often stony clay; sites covered with snow for long periods

152 **Holly fern – P. lonchitis**
Plant up to 50 cm high, with thick rhizome and leaves dying off in July of the following year

Polystichum lonchitis (L.) Roth
Holly fern

Aspidiáceae – Spleenwort family

Leaves: 20–50 cm long standing stiffly upright; stem up to 8 cm long, with large brown, hair-like scales; leaf proper lanceolate in outline, up to 45 cm long, pointed, narrowing towards base and simply pinnate; pinnae are leathery, stiff, glossy on top, having fine scales underneath, short-stalked, generally touching each other, bending upwards in a sickle shape, with spiny teeth and with a sharp, enlarged spine pointing forward at the base

Sporangiophore: Leaves (underside of pinnae)

Sori: Underneath pinnae, arranged in 2 rows, often touching one another and up to 1.5 mm in diameter; indusium round; spores ripe in late summer and autumn

Habitat: In mountain and alpine zones on scree, rock falls, and in mountain woodland and rock crevices on well-drained soil which is base-poor; sites generally covered in snow for long periods; plants live in light or half-shade; northern limit of penetration passes through Iceland

153 **Hard shield fern – P. aculeátum**
Plants up to 70 cm high, hardy and with thick rhizome; leaves remain green throughout the winter

Polystichum aculeátum Roth
Hard shield fern

Aspidiáceae – Spleenwort family

Leaves: 20–70 cm long, standing rigidly erect; stalk up to 20 cm long and like leaf spindle covered with numerous dark brown, glossy fine scales; leaf proper 15–60 cm long, lanceolate to narrow oval in outline, coming to a sharp point, narrowing towards base, coarse in texture, on the underside only covered with hairy, whitish scales; doubly pinnate; first-order pinnae in the upper portion point distinctly upwards, are up to 10 cm long, narrowly lanceolate, ending in a fine point, and up to 40 in number on each side; second-order pinnae are small-stemmed or sessile, lanceolate and serrated, with awns at tip; bottom second-order pinna is distinctly bigger than others

Sporangiophore: Leaves (underside of pinnae)

Sori: Large, very often touching each other; spores ripe in late summer and autumn

Habitat: In mountain and subalpine zones on shady scree, woods in ravines, and shady spruce woods on moist soil which is wet, more or less base-rich and rich in nutrients, lime-poor or lime-rich soil

154 **Soft shield fern – P. setiferum**
Perennial plants up to 100 cm high, with a thick rhizome; found in areas with mild winters

Polystichum setiferum (Forskal) Moore
Soft shield fern

Aspidiáceae – Spleenwort family

Leaves: 30–100 cm long, soft and leathery in feel and doubly pinnate; stem up to 30 cm long and thick with fine yellow-brown scales; first-order pinnae up to 30–40 pairs, covered underneath with yellow-brown hair-like scales, smooth above, standing out at right angles or pointing slightly forward, up to 15 cm long and finely pointed; second-order pinnae mounted on a thin short stalk, narrow oval, mostly pointing a little forward, with a pronounced point, at base often cut through to middle vein and at further edge generally clearly awned

Sporangiophore: Leaves (underside of pinnae)

Sori: Smaller than in other species, but distributed in great numbers over whole underside of pinnae (usually in 2 rows) and hardly or not at all touching each other; spores ripe in summer

Habitat: In hill and mountain zones in mixed deciduous forests, beech and oak woods, and on banks, on crumbly, lime-free, moderately acidic loam

155 **Green spleenwort – A. viride** Plants up to 30 cm high, with horizontal branching rhizome and rarely remaining green throughout the winter

Asplénium viride Hudson
Green spleenwort

Aspleniáceae – Spleenwort family

Leaves: 10–30 cm long, forming large clumps, recumbent or slightly erect; simply pinnate; stem covered scantily with a few fine scales, generally greenish (in lower parts also somewhat reddish-brown); leaf proper up to 15 cm long and 10–15 mm wide, narrowly lanceolate in outline, yellow-green to green and generally with up to 15 pinnae; these are oval to triangular in outline, narrowing to a wedge-shaped base, irregularly toothed or notched and with rounded teeth/segments

Sporangiophore: Leaves

Sori: 4–8 per pinna, longish, touching when spores are ripe, and mostly distributed below middle; when ripe indusia shrivel, then generally fall off; spores ripen in summer

Habitat: In mountain and alpine zones on walls, rocks, boulders, and next to tree-trunks on fresh to damp and generally limy soils in humid places; also on lime-free subsoils when irrigated by hard (lime-bearing) water

156 **Maidenhair spleenwort – A. trichómanes**
Perennial plants up to 30 cm high, with thick rhizomes and staying green throughout the winter

Asplénium trichómanes L.
Maidenhair spleenwort

Aspleniáceae – Spleenwort family

Leaves: 10–30 cm long, numerous and in thick bundles, recumbent or slightly erect; simply pinnate; stalk and central vein winged on both sides and glossy dark brown throughout; leaves proper up to 25 cm long, up to 2 cm wide, narrowly lanceolate in outline and coloured dark green; pinnae on both sides, up to 40 in number, oval, with unmatched bases, round at tip and bluntly toothed

Sporangiophore: Leaves (underside of pinnae)

Sori: Up to 2 mm long, and lying outwards from middle vein, almost reaching the edge; indusium visible when spores are ripe in late summer

Habitat: From hill to subalpine zones on walls and scree and in rock crevices on lime-poor or lime-rich, dry or moist soils in sunny and shady places; found in different kinds of rock crevices; in sunny habitats pinnae are smaller; various forms are recognised

157 **False spleenwort – A. adulterínum**
Plants up to 20 cm high, with rhizome from horizontal to vertical, and staying green through winter

Asplénium adulterínum Milde
False spleenwort

Aspleniáceae – Spleenwort family

Leaves: 5–20 cm long, standing in bundles in large numbers, rigidly upright and simply pinnate; stalk and most of leaf spindle glossy dark brown and without wings; top 1–3 cm of the spindle are green; leaf proper up to 18 cm long, narrowly lanceolate in outline, and with up to 30 pinnae on each side; these are broad oval to rhomboidal, with a distinctly vaulted top surface, rounded at tip and with blunt teeth

Sporangiophore: Leaves (underside of pinnae)

Sori: 6–8 per pinna, longish, when ripe generally not touching the edge but often touching each other; indusia folded back when spores are ripe in summer

Habitat: From hill to subalpine zones on rocks and scree (only on serpentine stone); Asplenium serpentini and Asplenium adulterinum form the only typical spleenworts on serpentine stone; northern penetration as far as Norway and southern Sweden

158 **French spleenwort – A. foreziénse**
Plants up to 25 cm high, with ascending or upright rhizome and staying green throughout the winter

Asplénium foreziénse Le Grand
French spleenwort

Aspleniáceae – Spleenwort family

Leaves: 5–25 cm long, tufty and singly or doubly pinnate; stalk generally shorter than leaf proper and greenish to red-brown; leaf proper up to 15 cm long, lanceolate to oval, 1–3 cm wide, narrowing towards base and usually rounded with a short point at tip; first-order pinnae narrow oval, deeply toothed or pinnately cut, with lowest pinnae often sloping backwards; second-order pinnae round with small teeth, with distinct awned point

Sporangiophore: Leaves (underside of pinnae)

Sori: Generally close to central vein and up to 1 cm long; covering clearly visible when spores are ripe; spores ripe throughout the year

Habitat: In hill and mountain zones in rock crevices and on walls; on silicate rocks; centre of distribution in southwest France; northwards as far as Belgium; southwards as far as Corsica, Sardinia, Spain, Algeria, and Morocco; in the Vosges region, Alsace, and the southern foothills of the Alps as far as the Tyrol

159 **Beech fern – P. connéctilis**
Plants up to 30 cm high, with thin creeping rhizome; leaves do not survive the winter

Phegópteris connéctilis Watt
Lastréa phegópteris Bory
Beech fern

Thelypteridaceae – Beech fern family

Leaves: 10–30 cm long, sticking up from rhizome in several large bundles and singly or doubly pinnate; leaf stem thin, greenish to yellowish and dark brown at base; leaf proper is arrow-shaped or triangular, up to 20 cm long, soft to the touch, light green in colour and hairy on both sides; lowest pair of pinnae generally slant backwards, succeeding ones stand out at right angles or point towards tip, their ends pointing upwards; first-order pinnae are narrow lanceolate, sessile, round or blunt-ended, and pinnate or toothed

Sporangiophore: Leaves

Sori: Generally close to edge of the segments and always without indusium; spores ripe in summer

Habitat: In hill to subalpine zones by tree-trunks, in oak and spruce woods rich in vegetation, mixed deciduous woods, spinneys , on good to moist, loose soil more or less base-rich and rich in nutrients, also moderately acidic and stony loam/clay soils; plants grow in shade

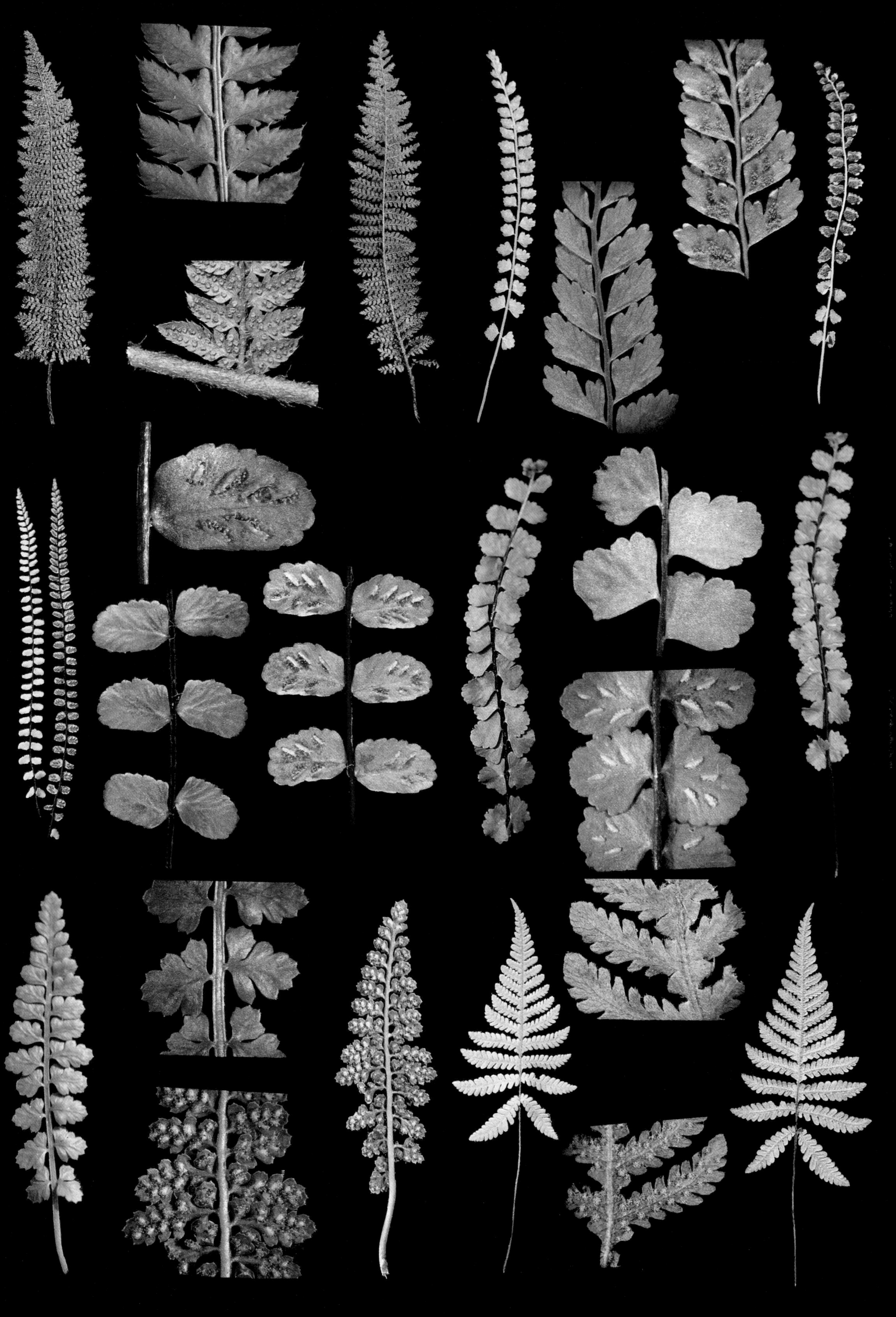

160 **Oak fern – G. dryópteris**
Perennial plant up to 50 cm high, with thick rhizome spreading underground

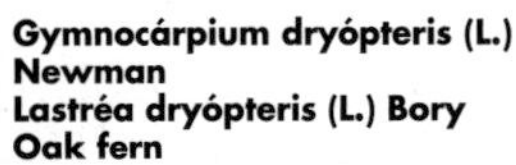

Gymnocárpium dryópteris (L.) Newman
Lastréa dryópteris (L.) Bory
Oak fern

Aspidiáceae – Spleenwort family

Leaves: 10–50 cm long, springing from rhizome in several large clumps, glandless and doubly pinnate; stalk thin, yellow-green and 2–3 times as long as leaf proper, which lies almost horizontal; this is triangular, up to 25 cm long, lowest pair of pinnae being smaller than the rest; first-order pinnae have stalks in lower part of leaf but are sessile in upper part and are wide lanceolate to narrow oval; second-order pinnae are often incised as far as middle vein; sections are entire, crenate, or with deeply toothed edge (blunt teeth), and rounded at tip

Sporangiophore: Leaves

Sori: Round, mostly near the edge and without indusium; spores ripen in summer

Habitat: Fairly frequent in groups in mountain and subalpine (rarely hill) zones, in woods rich in vegetation: mixed woods with beech, fir, and spruce, high undergrowth, on walls, scree, and banks on well-drained to damp, cool soils rich in nutrients

161 **Limestone fern – G. robertiánum**
Plant up to 55 cm high, with thick rhizome spreading underground and growing in shady places

Gymnocárpium robertiánum Newman
Limestone fern

Aspidiáceae – Spleenwort family

Leaves: 15–55 cm long, springing from rhizome in several large clumps, with yellowish glands and singly or doubly pinnate; stalk thin, yellowish-green and covered with scales, particularly in upper region, 2 to 3 times as long as leaf proper; this is triangular to rhomboid, generally upright and 10–25 cm long; lowest pinnae are disinctly larger than those above; first-order pinnae have stalks in lower part of leaf but are sessile in upper part; in outline they are wide lanceolate to oval; second-order pinnae often incised as far as middle vein; segments have a smooth, deeply toothed or notched edge and are rounded at tip

Sporangiophore: Leaves

Sori: Round, mostly near the edge and without indusium

Habitat: In mountain and subalpine (rarely hill) zones on limestone scree, on walls and rocks, and in woods containing rocky detritus on well-drained to damp soil containing lime

162 **Black spleenwort – A. adiantum-nigrum**
Plants up to 50 cm high, with ascending rhizome, mostly in thick bunches; lasts throughout the winter

Asplénium adiántum-nigrum L.
Black spleenwort

Aspleniáceae – Spleenwort family

Leaves: 10–50 cm long, dark green and doubly or trebly pinnate; stalk is chestnut brown to dark purple and is often longer than leaf proper; this is up to 30 cm long, triangular or oval, pointed and glossy; first-order pinnae are lanceolate to oval, gradually coming to a point, generally with point directed forwards, and generally narrowing to a wide wedge shape at the base; second-order pinnae are narrow oval, wedge-shaped, round or flat at tip and in the top part distinctly toothed

Sporangiophore: Leaves (underside of pinnae)

Sori: 2–4 mm long, reaching the edges only at ripening time, and then covering the whole surface; at ripening time the indusia of the spores are generally covered by the sporangia; spores ripen in summer

Habitat: In hill and mountain zones on sunny rocks, walls, on scree, and in rocky woodland, on acid, well-warmed rocks; plants grow in light or half-shade, preferring locations with mild winters; plants are distributed worldwide

163 **Donkey spleenwort – A. onópteris**
Plants up to 50 cm high, with ascending rhizome; lasts throughout the winter

Asplénium onópteris L.
Donkey spleenwort

Aspleniáceae – Spleenwort family

Leaves: 10–50 cm long, mid to dark green, and doubly or trebly pinnate; stalk is dark brown on top only in lower part; underneath stalk is dark brown to purplish-black for a long way up; leaf proper is up to 30 cm long, somewhat glossy, triangular to oval in outline and rounded or blunt-ended at tip; first-order pinnae in upper part of leaf point distinctly towards leaf tip, being lanceolate to narrow oval; at base they are wedge-shaped and come to a long point; second- order pinnae are narrow oval, but wedge-shaped at base and, especially in the top half, distinctly toothed or lobed, with pointed sections

Sporangiophore: Leaves (underside of pinnae)

Sori: 2–5 mm long, slanting toward the outer edge; when ripe covering most of segment; spores ripe in winter

Habitat: In hill zone on walls and rocks, in light, stony woods, copses, and on scree on dry and acid soil

164 **Wedge-leaved spleenwort - A. cuneifólium**
Plant up to 40 cm high, with ascending rhizome and tender foliage, not retained in winter

Asplénium cuneifólium Viv.
A. serpentíni Tausch
Wedge-leaved or serpentine spleenwort

Aspleniáceae - Spleenwort family

Leaves: 10–40 cm long, light to mid green and mostly doubly pinnate; above, stalk is not green until it reaches the leaf proper, below it is generally brown throughout; leaf proper is up to 30 cm long, matt, triangular to oval in outline, coming to a narrow point, with point directed forwards and a slightly leathery feel; first-order pinnae are oval, bluntly pointed or rounded and narrowing to a wedge shape at the base; second-order pinnae are oval, rounded at the end, narrowing to a wedge shape, and serrated in the upper part

Sporangiophore: Leaves (underside of pinnae)

Sori: 3–5 mm long, slanting toward the outer edge, but not always reaching it; when ripe covering most of the segment; spores ripen in summer

Habitat: In mountain and subalpine zones on rocks and boulders (only of serpentine) and on stony slopes; this middle and south European plant's range extends south to Corsica and the mountains of the Balkan peninsula

165 **Wall rue – A. rúta-murária** Plant up to 25 cm high, with a short rhizome often branching many times; leaves in bundles, retained during winter

Asplénium rúta-murária L.
Wall rue

Aspleniáceae – Spleenwort family

Leaves: 5–25 cm long, dark green, matt, coarse, and doubly or trebly pinnate; stalk is thin, green, as long as or longer than leaf proper, and at the base dark brown and covered in fine scales; when young stalks are covered in short glands; leaves are lanceolate to oval or triangular, coarse, without glandular hairs, and grey-green; pinnae mostly have distinct stems, and are generally lanceolate, oval or rhomboidal, having fine teeth or being divided up to the middle into 3 parts

Sporangiophore: Leaves (underside of pinnae)

Sori: Up to 1.5 mm long, linear, with 2–5 per pinna, sloping upwards and when ripe often clearly reaching the side of the pinna, particularly at the side; indusium toothed and often covered by the sporangia when ripe; spores ripen throughout the year

Habitat: From hill to alpine zones, frequently on rocks and walls, especially on lime in both sunny, dry places and also in shady, damp ones; species shows great variey

166 **Mountain bladder fern – C. montána** Plants up to 40 cm high, mostly with a horizontal rhizome; leaves rise singly from the rhizome

Cystópteris montána Desv.
Mountain bladder fern

Athyriáceae – Spleenwort family

Leaves: 10–40 cm long, trebly or quadruply pinnate, growing from the rhizome singly and several centimetres apart; stalk 5–30 cm long, thin, longer than the leaf proper and dark brown beneath (otherwise yellowish to green); leaf proper up to 15 cm long, triangular in outline and pointed; first-order pinnae have stems in the lower part of the leaf, being narrow triangular to oval, pointed, and less deeply divided towards the point; the lowest pair of pinnae is larger than the rest; second- and third-order pinnae are pinnately cut almost to the middle vein; the segments are narrow, rounded, and with teeth in groups of two or more

Sporangiophore: Leaves (underside of pinnae)

Sori: Small, and scarcely touching; indusium smooth or covered in small glands; spores ripen in summer

Habitat: Usually in the subalpine (rarely mountain) zone in rock crevices, in deciduous and coniferous woods, in alder thickets and on limestone scree, on damp soil in shady places; range extends northwards to South Greenland, south to the mountains of central Spain

167 **Brittle bladder fern – Cystópteris régia** Plants up to 30 cm high with short horizonal rhizomes

Cystópteris régia Desv.
Cystópteris críspa
Brittle bladder fern

Athyriáceae – Spleenwort family

Leaves: 5–30 cm long, arranged in a rosette (as the rhizome is short), very delicate in structure, often rather transparent and doubly or trebly pinnate; stalk generally shorter than the leaf proper; this is ensiform, pointed, and with one pair of pinnae much shorter than those above; first-order pinnae narrow oval to longish and rounded at the end, being wedge-shaped at the base and further pinnate; second-order pinnae have deeply incised segments, rounded at the end or bluntly pointed, and clearly rimmed or doubly toothed

Sporangiophore: Leaves (underside of pinnae)

Sori: When ripe touching and often covering the whole surface of each segment; spores ripen in late summer and autumn

Habitat: In subalpine and alpine zones in damp rock crevices and among boulders; only found on limestone; northward range extends to England and Sweden, southward to the Pyrenees; rarely found in hill zone and if so only by accidental distribution or through wind dispersal

168 **Sudeten bladder fern – C. sudética** Plants up to 30 cm high, with long horizontal, spreading rhizomes

Cystópteris sudética A.Br. and Milde
Sudeten bladderfern

Athyriáceae – Spleenwort family

Leaves: 10–30 cm long, trebly or quadruply pinnate, growing from the rhizome singly and several centimetres apart; stalk 5–20 cm long, thin, covered in scales resembling small glands, and longer than the leaf proper; this is triangular to wide oval, light or mid-green and pointed; first-order pinnae have stems a long way up the leaf, and are lanceolate to narrow oval in outline, pointed, and except in the topmost part themselves pinnate; second-order pinnae are themselves pinnately cut; segments are pinnately cut or incised or have many teeth

Sporangiophore: Leaves (underside of pinnae)

Sori: Small and scarcely touching; indusium covered with many glands; spores ripen in summer

Habitat: In the mountain zone in mixed deciduous forests, mixed forests with plenty of rocks on the ground, and in rock crevices, on damp, shady, lime soil rich in humus; plant found by itself in groups, or with other bladder ferns the limestone fern

169 **Parsley fern – C. críspa** Plants up to 30 cm high, with leaves arranged in bunches and spreading rhizomes

Cryptográmma críspa (L.) R.Br.
Parsley fern

Cryptogrammáceae – Maidenhair family

Leaves: 15–30 cm long, generally trebly pinnate and divided between leaves carrying spores and those without; stalk smooth, yellowish to greenish, and generally longer than the leaf proper; this is triangular to oval in shape and generally has 5–9 pairs of first-order pinnae, generally alternae; second-order pinnae oval, with stems; third-order pinnae pinnately cut, with a wedge-shaped base; in the case of leaves carrying spores the third-order pinnae cylindrical and rolled up to enclose the sori when the spores are rips; leaves not carrying spores occasionally have spore-carrying pinnae at the end

Sporangiophore: Leaves with segments rolled up in a cylinder

Sori: Roundish to oval, without indusium, initially enclosed, later free; spores ripen in late summer and autumn

Habitat: In subalpine and alpine zones in piles of coarse rubble and on drystone walls, on silicate rock; plants often found in large groups

170 **Brittle bladder fern – C. frágilis** Plants up to 40 cm high, with short, mostly horizontal and branching rhizomes

Cystópteris frágilis (L.) Bernhardi
Brittle bladder fern

Athyriáceae – Spleenwort family

Leaves: 10–40 cm long, doubly or trebly pinnate and arranged in a tight bunch; stalk often as long as the leaf proper, greenish or yellowish in the upper part, otherwise glossy red-brown and covered with a thick layer of fine scales beneath; leaf proper 10–30 cm long, lanceolate to oval in outline, and pointed; first-order pinnae more widely spaced at the base of the leaf than further up, and almost opposite; further up they lie closer together and are alternate, often with a stalk and pointed; second-order pinnae oval to lanceolate, pinnately cut or further pinnate

Sporangiophore: Leaves (underside of pinnae)

Sori: Touching when spores are ripe: they then cover the whole of the underside of the segments; spores ripen in summer

Habitat: From hill zone as far as the alpine zone (given snow cover as far as 3000 m) on rocks and walls, by old springs and in rocky forests, on well-drained to damp, generally shady, lime-rich soil or other base-rich soils; plants found in half-shade

171 **Spring spleenwort – A. fontánum** Plants up to 25 cm high, with short and spreading rhizomes; green throughout the winter only in sheltered places

Asplénium fontánum Bernhardi
Spring spleenwort

Aspleniáceae – Spleenwort family

Leaves: 10–25 cm long, doubly pinnate and growing in thick bunches; in the upper part the stem is green-brown, in the lower part red-brown, having dark-coloured hairy fine scales (later in the year the stalk is smooth), shorter than the leaf proper; this is 20 cm long, narrow lanceolate in outline, narrowing at the base, and dark green in colour; the lowest pinnae are oval to round, the upper ones mostly lanceolate to narrow oval and generally alternate; the lowest pinnae point distinctly downwards and are shorter than those above; second-order pinnae are pinnately cut, and segments have thorny points

Sporangiophore: Leaves (under the pinnae)

Sori: Up to 1 cm long, approching the middle vein; indusium not apparent when spores are ripe in summer and autumn

Habitat: In hill and mountain (rarely subalpine) zones on walls and on shady lime-containing rocks; widespread

172 **Bracken – P. aquilínum**
Plants up to 3 m high, forming thickets, with thick, hairy, branching rhizomes

Pterídium aquilínum (L.) Kuhn
Bracken

Pteridiáceae – Bracken family

Leaves: 20 cm to 3 m long, doubly or trebly pinnate and with long stems; stem up to 1 m long, smooth, generally straw-coloured or greenish and forming clumps of 10–20 (lying horizontally!); leaf proper upright or bending over, up to 1.5 m long and 1 m wide, triangular, leathery, glossy and slightly hairy or smooth above, and generally with a thick coating of hairs beneath; lower first-order pinnae are opposite, the upper ones mostly alternate and with stems; second-order pinnae sessile and pinnate or pinnately cut; third-order pinnae or segments opposite, arranged like a comb, those nearest the base often with wavy edges

Sporangiophore: Leaves (edges of underside of pinnae)

Sori: Situated on the edge of the pinnae, longish, with two indusia and partly covered by the edges of the pinnae; both indusia are hairy at the edge; spores ripen in summer

Habitat: From the hill zone to the sub-alpine zone in sparse woodland, on neglected rough pasture and heath, at the edge of woods and on banks, on moderately dry to well-drained soil

173 **Forked spleenwort – A. septentrionále**
Plants up to 15 cm high, with forked horizontal rhizome

Asplénium septentrionále (L.) Hoffm.

Forked spleenwort

Aspleniáceae – Spleenwort family

Leaves: 5–15 cm long, smooth, growing plentifully upwards in thick blades of 2–5 divisions; the stalk is up to 10 cm long, green, but brown at the base; the leaf proper is up to 5 cm long, with very narrow lanceolate divisions, leathery, smooth, and bluish to mid-green in colour; the divisions are up to 3 cm long, forked, mostly with thickened edges, and entire or with several long narrow teeth at the apex

Sporangiophore: Leaves (underside of divisions)

Sori: Longish, with 1–5 on each division, covering all the underside; the indusium is withdrawn when the spores are ripe, forming a seam at the edge of each division; spores ripen in late summer

Habitat: Found frequently from the hill zone to the alpine zone in rock crevices, in cracks in walls, and on scree on lime-free soil in sunny places exposed to light; found on erratic blocks as well as in the mountains

174 **Swimming fern – S. natans**
Plants with a branched or unbranched stem, floating on water; annual

Salvínia natans (L.) All.
Swimming fern

Salviniáceae – Salvinia family

Leaves: Arranged in whorls; the middle leaf of the three in the whorl forms a false root, while the two side leaves are those by which the plant swims; these are 5–15 mm long, oval to circular, cordate at the base, rather hairy beneath and on top covered with wart-like lumps which each have from 3–4 to 9 small brown hairs (which repel water); the root-like leaves (false roots) reach down into the water, and are split into up to 7 long divisions; there are no true roots

Sporangiophore: Spore cases in groups at the base of the false root

Sporocarps: Up to 8 spore cases are formed into balls about the size of peas

Habitat: In the hill zone in slow-moving or still lime-free water; the northern boundary of distribution lies through Holland and North Germany, and the southern through North Africa and Asia Minor; plants often distributed by wind and are not infrequently found further afield

175 **Kariba weed – S. molésta**
Tropical plants up to 10 cm long, floating on the water surface; perennial

Salvínia molésta Mitchell
Kariba weed

Salviniáceae – Salvinia family

Leaves: 1–2 cm long, oval to round, with several on one stalk up to 10 cm long, rounded above but narrowing to a wedge shape at the base, and light green in colour; they do not lie flat on the water; the outer side of the leaf is smooth and veined in white, the inner side spongy and white; there are no true roots; the third leaf is divided into fine root-like points

Sporangiophore: Spore cases lie at the top end of the false root

Sporocarps: Having a soft, double-layered wall, the spore cases either have numerous microsporangia on a stalk-like receptacle (= microsori) or have less numerous stalked megasporangia (= megasori); megaspores are globular or tetrahedral and smooth; microspores are smooth, globular or tetrahedral, and embedded in a hard frothy medium

Habitat: The plant is generally cultivated in botanical gardens as **Salvínia auriculáta Aubl.**; it has become naturalised here and there in warmer ponds

176 **Pepperwort – M. quadrifólia**
A herb-like perennial marsh plant up to 15 cm high; the spore cases are attached to the first shoots

Marsílea quadrifólia L.
Pepperwort

Marsileáceae – Pepperwort family

Leaves: 5–15 cm long, smooth and with very long stems; the leaf proper is divided into 4 leaflets as far as the base, looking like clover; new leaves are rolled up, opening out horizontally; the leaflets are inverted triangles with the outside edge rounded, entire, or somewhat wavy, covered with fine white hairs and brownish-green to matt green in colour

Sporangiophore: At the base of the leaf stalk on short stems

Sporocarps: Bean-shaped, 3–5 mm in diameter, hairy, with 2 small teeth on the edge, opening in pairs and each containing up to 24 compartments; each compartment contains micro- and macrosporangia; spores ripen in autumn

Habitat: In the hill zone in marshes, bogs, loamy hollows and dried-up pools in warm areas, generally with a lime-free subsoil; grows into clumps when present; range extends northwards to South Germany, but main habitat is Mediterranean area; has been introduced to North America; not very frequently found

177 **Azolla – A. filiculoídes**
Plants 1–0 cm long, moss-like in form, floating, with true roots

Azólla filiculoídes Lam.
Azolla

Azolláceae – Azolla family

Leaves: On the upper side of the stem, arranged very close to each other alternately in 2 rows, and folded over into 2 lobes; the larger, thick, upper lobes overlap each other like roof tiles and are glossy, often both green and brown in colour, floating on the surface of the water; there are hollows on both sides; the smaller lower lobes are submerged

Sporangiophore: Lower lobes of the first leaves on the stem

Sporocarps: Either with several microsporangia on stems, or with a single megasporangium without a stem

Habitat: In the hill zone in calm, sheltered ponds or in slowly flowing water rich in nutrients but often with a low lime content; plants found in light and half-shade; originally a native of the warm temperate and subtropical parts of America; dispersal usually by waterfowl; in favourable conditions shoots develop very quickly

178 **May lily – M. bifólium**
Plants 5–20 cm high, hardy, with bristles on the upper part of the stems; blooming in May

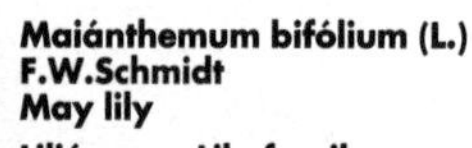

Maiánthemum bifólium (L.) F.W.Schmidt
May lily

Liliáceae – Lily family

Leaves: 2 on flowering stems, on non-blooming plants only one; all 4–8 cm long, 3–5 cm wide, cordate, with stalks, initially with flexible upright hairs beneath and arranged alternately

Inflorescence: Racemes 2–4 cm long, cylindrical, with 10–15 flowers; bracts like scales

Flowers: 4 petals, up to 3 mm long, oval, not fused, standing out horizontally or somewhat bent back, with stems 3–5 mm long and scented; 4 stamens; ovary superior and with a pistil; pollinated by insects

Fruits: Red berries, distributed by animals

Habitat: Hill and mountain (up to 1800 m) zones in deciduous and mixed woodland with a small range of species, and in mountain meadows on well-drained to moderately dry soil poor in nutrients, with or without lime, also on rather acid clay and well-bound sand; as plants growing on detritus or rotted matter and humus they are rooted up to 15 cm deep; with creeping roots plants soon cover a wide area; they grow in shade

179 **Summer snowflake – L. aéstivum**
Plants 35–60 cm high, generally growing in groups and blooming in March and April

Leucójum aéstivum L.
Summer snowflake

Amaryllidáceae – Daffodil family

Leaves: 30–50 cm long, up to 2 cm wide, blue-green in colour; 3 to 4 to each plant

Inflorescence: 3 to 7 flowers per stem

Flowers: 6 petals, 10–18 mm long, not fused, all of equal length, bunched in the form of a bell, and with a yellow or greenish fleck at the end; flowers nodding and lack a calyx; the longest flower stalks are longer than the bract; 6 stamens; ovary inferior and with a pistil; fertilised by insects

Fruits: Fleshy capsules

Habitat: Hill zone in moist or wet meadows or willow thickets, on wet (sometimes flooded) soil rich in nutrients (often manure) and humus, and on clay soils; native to the water meadows of southern and south-eastern Europe, and naturalised in Central Europe, particularly in the northern Vosges, the southern Black Forest, and the area in between

180 **Spring snowflake – L. vérnum**
Plants 30–60 cm high, generally growing in groups and blooming from February to April

Leucoóum vérnum L.
Spring snowflake

Amaryllidáceae – Daffodil family

Leaves: 20–30 cm long, up to 1 cm wide, linear, 3–5 on each plant

Inflorescence: 1 flower, rarely two, on each plant

Flowers: 6 petals, 20–25 mm long, not fused, all of equal length, bunched in the form of a bell, and with a greenish-yellow fleck at the end; flowers nodding and lack a calyx; flower stems no longer than the bract; 6 stamens; ovary inferior and with a pistil; fertilised by insects (bees and butterflies)

Fruits: Fleshy capsules

Habitat: Hill and mountain (rarely subalpine) zones in moist deciduous and mixed forests, meadows, moutain meadows, orchards and thickets, and on river-banks on clay soil rich in nutrients, deep, and loose; plants grow on rubbish; in natural habitats they are indicators of moisture; plants grow in half-shade or light; also grown in gardens as ornamental plants

181 **Snowdrop – G. nivális**
Plants 10–20 cm high, generally growing in groups and blooming from the end of March until June

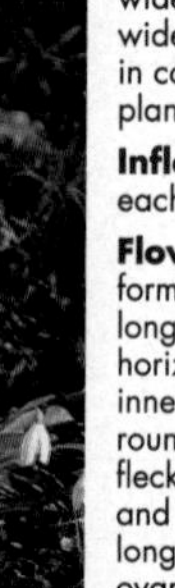

Galánthus nivális L.
Snowdrop

Amaryllidáceae – Daffodil family

Leaves: 10–25 cm long, up to 1 cm wide, broad-linear, generally widening towards the end, blue-green in colour and generally 2–3 on each plant

Inflorescence: 1 flower, rarely 2, on each plant

Flowers: 6 petals bunched in the form of a bell: outer petals 14–18 mm long, oval, often standing out horizontally and pure white in colour; inner petals about half as long, rounded at the tip, and with a green fleck on the outside; plants nodding and lack a calyx; flower stalks no longer than the bract; 6 stamens; ovary inferior and with a pistil; fertilised by insects

Fruits: Fleshy capsules (distributed by ants)

Habitat: In hill and mountain zones in gardens, thickets, and moist deciduous woods and willow thickets on moist soil rich in nutrients, also on fairly acidic, loose, humus-rich, deep clay; plants grow on rubbish or in shade; they attract bees

182 **Alpine leek – A. victoriális**
Plants 30–60 cm high, with round stalks; in bloom from June to August

Allium victoriális L.
Alpine leek

Liliáceae – Lily family

Leaves: 10–20 cm long, 2–5 cm wide, lanceolate to narrow oval, with stalks about 1–2 cm long, often tube-like and hollow; 1–3 on each plant; they smell like leeks

Inflorescence: Spherical heads with flowers clustered thickly, enclosed in a sheath before flowering

Flowers: 6 petals, 3.5–6 mm long, not fused, or only joining at the base, all of equal length, blunt or pointed at the end and whitish or yellow-green in colour; styles 2 –3 times as long as the petals; 6 stamens standing out from the petal sheath; anthers gradually widening towards the base; ovary superior with a pistil and three stigmata

Fruits: Capsules

Habitat: In subalpine and alpine (rarely mountain) zones in stony ground, thickets, scrub, and rocky slopes on moderately dry to moist soil. generally base-rich and lime-poor, and on clay soil

183 **Wild garlic – A. ursínum**
Plants 20–50 cm high, with blunt-edged stalks, smelling like garlic and blooming from April to May

Allium ursínum L.
Wild garlic, Ramsons

Liliáceae – Lily family

Leaves: 5–20 cm long, 2–5 cm wide, broadly lanceolate to oval, with long stems and generally 2 on each plant; all leaves radical.

Inflorescence: Moderately domed or flat, up to 20 florets in each head; sheath generally not standing out from the flower and soon falling away.

Flowers: 6 petals, 8–12 mm long, blunt or pointed at the end and tilted upwards; styles 1–2 times as long as petals; 6 stamens, about half the length of the petals; stamens only slightly wider at the base; ovary superior with a pistil and three stigmata; fertilised by insects

Habitat: In hill and mountain zones in mixed deciduous woodland and riverside woods with good ground cover, hedges, hollows, plantations, and at the foot of slopes; soil well-drained to moist, often saturated, clay and loam rich in humus and nutrients; often found covering a large area; on waste ground plants indicate wet soil or ground-water; plants grow in shade; a symbol of fertility

184 **Branched anthericum – A. ramósum** Plants 20–80 cm high, perennials with short rhizomes, blooming from June to August

Anthéricum ramósum L.
Branched anthericum

Liliáceae – Lily family

Leaves: Like grass, being narrow, linear, 15–25 cm long, 4–6 mm wide, coming gradually to a point, flat or grooved and growing from the ground; occasionally small leaves are found on the inflorescence

Inflorescence: Branching, with many flowers and lateral racemes

Flowers: 6 petals, not fused, 8–14 mm long, the inner petals distinctly wider than the outer ones and narrow-oval to oval, spread out, with 3–7 veins and pointed or rounded; 6 stamens, shorter than the petals; ovary superior with a pistil; fertilised by bees

Fruits: Spherical capsules 5–7 mm in diameter with 3 compartments and up to 8 seeds in each

Habitat: In hill and mountain (rarely subalpine) zones, frequently found in meadows, well drained grassland, at the edge of thickets and woodland and on banks on warm, dry, lime- and humus-containing sand, rocky or loess soil; deep-rooted; range extends northwards to Belgium, northern Germany, and southern Sweden

185 **Star of Bethlehem – O. umbellátum** Plants 10–35 cm high, perennials, generally with numerous secondary bulbs; blooming in April and May

Ornithógalum umbellátum L.
Star of Bethlehem

Liliáceae – Lily family

Leaves: Grass-like, narrow, linear, 1030 cm long, 3–7 mm wide, fleshy, with long pale stripe down the midribs, grooved, and growing from the ground

Inflorescence: Corymbs with up to 15 flowers; the lower stems noticeably elongated and generally extending above the end of the inflorescence; lower bracts up to 4 cm long

Flowers: 6 petals, not fused, up to 20 mm long, standing out in the form of a star, white inside and with a wide green stripe down the middle outside; 6 stamens, with narrow triangular filaments; ovary superior with a short pistil; fertilisation by insects, also self-fertile

Fruits: Capsules with 3 compartments and many seeds in each

Habitat: In hill and mountain (rarely subalpine) zones in vineyards, parks, orchards, meadows and fields on well-drained deep soils rich in nutrients and humus

186 **St Bruno's lily – P. liliástrum** Plants 20–50 cm high, with thickened roots (not like a turnip); blooming in June and July

Paradísea liliástrum (L.) Bertol.
St Bruno's lily

Liliáceae – Lily family

Leaves: Grass-like, narrow, linear, 20–40 cm long, up to 10 mm wide, gradually coming to a point and all growing from the ground; small leaves may be found on the inflorescence

Inflorescence: Racemes of 3–10 flowers growing only on one side; bracts are small, pointed, and longer than the flower stem

Flowers: 6 petals, not fused, lanceolate to oval, up to 6 cm long, coloured white, with 3 veins meeting at the point; 6 stamens, shorter than the petals; ovary superior; the pistil has a thickened stigma

Fruits: Leathery capsules with 3 compartments and many seeds in each

Habitat: In the mountain and subalpine (rarely alpine) zones in warm dry hills and meadows and on rocks on well-drained to damp, deep soil, which is poor or rich in lime and also manured; found in the Alps, particularly the central and southern Alps; found only occasionally in the northern Alps; perennial plants with short rhizomes

187 **St Bernard's lily – A. liliágo** Plants 20–70 cm in high with short rhizomes, blooming from May to June

Anthéricum liliágo L.
St Bernard's lily

Liliáceae – Lily family

Leaves: Grass-like, narrow, linear, 15 – 40 cm long, 4 – 8 mm wide, pointed, flat or grooved and rising from ground level; occasionally short leaves are found on the flower stem

Inflorescence: Simple raceme with flowers not only on one side, with bracts shorter than the flower stem

Flowers: 6 petals, not fused, 15–30 mm long, all of similar shape, narrow oval, with the greatest width towards the end, coming to a rounded point with 3 veins meeting at the end; 6 stamens shorter than the petals; ovary superior with a pistil which is often bent over; fertilisation by bees

Fruits: Capsules 10–15 mm long, pointed and divided into 3 compartments, each containing 4–8 seeds

Habitat: In hill and mountain (rarely subalpine) zones in meadows, well-drained and dry grassland, sparse oak and pine woods, at the edge of woods and thickets, and on rocky slopes, on dry soil usually poor in lime

188 **Lily of the valley – C. majális** Plants 10–20 cm high, perennials, with thin, wide spreading rhizomes; blooming in May

Convallaria majalis L.
Lily of the valley

Liliáceae – Lily family

Leaves: Lowest leaves are formed like scales; the 2 (seldom 3) upper ones are narrow oval to wide lanceolate, 10–20 cm long, up to 5 cm wide and coming to an obtuse point at the end

Inflorescence: Raceme with flowers on only one side, with long stem and 5–10 flowers, each with its own stem; bracts shorter than the flower stem and lanceolate

Flowers: Bell-shaped, nodding and sweet-scented; 6 petals, fused, 4–7 mm long, white, and with small points bent outwards; 6 sepals, not coming beyond the edge of the bell; ovary superior with a short, thick pistil; fertilisation by insects

Fruits: Red berries with 3 compartments, each with up to 6 seeds, which are distributed by animals

Habitat: Found from the hill zone up to the subalpine zone, common, forms clumps in oak and beech woods, mountain meadows, scree, and gardens (cultivated) on well-drained to fairly dry, loamy, sand and clay soil which is warm in summer, deep, loose, and rich in lime and humus

189 **Angular Solomon's seal – P. odorátum** Plants 15–50 cm high, with stems inclined, perennials blooming in May and June

Polygónatum odorátum (Mill.) Druce
P. officinále All.
Angular Solomon's seal

Liliáceae – Lily family

Leaves: 10–15 cm long, broad lanceolate to oval, sessile, dark green above, generally rather grey-green beneath, coming to an obtuse point at the end and alternate

Inflorescence: Flowers solitary, rarely in pairs

Flowers: 6 petals fused into a tube, white and green at the free end; 6 stamens, fused with the petals past their middle and after blooming often protruding out of the flower; filaments hairless; ovary superior with a pistil which is not thickened at the end; fertilised by bumble-bees

Fruits: Blue-black berries in 3 sections; distribution by animals

Habitat: From the hill zone to the subalpinhe zone in dry meadows, sparse undergrowth in oak woods, sparse woodlands, and in rocky places on dry, loose, stony and loamy soil which generally contains lime and humus; plants are found in light and half-shade; poisonous; grows well in pine woods

190 **Lesser butterfly orchid – P. bifólia**
Plants 20–50 cm high, with undivided, turnip-like tubers; flowering in May and June

Platánthera bifólia (L.) Rich
Lesser butterfly orchid

Orchidáceae – Orchid family

Leaves: Two, 5–20 cm long, up to 5 cm wide, oval and with parallel veins; leaves arise just above the base of the stem; there are up to 4 pointed lanceolate leaves on the stem

Inflorescence: Erect raceme with many flowers

Flowers: Whitish or yellow-green; 6 petals; outer 3 lanceolate, up to 15 mm long, pointed at the end or blunt, and standing away from the stalk; 2 inner petals shorter and less wide than the outer ones, sticking up and bent towards one another; lip points outwards, shaped like a band, up to 2 cm long, flat, undivided, and generally blunt; the spur is threadlike, gradually thinning towards the point; 1 stamen fused with the pistil into a column (gynostegium); filaments lie parallel, close to each other; ovary inferior

Fruit: Capsules cracking along their length, with numerous seeds

Habitat: In hill and mountain zones in sparse deciduous and coniferous forests, rough pasture, and near bushes, on moderately dry to damp soil

191 **Long-leaved helleborine – C. longifólia**
Plants 15–50 cm high, covered with leaves all the way up and flowering in May and June

Cephalánthera longifólia (L.) Fritsch
Long-leaved helleborine

Orchidáceae – Orchid family

Leaves: Linear to lanceolate, up to 12 cm long, folded, with parallel veins, coming to a long point, dark green, and alternately arranged in two lines in a single plane

Inflorescence: Loose spikes of flowers, from a few to 14

Flowers: All sloping outwards; 6 petals, bent together to form a bell, white, pointed, and generally concealing the lip; this is not undivided, is somewhat shorter than the petals, and has a deep incision between the two halves; there is no spur; 1 stamen; ovary inferior and smooth or with single glandular hairs; bracts lanceolate, often not reaching as far as the ovary

Fruits: Capsules containing many seeds

Habitat: In hill and mountain zones in mountain meadows, forest clearings, sparse oak and beech woods, and near bushes, on dry, base-rich soil containing lime and humus; soil may be loose stone or loam in sunny places; plants found in half-shade; distribution extends northwards to Ireland and southern Scandinavia

192 **Common cotton-grass – E. angustifólium**
Plants 20–55 cm high, perennial, with creeping roots underground; flowering from April to June

Erióphorum angustifólium Honckeny
Common cotton-grass

Cyperáceae – Sedge family

Leaves: Linear, 20–50 cm long, 3–5 mm wide, furrowed and comimg to a triangular point; the uppermost leaf has a funnel-shaped sheath and a short ligule

Inflorescence: Several spikes, bending over after flowering; spikes have smooth stems and a number of flowers; bracts 1 or more

Fruits: 1 triangular nut

Habitat: From hill to alpine zones in bogs, peat moors, marshy meadows, in ditches and on river banks, on wet, sometimes flooded, lime-free to acid soil poor in nutrients; appears first when ground becomes boggy; root reaches 50 cm deep; northern range extends to Iceland, Spitzbergen, southern range to northern Spain and southern Italy

193 **Scheuchzer's cotton-grass – E. scheúchzeri**
Plants 10–40 cm high, with long roots creeping underground; flowering from June to August

Erióphorum scheúchzeri Hoppe
Scheuchzer's cotton-grass

Cyperáceae – Sedge family

Leaves: Linear, 10–30 cm long, up to 5 mm wide, with smooth edges and flat or grooved; the top leaf is a sheath with an atrophied blade

Inflorescence: Solitary, upright, spherical spikes at the end of the stalk; bracts have grey points and narrow white edges

Flowers: Arranged in a double spiral; when in flower the petals are bristles in the form of long white hairs, the general shape of a fluffy, egg-shaped head 2–4 cm in diameter; 3 stamens; 1 ovary in 3 sections; 3 stigmata

Fruits: Triangular solitary nuts

Habitat: In the alpine (rarely subalpine) zone, in sizeable concentrations by ponds, small lakes, and bogs; soil is wet, poor in nutrients and flooded from time to time; northern range of this arctic/alpine sedge extends to Iceland, Spitzbergen, Finland, and arctic Russia; to the south it is found in isolated locations in the Pyrenees, the Alps, the northern Apennines, and the Carpathian mountains

194 **Crocus – C. albiflórus**
Plants up to 15 cm high, perennial, having corms covered with fibres; flowering from March to June

Crocus albiflórus Kit. ex Schult.
Crocus

Iridáceae – Iris family

Leaves: Grasslike, 5–15 cm long, up to 3 mm wide, curving downwards, with a white vein in the middle; springing from ground level

Inflorescence: 1 or 2 flowers on each plant, seldom more; flowers spring from ground level

Flowers: 6 petals fused into a long tube, their free parts forming a funnel 2–4 cm long and white to purple in colour, often striped; the outer 3 petals usually somewhat larger; the tubular throat is hairy; 3 stamens; ovary inferior, with a pistil and 3 stigmata with crimped edges

Fruits: Capsules in 3 compartments containing many seeds, distributed by ants

Habitat: Mostly in mountain and subalpine zones in meadows and pastures on well-drained clay and loamy soil rich in humus and nutrients and base-rich, moderately acid to neutral and deep; fairly common, in large concentrations; southern range extends to the Apennines, the mountains of the Balkan Peninsula, and Sicily

195 **Frog-bit – H. mórsus-ránae**
Water plants 5–15 cm long, with roots in bunches at the nodes; flowering from June to August

Hydrócharis mórsus-ránae L.
Frog-bit

Hydrocharitáceae – Frog-bit family

Leaves: Round, cordate at the base, floating on the water (without any immersed leaves!),

15–60 mm in diameter with stalks 5–10 cm long

Inflorescence: 3 flower buds at most enclosed in a common spathe

Flowers: Single-sexed; plants are dioecious; perianth usually of 3 outer and 3 inner white petals; the outer ones are oval and 3–5 mm long, the inner ones are rounder and have a yellow base; male flowers have 12 stamens; female flowers have only 1 flower bud in the sessile spathe; ovary of 6 carpels; many ovules; 6 stigmata, each divided in two

Fruits: Follicles ripening under water

Habitat: In the hill zone in stagnant water, ponds, pools, and slowly flowing water rich in nutrients, lime-poor, protected from the wind and well heated in summer; plants found in light and half-shade

196 **Water plantain – A. plantágo-aquática**
Plants 15–100 cm high, perennials, with thick rhizomes; flowering from June to August

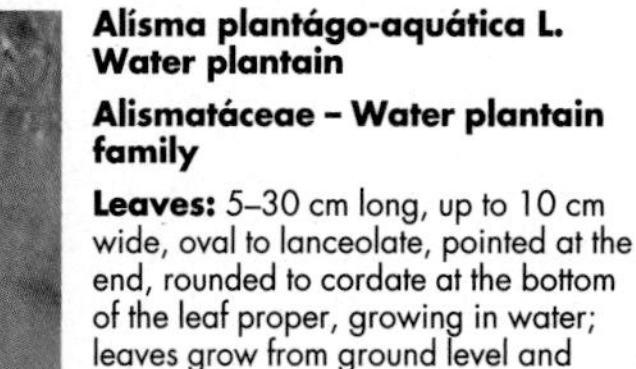

Alísma plantágo-aquática L.
Water plantain

Alismatáceae – Water plantain family

Leaves: 5–30 cm long, up to 10 cm wide, oval to lanceolate, pointed at the end, rounded to cordate at the bottom of the leaf proper, growing in water; leaves grow from ground level and have stalks

Inflorescence: Panicle with many flowers; side branches arranged in whorls

Flowers: 6 perianth segments usually with toothed edges; the 3 outer greenish and similar to sepals; the inner ones 2–3 times as long as the outer ones, round (2–3 mm in diameter), like petals, coloured white or pink, generally with toothed edges; 6 stamens, 2 by each of the inner petals; numerous superior carpels in a ring and not fused; styles longer than carpels and upright

Fruits: Achenes (each with one seed) usually with a groove along the seam between the 2 parts of the shell

Habitat: In hill and mountain zones in still or slowly flowing water 20–50 cm deep, in lakes and ponds rich in nutrients, on rather acidic sandy or muddy soil rich in humus

197 **Perfoliate penny cress – T. perfoliátum**
Plants 5–20 cm high, annuals or biennials, with smooth stems and thin roots; flowering in April/May

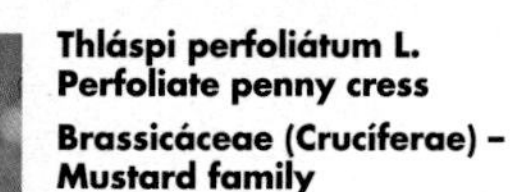

Thláspi perfoliátum L.
Perfoliate penny cress

Brassicáceae (Crucíferae) – Mustard family

Leaves: Up to 4 cm long, narrow oval, pointed, entire or rarely somewhat toothed, blue-green, smooth, the bottom leaves with stems, those above sessile and amplexicaul with rounded, blunt, or pointed lobes

Inflorescence: Raceme, becoming distinctly elongated after flowering

Flowers: 4 sepals, greenish, up to 18 mm long and smooth; 4 petals, white, up to 25 mm long and smooth; 6 stamens, the outermost two being shorter than the 4 inner ones; yellow stamens; superior ovary of 2 ovules set together

Fruits: Pods 3–6 mm long, narrow inverse heart-shape with wings narrowing towards the base

Habitat: In hill and mountain zones on dry hills, tops of walls, in fields and vineyards, on dry to moderately well-drained loamy soil rich in nutrients, lime and base-rich but containing little humus; shallow roots; originally a Mediterranean plant

198 **Alpine penny cress – T. sylvium**
Plants 5–15 cm high, usually with elongated, wide-spreading roots; flowering from June to August

Thláspi sylvium Gaudin T. alpinum Crantz ssp. silvium P. Fournier
Alpine penny cress

Brassicáceae (Crucíferae) – Mustard family

Leaves: Rising from ground level, 1–5 cm long, narrow spathulate and rounded at tip; leaves on stem lanceolate to narrow oval, pointed, sessile, and usually amplexicaul

Inflorescence: Racemes like composite flowers at the end of the stalk

Flowers: 4 sepals, narrow oval, 2–3 mm long, greenish and often with whitish edges; 4 petals, oval, 5–12 mm long and white; 6 stamens (the 2 on the outside short and the inner ones longer); anther golden; superior ovary of 2 ovules set together

Fruits: Pods with narrow wings, somewhat obcordate and narrowing to the base like a wedge; beak on the pods is generally over 2 mm long

Habitat: In subalpine and alpine zones on stony fields and scree (on silica rock); these alpine plants have been found, among other places, in the Valais (e.g. near Zermatt), in the Ticino, and in Piedmont

199 **Gipsywort – L. europaéus**
Plants 15–70 cm high, with thick spreading roots; flowering from July to September

Lycopus europaéus L.
Gipsywort

Lamiáceae (Labiátae) – Mint family

Leaves: Narrow oval, 2–8 cm long, up to 4 cm wide, sessile or with short leaves, pointed, with deep, coarse serrations, mostly with teeth pointing toward the tip of the leaf; mid to dark green above and grey-green beneath; the lower leaves of plants in wet locations are more pinnately cut

Inflorescence: Flowers arranged in axillary whorls about the stem

Flowers: 5 fused sepals, 2–4 mm long and with 5 stiff teeth with long points; 4 petals fused into a tube; points of corolla 3–6 mm long, white, the 3 lower ones with red spots; 2 stamens, protruding beyond the corolla and free-standing; ovary superior; flowers fertilised by flies

Fruits: Tetrahedral nutlets; dispersal by water and waterfowl

Habitat: In hill and mountain zones in marshes, reed beds, alder thickets, and on banks on wet, periodically flooded, sandy, clay, or peat soil rich in nutrients, base-rich, and containing rotted humus

200 **Garlic mustard – A. petioláta**
Plants up to 90 cm high, smelling like garlic and flowering from April to June

Alliária petioláta Scop.
Garlic mustard

Brassicáceae (Crucíferae) – Mustard family

Leaves: Basal leaves kidney- or heart-shaped, with long stems, 10–15 cm long, with bluntly toothed or notched, and smooth (but somewhat hairy) stem; upper leaves short-stemmed, cordate, pointed and generally with pointed teeth

Inflorescence: Racemes at ends of the main stem and branches

Flowers: 4 sepals, greenish, smooth, 2–4 mm long, and not protruding; 4 petals, white, 3–6 mm long, narrow oval and rounded; 6 stamens (2 outer ones shorter than the 4 inner ones); ovary superior, made up of 2 ovules with 3 veins; fertilisation by bees or self-fertile plants

Fruits: Pods 2–7 mm long

Habitat: In hill and mountain zones among bushes, in gardens and parks run wild, spinneys, and on the edge of woods on well-drained, loose, loamy soil preferably rich in nitrogen and humus, mainly in humid conditions; grows as a weed in riverside woodland; used as a salad plant in earlier times

201 **Scurvy grass – C. officinális**
Plants 15–50 cm high, generally biennials, smooth with thin tap-roots; flowering from June till August

Cochleária officinális L.
Scurvy grass

Brassicáceae (Crucíferae) – Mustard family

Leaves: Basal leaves in a rosette, long-stalked, round to kidney-shaped, up to 2 cm long without the stalk, entire or curved; lower leaves on the stalk oval or wedge-shaped and sessile, upper leaves with 2 points and amplexicaul

Inflorescence: Branched racemes at the end of the stalk, with many flowers

Flowers: 4 sepals, narrow oval, 2–3 mm long, the outsides often brownish at the point; 4 petals, oval, white, 3–6 mm long, narrowing to a wedge shape at the base and rounded at the end; 6 stamens, rather longer than the petals; superior ovary of 2 ovules; fertilised by insects

Fruits: Pods 3–7 mm long, with distinct middle vein and containing 4–8 seeds

Habitat: In hill and mountain zones by rubbish dumps and on salt marshes along the coast, on salty soil; earlier cultivated as a medicinal plant and now run wild

202 **Alpine rock cress – A. alpina**
Plants 10–25 cm high, perennials, with basal rosettes of leaves; flowering from March to October

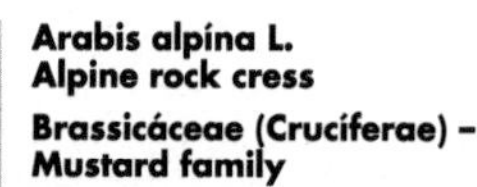

Arabis alpína L.
Alpine rock cress

Brassicáceae (Cruciferae) – Mustard family

Leaves: Basal leaves broad oval to oval, with short stems, pointed, deeply toothed and with numerous hairs; stem leaves lanceolate to oval, pointed, deeply toothed (with teeth pointing upwards) and with 2 lobes surrounding the stem

Inflorescence: Raceme with many flowers, growing longer as flowering approaches

Flowers: 4 sepals, greenish, 2–4 mm long and smooth or having single hairs; 4 petals, white, obovate, clawed, and 5–10 mm long; 6 stamens; ovary of 2 fused ovules and superior; fertilisation by insects or self-fertile plants

Fruits: Pods 2–6 cm long, up to 2 mm wide, smooth, without distinct veins, with protruding stalk, and containing many seeds

Habitat: From mountain to alpine zones in scree, boulders, rock crevices deep in ravines, or alluvial gravel on well-watered stony soil, usually containing lime but little humus, and undisturbed

203 **Haller's rock cress – C. hálleri**
Plants 20–40 cm high, with root runners above ground; flowering from April to June

Cardaminópsis hálleri (L.) Hayek
Haller's rock cress

Brassicáceae (Cruciferae) – Mustard family

Leaves: Basal and lower stem leaves narrow oval, stalked, most pinnately cut, with a large, oval, toothed end section and 1–3 lateral sections on each side; leaves on upper stem lanceolate, toothed or entire, narrowing towards the base and sessile

Inflorescence: Racemes

Flowers: 4 sepals, narrow oval to lanceolate, yellowish, 2–3 mm long, with small single hairs; 4 petals, white, occasionally pale lilac, 3–6 mm long, wide oval and rounded or square-trimmed ends; 6 stamens (outer 2 short and inner ones longer); superior ovary: both ovules have indistinct veins

Fruits: Pods on upright or horizontal stalks

Habitat: In mountain and subalpine zones on river banks, mountain meadows and damp rocks on well-drained to fairly moist soil poor in lime but rich in nutrients; one of the first plants to colonise a fresh grave

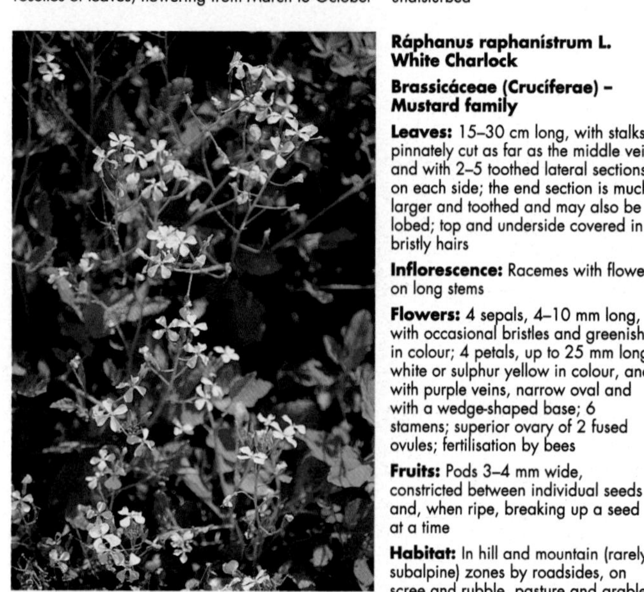

204 **White Charlock – R. raphanistrum** Plants 15–60 cm high, annual weeds that have accompanied cultivation worldwide; flowering May to Oct.

Ráphanus raphanístrum L.
White Charlock

Brassicáceae (Cruciferae) – Mustard family

Leaves: 15–30 cm long, with stalks, pinnately cut as far as the middle vein and with 2–5 toothed lateral sections on each side; the end section is much larger and toothed and may also be lobed; top and underside covered in bristly hairs

Inflorescence: Racemes with flowers on long stems

Flowers: 4 sepals, 4–10 mm long, with occasional bristles and greenish in colour; 4 petals, up to 25 mm long, white or sulphur yellow in colour, and with purple veins, narrow oval and with a wedge-shaped base; 6 stamens; superior ovary of 2 fused ovules; fertilisation by bees

Fruits: Pods 3–4 mm wide, constricted between individual seeds and, when ripe, breaking up a seed at a time

Habitat: In hill and mountain (rarely subalpine) zones by roadsides, on scree and rubble, pasture and arable fields on well-drained to damp alkaline soil, lime-poor but rich in nutrients

205 **Sand rock cress – C. arenósa** Plants 10–25 cm high, often biennials, without a sterile rosette of leaves at the base; flowering April to Oct.

Cardaminópsis arenósa (L.) Hayek
Sand rock cress

Brassicáceae (Cruciferae) – Mustard family

Leaves: Basal leaves and those on the lower stem stalked, most pinnately cut to the middle vein, with 1–6 leaflets on either side and a long teminal leaf; leaves on the upper stem are narrow lanceolate, generally sessile and either toothed or entire

Inflorescence: Racemes

Flowers: 4 sepals, up to 3.5 mm long, green, very often with hairs; 4 petals, 5–10 mm long, and white or pale lilac; often faintly veined in deep purple; 6 stamens; superior ovary of 2 fused ovules; fertilisation by insects

Fruits: Pods 20–40 mm long, generally at a distinct upward angle to the fruit stalks

Habitat: In hill and mountain zones on rubbish dumps, railway sidings, by ditches, at the edge of woodland, on sand and pebble paths, grass meadows, on moderately damp to dry stony or sandy soil which is loose, contains lime and humus and is base-rich; typically one of the first plants to appear when ground is cleared

206 **Seven-leaved toothwort – D. heptaphylla**
Plants 20–60 cm high, with smooth stems and rhizomes; perennials flowering in April and May

Dentária heptaphylla Vill.
Cardámine heptaphylla (Vill.) O.E.Schulz
Seven-leaved toothwort

Brassicáceae (Cruciferae) – Mustard family

Leaves: Situated alternately on the stem, unequally pinnate, dark green, with stems and with 5, 7, or rarely 9 leaflets; these are lanceolate to narrow oval, coming to a fine point, toothed and without stems

Inflorescence: Racemes, nodding before flowering starts

Flowers: 4 sepals, up to 9 mm long, smooth and greenish-yellow; 4 petals white or pale lilac, 10–20 mm long, and narrow obovate; 6 stamens with yellow anthers; superior ovary of 2 ovules

Fruits: Pods up to 7 cm long and 2–5 mm thick, with beak 3–10 mm long

Habitat: In hill and mountain (rarely subalpine) zones in beech and mixed beech/oak woods with good ground cover, on well-drained to damp, loose clay and peat soil rich in nutrients, generally containing lime and humus and rather stony; found in shade; sometimes found in mountain forests; typically found on waste ground

207 **Large bittercress – C. amára** Plants 10–40 cm high, perennials, with horizontal creeping rhizomes from which roots and shoots emerge; flowers May/ June

Cardámine amára L.
Large bittercress

Brassicáceae (Cruciferae) – Mustard family

Leaves: Situated on the stalk, unequally pinnate, with stems and up to 12 cm long; 5, 7, 9, or 11 leaflets, lanceolate to oval and entire or crenate; end leaflet is round to oval; basal leaves are not arranged in a rosette and are pinnate, with stems

Inflorescence: Panicles

Flowers: On long stems; 4 sepals, greenish, 2–5 mm long and smooth; 4 petals, white, oval, distinctly veined, and up to 10 mm long; 6 stamens, narrow triangular, greenish and white towards the end; anthers purple or violet; superior ovary of 2 ovules set together; fertilisation by insects

Fruits: Pods 15–40 mm long, up to 2 mm wide, with many seeds and opening suddenly, flinging the seeds some distance

Habitat: From hill to subalpine zones next to springs and brooks, and in alder thickets, on wet, often flooded soil which is cool, rich in nutrients and humus and base-rich; in earlier times used as a medicinal and salad plant; found especially in shady places

208 **Watercress – N. officinále**
Plants 20–90 cm high, with creeping or upright stems; flowering from June to September

Nastúrtium officinále R.Br.
Watercress

Brassicáceae (Cruciferae) – Mustard family

Leaves: Unevenly pinnate, with stalks and 2–4 oval leaflets on each side; these are narrow oval and entire, crenate, or irregularly toothed and usually rounded at the end; the end leaflet is larger, oval to round, with irregular blunt teeth; the upper side of the leaf stalk is covered with short hairs; stays green in autumn and winter

Inflorescence: Racemes

Flowers: 4 sepals, greenish, smooth and 2–3 mm long; 4 petals, white, round, 3–5 mm long and, like the filaments, turning lilac after flowering; anthers yellow (not purple as with Cardámine amára!); 6 stamens; ovary superior; fertilised by insects or self-fertile plants

Fruits: Pods 12–20 mm long, up to 2.5 mm wide and smooth; dispersal by waterfowl

Habitat: In hill and mountain (rarely subalpine) zones in streams, by springs, in flushes and reed beds with running, fairly cool water rich in nutrients

209 **Sweet woodruff – A. odoráta**
Plants 10–30 cm high, perennials, with branching rhizomes; flowering from April to June

Aspérula odoráta L.
Gálium odorátum (L.) Scop.
Sweet woodruff

Rubiáceae – Bedstraw family

Leaves: In whorls of 6–9 leaves each; each leaf lanceolate, with the widest part near the end, 20–45 mm long, with a clearly marked main vein, forward-pointing prickles on the edges and the main vein, covered with rough hairs

Inflorescence: Long-stalked, terminal with 1–4 lateral cymes, resembling an umbel in form; cymes contain only a few flowers

Flowers: Stems 0.5–3 mm long; sepals usually reduced to an indistinct ring; 4 petals, fused to half-way, white, spreading out flat or in a funnel shape, and either rounded or pointed at the end; 4 stamens, alternating with the petals; ovary inferior, divided into 2 compartments, with a pistil divided in two; fertilised by insects

Fruits: Wrinkled, with hooked black-tipped bristles, 2–3 mm long and divided into two

Habitat: In hill and mountain (rarely subalpine) zones in beech and mixed deciduous forests with ground cover, on well-drained soil rich in humus and nutrients and base-rich

210 **Hedge bedstraw – G. mollúgo**
Plants 20–140 cm high, recumbent, climbing or erect; flowering from May to September

Gálium mollúgo L.
Gálium elátum Thuill.
Hedge bedstraw

Rubiáceae – Bedstraw family

Leaves: Arranged in several whorls; each leaf 10–30 mm long, up to 7 mm wide, oval to broad lanceolate, entire, coming to a sharp point, very thin but slightly thickened at the edges

Inflorescence: Pyramidal, loose thyrses with many flowers at the nodes of the topmost leaves; the outermost branches are without bracts

Flowers: Stalks 3–4 mm long; sepals very small; 4 petals at most, fused, white, 1–2 mm long and with sharply pointed ends; stamens equal in number to petals; filaments up to 3 mm long; ovary inferior and in 2 parts; fertilisation by insects

Fruits: Wrinkled and brownish, without hairs or hooks,with a more or less smooth surface

Habitat: In hill and mountain zones in sparse deciduous forest, fertile meadows, and bushes on moist to intermittently moist soil

211 **Dwarf bedstraw – G. púmilum**
Plants 10–30 cm high, generally without non-flowering shoots; blooming from May to July

Gálium púmilum Murray
Gálium silvéstre Poll.
Dwarf bedstraw

Rubiáceae – Bedstraw family

Leaves: In whorls of 5–9 leaves each; each leaf 1–2 cm long, lanceolate, often rather sickle-shaped, with a sharp point, with more or less thickened edges and short single hairs

Inflorescence: Thyrses with many flowers; the side shoots spring from the leaf nodes

Flowers: On stems; sepals very small, green; 4 fused petals, white, 1–2 mm long and pointed; the same number of sepals as petals; ovary inferior and in two parts; fertilisation by insects

Fruits: Wrinkled, more or less smooth and with straight stalks

Habitat: Frequently found in hill and mountain zones on siliceous or low-grade pasture or meadows, among thin bushes, in open woodland and dry willow beds, and on dry grassland on moderately well-drained to intermittently wet soil tending to sand or loam, which is base-rich, lime-poor, moderately acid, with rotted humus; the plant is an indicator of acidification; it will not endure excess manure or wetness; plants grow in light and half-shade

212 **Alpine poppy – P. alpínum**
Plants 5–25 cm high, perennials, with a rhizomes producing to many shoots; flowering in July

Papáver alpínum L.
Alpine poppy

Papaveráceae – Poppy family

Leaves: Forming a basal rosette; single leaves are smooth or covered with short hairs, singly or doubly pinnately cut as far as the middle vein, grey-green in colour, and with several lanceolate sections; flower stem lacks leaves

Inflorescence: Solitary flower on the stem

Flowers: 2 sepals covering the flower bud completely, oval, blunt-ended, entire, covered with brown hairs and dropping off when the flower opens; 4 petals, oval to round, 15–25 mm long, entire or wavy-edged, without stems and white; numerous stamens; filament light and thread-like; ovary superior with many ovules; fertilisation by insects

Fruits: Capsules 10–15 mm long, with projecting longitudinal lines underneath the 4–9 stigma-rays and covered with white hairs

Habitat: In the alpine zone on moraines, coarse limestone rubble, and alluvium on scree covered with snow for long periods; lives on rock debris and holds it together; rarely flooded out

213 **Yellow-white fumitory – C. ochroleúca**
Plants 10–40 cm high, smooth-leaved, with branching rhizomes; flowering from June to Sept.

Corydalis ochroleúca Koch
Yellow-white fumitory

Fumariaceae – Fumitory family

Leaves: 10–25 cm long, oval, doubly or trebly pinnate and blue-green; sections are oval, generally narrowing to a wedge shape and with unequally notched or incised sections; stems have a narrow wing-like edge toward the base

Inflorescence: The end raceme is longer than the lateral ones

Flowers: Zygomorphic, 10–15 mm long, white to pale yellow and yellow at the end; sepals are toothed and 2–3 mm long; 2 outer petals, the top lip with a backward-pointing spur, widening out at the front and bending upwards, the lower lip widening out at the front and bending downwards; the spur usually bends downwards; the inner 2 petals are similar in form and light or dark yellow in colour; 4 stamens (2 inner and 2 outer); superior ovary with 2 ovules; 2 stigmata on one pistil

Fruits: Podlike capsules

Habitat: In hill and mountain zones on rocks and scree, and on walls on damp, stony soil

215 **Hoary plantain – P. média**
Plants 15–40 cm high, common perennials, flowering from May to July

Plantágo média L.
Hoary plantain

Plantagináceae – Plantain family

Leaves: All in a basal rosette, generally lying on the ground, oval, or oval to broad oval, up to 15 cm long, occasionally with short teeth, with short stems, narrowing toward the base, pointed, with 5–9 veins and scattered to thick fine hairs

Inflorescence: Spikes carrying many flowers, 2 – 7 cm long and with flat hairy stalks

Flowers: 4 sepals, free-standing almost to the base; 4 petals fused together into a tube, 2–4 mm long, smooth and with white points; 4 stamens of equal length, lilac and fused to the tube formed by the petals; superior ovary of 2 ovules; pistil with a hairy stigma; fertilisation by insects or self-fertile plants

Fruits: Oval capsules splitting in two round the centre to release 3–8 seeds; dispersal by wind

Habitat: Hill and mountain (rarely subalpine) zones in dry meadows, semi-dry grassland, and by the roadside on well-drained to dry loamy soil

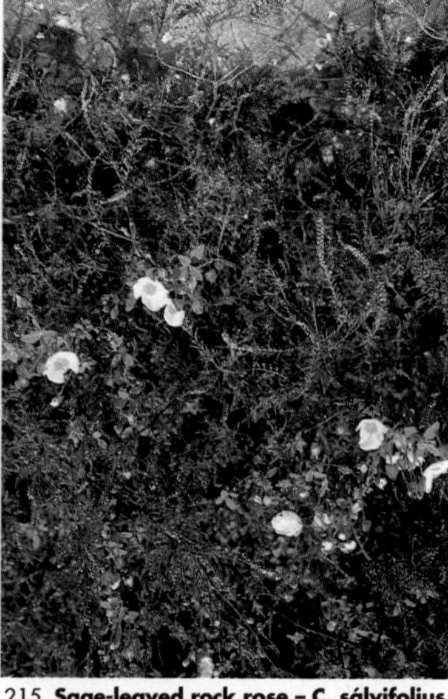

215 **Sage-leaved rock rose – C. sálvifolius**
Plants 15–70 cm high, with twigs covered in thick hairs; flowering in May

Cístus salvifólius L.
Sage-leaved rock rose L.

Cistáceae – Rock rose family

Leaves: Opposite, narrow oval, 1–4 cm long, pointed or rounded at the end, narrowing to a wedge or heart shape at the base, with clearly etched vein network, wrinkled and thickly covered with hairs on the underside; leaf stalk covered with bushy hairs 1.5 mm long

Inflorescence: Solitary, long-stemmed flowers rising from leaf nodes

Flowers: 5 sepals (outer 2 heart-shaped at base, inner 3 rounded), not fused together, green, but often reddish in colour and thickly covered with hairs; 5 petals, up to 2 cm long, white, not fused together, each with a black spot at the base; numerous stamens; superior ovary of 5 ovules with 5 compartments; pistil 3 mm long, with 5-fold stigma

Fruits: Capsules containing many seeds

Habitat: In hill zone on south-facing rock faces in sunlit situations, generally on dry, stony soil poor in nutrients and weakly to strongly acidic (e.g. by the Langensee)

216 **Grass of Parnassus – P. palústris**
Plants 5–30 cm high, perennials, completely smooth-leaved; flowering from July to September

Parnássia palústris L.
Grass of Parnassus

Saxifragáceae – Saxifrage family

Leaves: Rising from ground level, cordate, generally with rounded end, stem 5–10 cm long and generally blue-green in colour; there is occasionally 1 amplexicaul leaf on the flower stem

Inflorescence: Soliary flowers at the end of long stalks

Flowers: 5 sepals; 5 petals, oval, 2–3 times as long as the sepals, coloured white or pink, with numerous clearly incised parallel veins on the top side, arranged radially; 5 staminodes with numerous spherical yellow glands on stems 1–4 mm long; 5 stamens with light green or whitish anthers next to the sepals; superior or intermediate ovary of 4 ovules; fertilisation by insects

Fruits: Capsules containing 1 compartment

Habitat: Ranging from hill zone to alpine zone on marshes and wet moors and by springs, and in the alpine zone on water-bearing scree; preferred soil moist to wet. containing little humus

217 **Buckwheat – P. esculéntum**
Plants 15–70 cm high, annuals flowering from July to September

Polygonum fagopyrum L.
Fagopyrum vulgáre Hill.
Fagopyrum esculéntum
Buckwheat

Polygonáceae – Dock family

Leaves: Cordate to hastate, 3–8 cm long, generally longer than broad, pointed, with stems on the lower part of the stem but generally sessile on the upper part, entire, dark green above and grey-green beneath

Inflorescence: Usually spikes growing from the leaf nodes

Flowers: 5 petals, white to light red in colour and when ripe up to 5 mm long; 8 stamens with reddish anthers; superior ovary; 3 pistils each with a stigma at the end; fertilisation by insects

Fruits: Capsules 5–7 mm long, oval, gradually coming to a point away from the base and without burls or margins

Habitat: Wild in the mountain zone on rubbish dumps and waste ground and by roadsides, on well-drained to dry loamy/sandy soil rich in nutrients and humus and moderately acid; tends to seek warmth; cultivated in earlier times, but rarely nowadays; attracts bees

218 **Swallow wort – V. officinále**
Plants 20–120 cm high, with short rhizomes and hairy stems; flowering from June to August

Vincetóxicum officinále Moench
V. hirudinária Medicus
Swallow wort

Asclepiadáceae – Milkweed family

Leaves: Opposite, with short fluffy hairs on the stem, 5–12 cm long, the lower ones tending to be heart-shaped, those above long lanceolate with long points, green to blue-green in colout, with hairs on the veins beneath

Inflorescence: Cymous flower heads surrounded by lanceolate bracts, with long stems, springing from the leaf nodes on the upper part of the stalk

Flowers: 5 sepals, with narrow, sharp points up to 2 mm long; 5 petals, triangular, off-white to yellow-green, 2–4 mm long, with the edge of the point rolled inwards; perigynous zone fused and hemispherical; 5 stamens; filaments fused together with the pistil into a column (gynostegium); superior ovary; fertilisation by insects

Fruits: Follicles with skins; seeds have mass of hair; dispersal by wind

Habitat: From hill to subalpine zones in sparse oak or pine woods, at the edge of thickets and on scree

219 **Chickweed – S. média** Plants 5–35 cm high; they have accompanied cultivation since Palaeolithic times; flowering throughout the year

Stellária média L.
Chickweed

Caryophylláceae – Pink family

Leaves: Oval to narrow oval or cordate, 3–5 cm long, opposite, pointed at the end, somewhat rounded at the base, with a short hairy stem (the lower leaves having more distinct stems)

Inflorescence: Flowers in thick dichasia

Flowers: 5 sepals, not fused, 3–5 mm long and with white hairs; 5 petals, white and deeply bifid; up to 10 stamens, generally 3–5; superior ovary; fertilisation by insects or self-fertile plants

Fruit: Capsules opening with teeth; seeds up to 1.5 mm long, with wide, blunt burls.

Habitat: From hill to subalpine zone in gardens, vineyards, rubbish dumps, by the roadside, and on river banks on well-drained to moist, loose soils containing humus and often lime and nitrogen; an indicator of nitrogenous soil; present along with cultivation; mat-forming; mainly a weed on cultivated ground; a plant with worldwide distribution

220 **Creeping baby's breath – G. répens** Plants 10–25 cm high, creeping or climbing perennials flowering from May to August

**Gypsóphila répens L.
Creeping baby's breath**

Caryophylláceae – Pink family

Leaves: Very narrow lanceolate, 1–3 cm long, blue-green in colour in this specimen, opposite and without stipules

Inflorescence: Cymose corymbs or panicles

Flowers: 5 sepals, fused together, 2–4 mm long, smooth with 5 pointed teeth about half as long as the fused sepals; 5 petals, white or reddish in colour, 5–10 mm long, obovate, rounded, or slightly notched at the end, narrowing gradually into a claw and not fused; 10 sepals; superior ovary with 2 styles

Fruits: Capsules 3–5 mm opening by 4 teeth; seeds kidney-shaped, generally rather flattened, dark in colour, burled, and 1–2 mm in diameter

Habitat: In subalpine and alpine zones in rubbish dumps, beds and banks of streams, and in shingle by alpine streams, on moist or periodically flooded , loose, lime-rich soil

221 **Common pink – D. plumárius** Plants 20–50 cm high, perennials with branching rhizomes; flowering from June to August

**Diánthus plumárius L.
Common pink**

Caryophylláceae – Pink family

Leaves: Narrow lanceolate, 2–6 cm long, opposite, with leaf sheaths1–2 times the width of the leaf; without stipules

Inflorescence: Long-stemmed solitary flowers or sparse flower heads

Flowers: Usually 5 sepals, fused into a tube and with short scales; 5 petals, with the spread part up to 18 mm long, not fused, and with long claws, incised almost to the middle, white or pink in colour and generally dark red at the base of the spread; 10 stamens; superior ovary with 2 styles

Fruits: Capsules opening by 4 teeth; seeds shaped like a shield with thickened edges

Habitat: From hill to subalpine zones in rocks, crags, and thickets, on stony, lime-rich soil in sunny places; this pink is very variable and can be divided into many subspecies; many varieties and hybrids are cultivated in gardens, and these are sometimes naturalised

222 **Yellow soapwort – S. lútea** Plants 5–12 cm high, perennials, with branching rhizomes, covered with short hairs and flowering in July/August

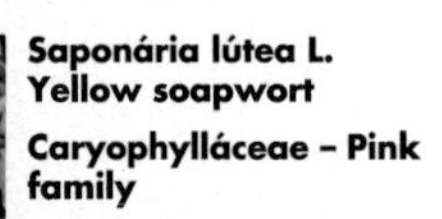

**Saponária lútea L.
Yellow soapwort**

Caryophylláceae – Pink family

Leaves: Linear or narrow lanceolate, opposite, 1–4 cm long, pointed, with a clearly protruding main vein on the underside, smooth on the surface, with hairs at the edges; basal rosette does not produce flowers

Inflorescence: Terminal corymbs, the flowers having short stems

Flowers: 5 sepals, fused, 6–12 mm long, often whitish towards the base, greenish above, covered with woolly hairs and with similar teeth; 5 petals, narrowly obovate, 10–15 mm long, light yellow to pale sulphur yellow, with a violet claw; 2 coronal scales about 1 mm long ; 10 stamens; ovary superior with free-standing styles

Fruits: Capsules 6–10 mm long

Habitat: In the alpine (rarely subalpine) zone on grassland, among rocks, and on scree, on loose, stony, well-drained to moderately dry, lime-rich soil

223 **Creeping baby's breath – G. répens** Plants 10–25 cm in height, creeping or climbing perennials flowering from May to August

**Gypsóphila répens L.
Creeping baby's breath**

Caryophyllaceae – Pink family

Leaves: Very narrow lanceolate, 1–3 cm long; in this specimen, found in a different location from number 220, the leaves were distinctly light green

Inflorescence: Cymose corymbs or panicles

Flowers: 5 sepals, fused together, 2–4 mm long, smooth with 5 pointed teeth about half as long as the fused sepals; 5 petals, white or reddish in colour, 5–10 mm long, obovate, rounded, or slightly notched at the end, narrowing gradually into a claw and not fused; 10 sepals; superior ovary with 2 styles

Fruits: Capsules 2–3 mm opening by 4 teeth; seeds kidney-shaped, generally rather flattened, dark in colour, burled, and 1–2 mm in diameter

Habitat: In subalpine and alpine zones in rubbish dumps, beds and banks of streams, and in shingle by alpine streams, on moist or periodically flooded , loose, lime-rich soil

224 **Sandwort – M. ciliáta** Plants 2–15 cm high, with recumbent or erect stems; flowering in July and August

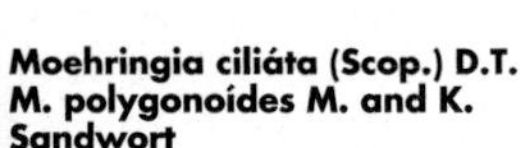

Moehringia ciliáta (Scop.) D.T. M. polygonoídes M. and K. Sandwort

Caryophylláceae – Pink family

Leaves: Narrow lanceolate, opposite and dark green, up to 1 cm long, generally rather fleshy, with 1–3 indistinct veins, generally smooth towards the top, but with short hairs towards the ground

Inflorescence: Often solitary flowers or 2–3 at end of stem

Flowers: Stalk 2–4 times as long as the small topmost leaves and greenish to dark red; 5 sepals, narrow oval, with blunt or pointed ends, 1–3 indistinct veins, and green in colour; 5 petals, longer than the sepals, white in colour, oval, not fused and entire; 10 stamens; superior ovary with 3 styles

Fruits: Capsules; seeds up to 1.5 mm long with a short fringed appendage

Habitat: In the alpine (rarely subalpine) zone in scree on well-drained to moist, loose, stony, lime-rich soil; a plant of the central and south European mountains

225 **White stonecrop – S. álbum** Plants 5–20 cm high, with numerous recumbent shoots, flowering from June to September

**Sédum álbum L.
White stonecrop**

Crassuláceae – Stonecrop family

Leaves: Linear and cylindrical, round in cross-section, blunt or rounded at the end, 5–15 mm long, without stalks, smooth, greenish, light yellow or reddish and alternate; the leaves are closer together on the sterile recumbent shoots than on the stems bearing flowers

Inflorescence: Umbel-like panicles with many flowers

Flowers: Usually 5 sepals, small and gradually rounded at the tip; usually 5 petals, up to 6 mm long, not fused, narrow lanceolate, blunt, often with red veins at the base and white or pale pink in colour; 10 stamens arranged in 2 rings; anther red and spherical; superior ovary of 5 carpels; fertilisation by insects

Fruits: Schizocarps bearing many seeds

Habitat: From the hill to the subalpine zone on rocks, gravel layers, the tops of walls, embankments, and rocky meadows on dry, stony, rocky soil poor in nutrients; mat-forming

226 **Alpine bistort – P. vivíparum**
Plants 5–30 cm high, often with with tortuous serpentine rhizomes; flowering from June to August

Polygonum vivíparum L.
Alpine bistort

Polygonáceae – Buckwheat family

Leaves: Lanceolate or narrow oval, 2–7 cm long, coming to a long point, with edges rolled downwards, with a rather deep whitish middle vein above, generally with short hairs beneath, lower leaves with stems and upper leaves sessile; sheathing stipules narrow triangular, long, and often brownish

Inflorescence: At end of stem, a loose and narrow spike with small bulbils in the lower part

Flowers: Single- or double-sexed; plants may be monoecious or dioecious; 5 petals, white and 2–4 mm long; ovary superior

Fruits: Triangular dark brown achenes

Habitat: In subalpine and alpine (rarely mountain) zones in sparse woodlands, meadows, rough pasture and grassland on damp to dry stony and loamy soil which is base-rich with lime and humus on top; plant also reproduces by means of tubers; preferably roots in humus with a tap-root; north of range extends to Iceland and Spitzbergen

227 **Livelong saxifrage – S. paniculáta**
Plants 5–45 cm high, perennials forming loose mats with basal rosettes; flowering from May till July

Saxifrága paniculáta Miller
Saxifrága aizáon Jacq.
Livelong saxifrage

Saxifragáceae – Saxifrage family

Leaves: Form part of numerous rosettes, 1–5 cm long, generally spathulate, blunt, or pointed at the end, with fine regular teeth, with horizontal hairs at the base, spreading horizontally or obliquely next to the flower stem, and on the upper side generally with a chalk gland at each serration; leaves on the flower stem similar to those in the rosette, but smaller and alternate

Inflorescence: Panicles with 1–5 flowers on each branch

Flowers: 5 sepals, with glands on the edge, up to 3 mm long, fitting closely to the petals and pointed or blunt at the end; 5 petals, oval to round, white with numerous red dots and up to 9 mm long; 10 stamens; inferior ovary of 2 carpels

Fruits: Capsules containing many seeds

Habitat: In subalpine and alpine zones on scree and rock faces, on moderately dry to dry stony soil which is base-rich and usually contains lime and humus and lacks fine earth

228 **Rock jasmine – A. villósa**
Perennial plants, building up thick mats and flowering in June and July

Andrósace villósa L.
Rock jasmine

Primuláceae – Primrose family

Leaves: Lanceolate, entire, 5–10 mm long, with numerous hairs at the edges and on the underside, particularly in the upper part; in addition to the rosettes spread on the surface there are hemispherical rosettes 2–5 cm above the ground (on red stalks!)

Inflorescence: Umbels with 2–8 blooms on red hairy stems

Flowers: 5 sepals, oval, pointed, hairy, often reddish towards the points, and deeply divided below this level; 5 petals fused in the lower part into a tube, in the upper part with slightly notched edges, white or reddish , with a yellow or red central spot; 5 stamens; filaments fused in the tube of the corolla; superior ovary

Fruits: Capsules up to 4 mm long, spherical or egg-shaped and often opening up to the middle with 5 teeth

Habitat: In subalpine and alpine zones in grassland and stony rock faces on soil which is lime-rich and often free of snow

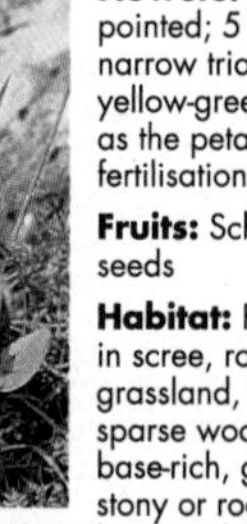

229 **Orpine or live forever – S. teléphium**
Plants 15 – 70 cm in height, perennial, smooth, with a turnip-like rhizome and blooming from June/Sept.

Sédum teléphium L. ssp. máximum (Hoffm.)
Orpine or live forever

Crassuláceae – Stonecrop Family

Leaves: Oval, 3–15 cm long, somewhat amplexicaul or sessile, usually opposite or 3 to a whorl, cordate or rounded at the base, somewhat fleshy and with irregular blunt teeth

Inflorescence: At end of stem, compact terminal and axillary cymes with many flowers

Flowers: 5 sepals, 1–2 mm long and pointed; 5 petals, 3–5 mm long, narrow triangular and white or light yellow-green; stamens twice as many as the petals; ovary superior; fertilisation by flies

Fruits: Schizocarps containing many seeds

Habitat: From hill to subalpine zone in scree, rock debris, in rocky grassland, fields, on walls and in sparse woods, on moderately dry, base-rich, generally lime-poor, loose stony or rocky soil containing little humus; plant often appears where others do not; in earlier times used as a medicinal plant

230 **Starry saxifrage – S. stelláris**
Plants 5–20 cm high, forming leaf rosettes from its spreading roots; flowering in July and August

Saxifrága stelláris L.
Starry saxifrage

Saxifragáceae – Saxifrage family

Leaves: Leaves in rosette oval and narrowing in a wedge shape or spathular, 1–5 cm long, with up to 9 deep teeth, without or with only short stems, whitish on the underside and rather fleshy; only bracteoles are found on the flower stem, which has glandular hairs

Inflorescence: Racemes containing 3–12 flowers; lateral flowers may be zygomorphic with shorter and longer petals

Flowers: 5 sepals, oval, greenish or reddish, 2–3 mm long and bent back; 5 petals or more rarely 6, narrow oval, pointed and with 3 yellow spots at the base; 10 stamens; superior ovary 2 carpels; fertilisation by insects

Fruits: Capsules containing many seeds

Habitat: In mountain and subalpine zones by cold springs, banks of streams, on damp rocks, in snowy valleys, and by rivulets on cool, wet, stony or sandy soil (or on pure sand) which is base-rich and fairly rich in nutrients, containing humus and often mossy; prefers cold water; variable species

231 **Mountain avens – D. octopétala** Dwarf shrub up to 10 cm high, spreading outwards up to 50 cm and forming mats; flowering in June and July

Dryas octopétala L.
Mountain avens

Rosáceae – Rose family

Leaves: Generally narrow oval, 1–3 cm long, on stalks, leaf proper heart-shaped, generally blunt at the end, smooth (rarely haired) and dark green above; underside white and with white, thickly matted hairs beneath, evergreen, with deeply notched or toothed edges rolled downwards

Inflorescence: Solitary flowers at end of stems

Flowers: 7–9 sepals, narrow lanceolate, up to 1 cm long, with dark glandular hairs and thickly set with white hairs on the edge; 7–9 petals (usually 8), oval, white, and up to 2 cm long; many stamens; numerous carpels on a vaulted perigynous zone; pistil grows in size after flowering and is covered with fine hairs

Fruits: Achenes with a beak 2–3 cm long, covered in feathery hairs

Habitat: In subalpine and alpine zones in sparse limestone or stony grassland, on rock faces, and on the dried-up shingle of rivers, on well-drained to moderately dry soil which is base-rich and contains more or less lime

232 **Wintergreen – O. secúnda**
Plants 5–25 cm high, with creeping branched rhizomes; flowering from July to September

Orthília secúnda (L.) House
Pyróla secúnda L.
Serrated wintergreen

Pyroláceae – Wintergreen family

Leaves: Present only on the bottom third of the stem, with stalks and glossy green in colour; the leaf proper is broad lanceolate to oval, pointed at the end and up to 4 cm long

Inflorescence: Raceme with up to 30 nodding flowers on one side only

Flowers: 5 sepals, green, forming a triangle; 5 petals, not fused, 2–4 mm long, bending together to form a bell and white or pale green in colour; 10 stamens; filaments somewhat bent but with pollen grains free; ovary superior with pistil; fertilised by insects

Fruits: 5 compartments with hanging capsules

Habitat: Distributed in mountain and subalpine zones, forming clumps in spruce, mixed spruce/oak, pine and pine/larch woods on moist to dry, infertile, neutral to acid sand and loamy soil with rotted humus; plants grow in half or full shade; medicinal plant; seldom found in beech woods

233 **Round-leaved saxifrage – S. rotundifólia**
Plants 20–50 cm high, with alternate leaves; flowering from June to September

Saxifrága rotundifólia L.
Round-leaved saxifrage

Saxifragáceae – Saxifrage family

Leaves: Basal leaves have long stalks, forming a rosette; the leaf proper is round to kidney-shaped, up to 5 cm in diameter, heart-shaped at the base, toothed, with sparse hairs on both sides, dark grey-green above and light grey-green beneath; leaf stalk is hairy; stem leaves also have stalks but are less complex and without stalks higher up the stem

Inflorescence: Panicles

Flowers: 5 sepals, lanceolate to oval and up to 5 mm long; 5 petals, narrow oval, white with red and yellow spots; 10 stamens; ovary superior with 2 carpels; fertilisation by flies

Fruits: Capsules containing many seeds

Habitat: In mountain and subalpine zones (rarely hill and alpine zones) among mountain shrubs, mountain mixed forests with shrubby undergrowth or good ground cover, banks of streams and between boulders, in shady places, on well-drained to wet soil which is alkaline, rich in nutrients, and generally contains lime and humus

234 **Liverwort – H. trilóba**
Plants 5–20 cm high with almost vertical rhizomes; flowering from March to May

Hepática trilóba Gilib.
H. nóbilis Schree
Liverwort

Ranunculáceae – Buttercup family

Leaves: Basal, staying throughout the winter and with long hairy stalks; the leaf proper is heart-shaped to 3-lobed, frequently with overlapping sections, mostly green above and reddish brown to violet beneath; the 3 stem leaves are oval, blunt-ended or pointed, 5–10 mm long, entire, close to the petals and forming a false calyx

Inflorescence: Solitary flowers on hairy, dark red stems

Flowers: False calyx consisting of 3 bracts; 5–10 petals, oval, smooth, and blue, red, or white; many stamens; numerous carpels with superior ovules; fertilisation by insects

Fruits: Achenes

Habitat: In hill and mountain zones in oak and beech woods, rarely coniferous forests, and on bushy slopes, on soil which is warm in summer, well-drained to dry, more or less base-rich and rich in nutrients, generally containing lime but neutral

235 **Wood sorrel – O. acetosélla**
Plants 5–15 cm high, perennials with runners spreading underground; flowering from April to June

Oxalis acetosélla L.
Wood sorrel

Oxalidáceae – Wood sorrel family

Leaves: Basal, with long stems and ternate; leaflets inversely heart-shaped, with scattered hairs, flexibly joined together, and folded up in bad weather and at night

Inflorescence: Solitary flowers, nodding before flowering

Flowers: 5 sepals, narrow oval, thin, 3–5 mm long and green (often with lighter edges); 5 petals, oval, with irregular outer edges, up to 15 mm long, white (rarely pink or blue), and with red veins; 10 stamens, the 5 outer being shorter than the 5 inner; ovary 5-celled and superior; fertilisation by insects or self-fertile plants

Fruits: Capsules with compartments, from which flat seeds are flung as far as 2 m

Habitat: From hill to subalpine zone, forming clumps in shady, coniferous, beech, and oak mixed forest with good ground cover, on well-drained to moist soil moderately rich in nutrients

236 **Bogbean – M. trifoliáta**
Plants 10–30 cm high, perennials with long creeping rhizomes; flowering from May to July

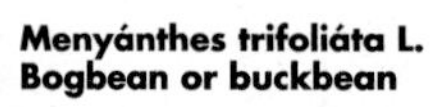

Menyánthes trifoliáta L.
Bogbean or buckbean

Gentianáceae – Gentian family

Leaves: Ternate, alternate, and with stalks widened at the end into sheaths; leaflets are wide lanceolate to oval, entire or with small blunt teeth and a network of veins

Inflorescence: Racemes

Flowers: 5 sepals, curved back in 5 parts; 5 petals, fused into a short funnel, white or reddish, bearded on the inside and with points rolled outwards; 5 stamens, fused to the tube of the corolla; superior ovary of 2 ovules; pistil with stigma in 2 parts; fertilisation mainly by bumble-bees

Fruits: Capsules containing many seeds bursting along fruit leaves; dispersal in water

Habitat: From hill to subalpine zone in bogs, marshes, and polders on wet or flooded peaty and muddy soils moderately rich in nutrients and bases, generally poor in lime and rather acid; appears early on newly drained polders

237 **Wild strawberry – F. vésca**
Plants 5–20 cm high, with runners above ground; flowering from April to July

Fragária vésca L.
Wild strawberry

Rosáceae – Rose family

Leaves: Basal leaves forming a rosette, ternate and with long stalks; stalk and stem hairy; leaf sections oval, deeply toothed, with a moderate number of hairs beneath and with a silky gloss; the end tooth is sometimes longer than its neighbours

Inflorescence: Solitary flowers or an umbel with several flowers

Flowers: Sepals doubled: the inner 5 are oval, with long points, the outer ones narrow lanceolate; 5 petals, round, white, and usually longer than the calyx; many stamens; many carpels each with an ovule and with a fleshy perigynous zone; pollination by insects

Fruits: Receptacles 1–2 cm in diameter, spherical and fleshy, red and covered with seeds projecting from the surface; dispersal by animals

Habitat: From hill to subalpine zones in forest clearings, by woodland paths, and at the edge of woods, on well-drained to moderately dry, stony, sandy or pure peat soils rich in nutrients and containing humus; plants grow in light or half-shade; medicinal plants

238 **Alpine buttercup – R. alpéstris**
Plants 5–15 cm high, smooth-leaved perennials with short rhizomes; flowering from June to August

Ranúnculus alpéstris L.
Alpine buttercup

Ranunculáceae – Buttercup family

Leaves: Basal, long-stemmed; the leaf proper is round to kidney-shaped, divided into 3–5 leaflets and glossy; leaflets incised to various depths and irregularly and bluntly toothed; where present, leaves on the stem are simple or ternate, with narrow lanceolate sections

Inflorescence: Generally solitary flowers on stems

Flowers: White and 10–25 mm in diameter; 5 sepals, oval, green, and fitting closely to the petals; 5 petals, oval and distinctly notched or inversely cordate; many stamens; many superior carpels on a smooth domed perigynous area

Fruits: Achenes up to 2 mm long, smooth, with long beaks

Habitat: In subalpine and alpine (rarely mountain) zones on rocks, in snow valleys, on snow slopes, open grassland, moist rough soil and scree, with long snow cover, containing lime and rich in loam; commoner in the northern Alps than in the central ones

239 **White cinquefoil – P. álba**
Plants 10–25 cm high, perennials flowering from April to June

Potentílla álba L.
White cinquefoil

Rosáceae – Rose family

Leaves: Basal leaves with 5 (rarely 7) sections and with recumbent or upright hairy stems; sections are wide lanceolate, 2–6 cm long, pointed, and broadest towards the end, with several small teeth towards the end, usually smooth above, hairy and light grey beneath; leaves directly on the stem have a simpler structure

Flowers: 1 or 2 flowers on stem, rarely more

Flowers: 10 sepals, the outer 5 narrow lanceolate and much smaller than the inner 5; 5 petals, white, inversely cordate, and protruding only a little beyond the sepals; many stamens; many carpels on a domed perigynum; fertilisation by insects

Fruits: Achenes 2–3 mm long; dispersal by insects

Habitat: In the hill zone in sparse oak and oak/pine woods, at the margins of woods and thickets, on moderately dry to dry loam or clay soil which is often poor in lime and sandy or stony; plants like warmth

240 **Snowdrop windflower – A. sylvéstris**
Plants 20–40 cm high, perennials with upright, hairy stems; flowering in April and May

Anémone sylvéstris L.
Snowdrop windflower

Ranunculáceae – Buttercup family

Leaves: Springing from ground level, in 3 or 5 sections, 5–10 cm in diameter, with loose hairs on both sides and pentagonal or heptagonal; sections are also on stems and deeply toothed at the ends; leaves on the stem are the same shape as basal leaves

Inflorescence: Usually solitary flowers

Flowers: Usually 5 or 6 petals, white, and hairy outside; numerous stamens; numerous carpels on a domed perigynum; pollination by wind

Fruits: Achenes, more or less flat, set closely together and covered with white hairs; hairs longer than the fruits

Habitat: In the hill (rarely mountain) zone on warm, dry hills, at the edge of woodland, in pine and oak woods, and scrub on sandy to loamy soil which is warm and dry in summer, lime-rich, containing humus, loose and deep; creeping roots; this plant of the Euro-Siberian plain reaches north to northern Germany, west to Lorraine, and south to northern Spain

241 **Wood anemone – A. nemorósa**
Plants 10–25 cm high, perennials with horizontally spreading rhizomes; flowering from March to May

Anémone nemorósa L.
Wood anemone

Ranunculáceae – Buttercup family

Leaves: At the time of flowering there are normally no basal leaves; 3 stem leaves on the top third of the stem, generally with stalks, rhomboid to pentagonal in outline, 2–6 cm long and ternate; the sections are deeply incised and also coarsely toothed

Inflorescence: Solitary flowers (rarely 2)

Flowers: 2–4 cm in diameter, with crinkly hairs on the stem; generally 6–8 (rarely up to 12) petals, narrow oval, rounded at the ends or slightly incised, white, often with a pink tinge; numerous stamens; numerous superior carpels on a domed perigynous area; fertilisation by insects

Fruits: Single achenes with short bristly hairs; dispersal by insects

Habitat: In hill and mountain (rarely subalpine) zones in orchards, shady meadows, deciduous and coniferous forests with good ground cover, and in hedgerows, on well-drained soils rich in nutrients, neutral to moderately acid and containing humus

242 **Plane-leaved buttercup – R. platanifólius**
Plants 40–130 cm high, branching, with strong rhizomes; flowering in June and July

Ranúnculus platanifólius L.
Plane-leaved buttercup

Ranunculáceae – Buttercup family

Leaves: Basal leaves on long stalks, those close to the ground divided into 3–7 sections and smooth; sections irregularly toothed, with teeth pointing forwards, and are wide lanceolate to narrow oval, coming to a fine point, narrowing toward the base in a wedge shape; sections without stems combine at the base; topmost leaves on the stem lack stalks; there sections are narrow lanceolate, coarsely toothed, coming to an entire point

Inflorescence: Stem branches out and bears many flowers

Flowers: Stems more or less parallel, slender and always smooth below the flower; 5 sepals, smooth, rather reddish and falling during or soon after flowering; 5 or more petals, narrow oval and white; numerous stamens; numerous carpels

Fruits: Achenes 2–4 mm long, with beaks 1.5 mm long

Habitat: Generally in the subalpine zone in woodland with shrubby undergrowth, thickets, high grass, and woods in ravines, on well-drained, alkaline soil rich in nutrients

243 **Aconite-leaved buttercup – R. aconitifólius**
Plants 20–100 cm high, branching, with strong rhizomes; flowering from May to July

Ranúnculus aconitifólius L.
Aconite-leaved buttercup

Ranunculáceae – Buttercup family

Leaves: Basal leaves on long stalks, those close to the ground divided into 3–7 sections and smooth; leaflets regularly toothed, often rhomboid, widest towards the ends, often with short stems and narrowing sharply at the base; topmost leaves on the stem lack stalks; their sections are rhomboid and regularly toothed almost as far as the tip

Inflorescence: Stem branches and bears many flowers

Flowers: Stems tend to spread rather than run parallel, are generally rather thick and covered in hairs below the flower; 5 sepals, smooth, rather reddish and falling during or soon after flowering; 5 or more petals, wide oval and white; numerous stamens; numerous carpels

Fruits: Achenes 2–4 mm long, with beaks 0.5 mm long, sometimes more than one on each

Habitat: In mountain and subalpine zones in moist meadows, water meadows, woods with shrubby undergrowth, by springs and on the banks of streams, on wet to moist soil rich in nutrients, generally poor in lime and containing humus

244 **Pallid cranesbill – G. rivuláre**
Plant 20–60 cm high, perennials with forked stems; flowering in July and August

Gerânium rivuláre Vill.
Pallid cranesbill

Geraniáceae – Geranium family

Leaves: Lower ones close to the ground in 5–7 sections, 5–10 cm wide, with long stalks; sections deeply incised and also toothed; the teeth and points are longer than they are wide; leaves on the upper part of the stem lack stalks

Inflorescence: In bunches, flowers in pairs; flower stems have backward-pointing hairs

Flowers: Above the topmost leaves; 5 sepals, narrow oval, 6–9 mm long, with points up to 2 mm long; 5 petals, not fused, 10–15 mm long, white with reddish to violet veins; 10 stamens; filaments widen at the base and are distinctly crinkled at the edges; ovary superior

Fruits: Divided into 5 lobes, with one seed in each; seeds flung from fruit when ripe

Habitat: In subalpine (rarely mountain or alpine) zone by the edges of streams, in thickets and scrub, in sparse larch and pine woods on well-drained, lime-poor, generally stony soil

245 **Narcissus-flowered anemone – A. narcissiflóra** Plants 15–40 cm high, perennials, usually with upright rhizomes; flowering from May to July

Anémone narcissiflóra L.
Narcissus-flowered anemone

Ranunculáceae – Buttercup family

Leaves: Basal leaves have long hairy stalks; leaves proper are roundish, 3–8 cm in diameter, divided into 3 or 5 sections and with hairs on the veins and the edges; sections often deeply incised and the narrow points also have teeth; leaves on stem also have stalks

Inflorescence: Umbels of 3–8 flowers; stems are hairy

Flowers: 15–30 mm in diameter; 5–7 petals, red outside before opening, later white and smooth; numerous stamens; numerous superior carpels on a slightly domed perigynal area

Fruits: Flat, smooth achenes without beaks

Habitat: In the subalpine (rarely mountain or alpine) zone in mountain meadows, on alpine rock faces and grassland, and near thickets on well-drained or water-retaining stony or loose loam or clay soil, usually containing lime; poisonous plants; not found on bare rock

246 **Alpine anemone – P. alpína**
Plants 15–50 cm high, with erect, hairy stems; flowering from May till August

Pulsatilla alpína (L.) Schrank
Anémone alpína L.

Ranunculáceae – Buttercup family

Leaves: At flowering the single basal leaf is scarcely developed; this later has a long stem; the leaf proper is triangular in outline, 10–30 cm wide and ternate; the long-stemmed sections are themselves ternate; second-order sections are pinnately cut; the 3 leaves on the stem (= bracts) are not different, being the same shape as the basal leaf, though rather smaller

Inflorescence: Solitary flowers

Flowers: Up to 6 cm in diameter; usually 6 petals, oval, rouded, or irregularly incised at the end, white (occasionally bluish or reddish), smooth inside and rather hairy outside; numerous stamens; numerous carpels; point of style smooth when ripe

Fruits: Achenes

Habitat: Mainly in the subalpine zone, in meadows and scrub on well-drained to dry, loose and stony or clay soil which is warm in summer, rich in bases and nutrients, containing lime and neutral to slightly acidic

247 **Rock cinquefoil – P. rupéstris**
Plants 15–30 cm high, usually with abundant hairy stems; flowering from May to July

Potentílla rupéstris L.
Rock cinquefoil

Rosáceae – Rose family

Leaves: Basal, with erect, hairy stems and simply pinnate; leaflets oval to round, up to 3 cm long, narrowing in a wedge shape towards the base, covered with hairs on both sides and doubly toothed; leaves on the stem are less complex

Inflorescence: Stem branches in its upper part; inflorescence cymose

Flowers: 5 outer, 5 inner sepals; outer ones are narrow lanceolate and shorter than the inner ones, which are oval; 5 petals, oval to round, not notched at the end, and white; numerous stamens, with smooth filaments; numerous carpels, superior, producing 1 seed each

Fruits: Smooth achenes; dispersal by wind

Habitat: In hill and mountain (rarely subalpine) zones on rocks, walls, furrows of fields, margins of woods and thickets, on moderately dry to dry stony or sandy loam or rock which is base-rich, often poor in lime and containing humus

248 **Meadowsweet – F. ulmária**
Plants up to 2 m high, perennials with short thick rhizomes; flowering from June to August

Filipéndula ulmária (L.) Maxim
Meadowsweet

Rosáceae – Rose family

Leaves: Basal and stem leaves are simply and unevenly pinnate, dark green above and light green or whitish (because of thick covering of hair) beneath; the end leaflet is longer than the others and lobed; small stipule-like leaves are present between the leaflets, which are oval, up to 6 cm long, and with fine double teeth

Inflorescence: Flowers numerous in irregular cymose panicles

Flowers: 5 or 6 sepals; 5 or 6 petals, whitish or pale yellow; numerous stamens, usually much longer than the petals;up to 15 carpels are on a conical perigynium; fertilisation by insects

Fruits: Achenes smooth, hard, and arranged spirally

Habitat: From hill to subalpine zone by ditches, river banks and springs, in marshes, bogs, wet woods, and among reeds, on wet to moist, sandy, loamy, or clay soil rich in nutrients, containing humus, acid or alkaline; plants grow in light or half-shade; often found in association with alders and ash trees

249 **Danewort – S. ébulus**
Plants up to 2 m high, with deep creeping rhizomes; flowering from June to August

Sambúcus bulus L.
Danewort

Caprifoliáceae – Honeysuckle family

Leaves: On short stalks, dark green and unequally pinnate with 5, 7, or 9 leaflets; these are long lanceolate, up to 15 cm long, sessile or on short stems, rounded at the base and often with asymmetric, regular teeth

Inflorescence: Flat or slightly vaulted panicles, appearing after the leaves

Flowers: 5 sepals, fused and reddish at the tip; 5 petals, narrow oval, with short points, with edges turning upwards, white and often pink towards the points on the lower side; 5 stamens, with white filaments and red anthers pointing outwards; ovary inferior; fertilisation by insects

Fruits: Berrylike drupes with 3–5 kernels; dispersal by birds

Habitat: In hill and mountain zones in forest clearings, meadows, by hedges and roadsides, and on rubbish dumps on wet to well-drained soil rich in bases and nutrients, generally containing lime and deep

250 **Furrowed saxifrage – S. exaráta**
Plants 2–10 cm high, forming mats and flowering from June to August

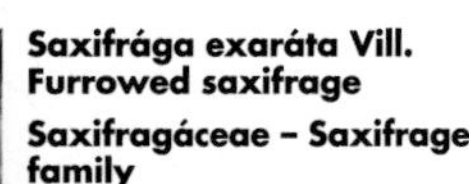

Saxifrága exaráta Vill.
Furrowed saxifrage

Saxifragáceae – Saxifrage family

Leaves: Arranged in a rosette, 5–20 mm long, in 3–7 blunt sections covered thickly with glands; in dry conditions forked between veins

Inflorescence: Racemes with 4–8 flowers

Flowers: 5 sepals, oval, blunt, with hairs on the edges, greenish but rather darker towards the tip; 5 petals, longer and up to twice as wide as the sepals, rounded at the ends, white, reddish, or rarely yellow and occasionally overlapping at the edges; 10 stamens, with reddish-yellow anthers; ovaries of 2 carpels, with 2 styles not fused at the base and not superior

Fruits: Capsules containing many seeds

Habitat: In subalpine and alpine zones on open grassland, static scree, and among bushes, on moderately well-drained, lime-poor, acid, infertile, stony or rocky soil ; occasionally penetrates the mountain zone; fairly common in the Alps on primordial rock

251 **Water violet – H. palústris**
Plants 20–80 cm long, branching under the water surface, and flowering from May to July

Hottónia palústris L.
Water violet

Primuláceae – Primrose family

Leaves: Arranged alternately or almost in whorls, submerged in the water, pinnately cut almost to the middle vein in the form of a comb, light green and covered with glands on the underside; sections up to 5 cm long and occasionally forked

Inflorescence: Racemose, flower heads set one above the other, and emerging above the water

Flowers: With stems, lying in the angle of narrow bracts; 5 sepals, narrow lanceolate, only fused at the base and 3–5 mm long; 5 petals, fused in the lower part and forming a yellow-green funnel, and with 5 oval, entire or slightly toothed, rounded, or notched lobes, 6–9 mm long and whitish or reddish in colour; 5 stamens, ovary superior

Fruits: Capsules up to 5 mm long, opening through slots

Habitat: In the hill zone in water fairly rich in nutrients and often poor in lime, set in peaty mud; found in shady places

252 **White campion – S. álba**
Plants 30–60 cm high, without bracts, with unisexual flowers; flowering from June to September

Siléne álba (Miller) Krause
White campion

Caryophylláceae – Pink family

Leaves: Wide lanceolate to oval, 3–8 times as long as wide. pointed or blunt at the tip and arranged alternately on the stem; a basal rosette is also present

Inflorescence: Flowers in loose panicles

Flowers: 5 sepals, fused together, green or pinkish, with red veins, distinctly hairy, up to 20 mm long and with 5 pointed teeth; the calyx of female flowers is enlarged; 5 petals, 20–35 mm long, white and deeply incised; 10 stamens; superior ovary with 5 styles; flowers open in the afternoon and at night and are scented; fertilised by moths

Fruits: Capsules up to 16 mm long, with vertical or slightly outward-pointing teeth

Habitat: In hill and mountain zones by roadsides, on rubbish dumps, in fields and hedgerows on dry, stony, sandy, or loamy soil rich in nutrients and humus or newly cultivated; tends to like warmth; tap-root goes 60 cm deep

253 **Bladder campion – S. vulgáris**
Plants 20–50 cm high, perennials with scanty hairs and some sterile shoots

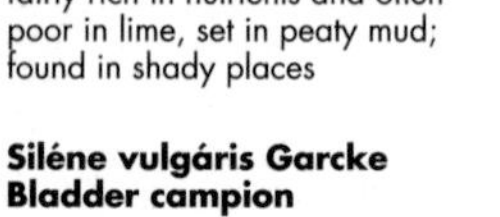

Siléne vulgáris Garcke
Bladder campion

Caryophylláceae – Pink family

Leaves: Lanceolate or oval, 2–12 cm long, opposite, sessile, pointed and blue-green

Inflorescence: Flowers in loose panicles

Flowers: Unisexual, male or female; 5 sepals, fused, 10–20 mm long, greenish-white or reddish, smooth, with red veins, swollen and with short points; 5 petals, fused, up to 25 mm long , deeply incised at the ends and with narrow, occasionally toothed, tips; 10 stamens, with dark anthers; ovary superior; fertilised by moths and bees

Fruits: Capsules up to 10 mm long, opening with 6 teeth

Habitat: From hill to subalpine zones in meadows, by roadsides and thickets, in grassland breaks, scree, and spinneys on moderately well-drained to dry humus-containing or newly cultivated soil moderately rich in nutrients and bases, and rather acid; appears early on newly cultivated soil; in earlier times used medicinally

254 **Nottingham catchfly – S. nútans**
Plants 20–50 cm high, covered fairly thickly in hairs and flowering from May to July

Siléne nútans L.
Nottingham catchfly

Caryophylláceae – Pink family

Leaves: Those on stem lanceolate or spathulate, 3–10 cm long, pointed, mid-green and opposite; stem sticky towards the top; sterile shoots have leaves in rosettes

Inflorescence: Loose panicles, with flowers directed to one side

Flowers: Nodding; 5 sepals, fused, 6–18 mm long, greenish or reddish, with dark veins, with glandular hairs and teeth; 5 petals, fused in the lower part, up to 25 mm long, white above and white, reddish or greenish beneath, deeply split, with narrow, rounded tips; 10 stamens; superior ovary; fertilised by moths

Fruits: Capsules containing many seeds, up to 16 mm long, opening by 6 spreading or reflexed teeth

Habitat: From hill to subalpine zones at the edges of woods, in sparse thickets or oak woods, in grassland on lime, scree and meadows, on soil containing humus which is dry, moderately rich in nutrients and bases and often poor in lime

255 **Wood sanicle – S. europaéa**
Plants 20–45 cm high, perennials with thick rhizomes; flowering from May to July

Sanícula europaéa L.
Wood sanicle

Apiáceae (Umbellíferae) – Carrot family

Leaves: Basal leaves with long stalks; leaves proper up to 10 cm wide, pentagonal, usually with 5 palmate lobes; sections deeply toothed or lobed, the teeth having awn-like tips; the stem leaves are similarly constructed but have only short stems or are sessile

Inflorescence: Umbels of few (often 3) few-flowered partial umbels, each 4–7 mm in diameter

Flowers: In each partial umbel there are many long-stemmed male flowers; hermaphrodite flowers 1 or a few to an umbel and sessile or with short stems; 5 sepals, free-standing in hermaphrodite flowers but often fused in male flowers, up to 1 mm long and pointed; 5 petals, usually white or pale yellow, rarely reddish; 5 stamens; inferior ovary

Fruits: Schizocarps

Habitat: In hill and mountain zones, frequent in beech woods, mixed deciduous woods, and rarely fir woods with good ground cover, on well-drained to moist, base-rich, lime-containing soil rich in nutrients and humus

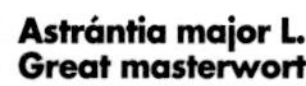

256 **Great masterwort – A. major**
Plants 20–90 cm high, with upright stems; flowering from June to August

Astrántia major L.
Great masterwort

Apiáceae (Umbellíferae) – Carrot family

Leaves: Basal leaves with long stalks, dark green; leave proper pentagonal to heptagonal, up to 20 cm wide and divided into 5 or 7 lobes almost to the base; lobes have one or two deep incisions, irregularly toothed and narrowing towards the base in a wedge shape; there is often an awn-like point on the tips of the teeth; stem leaves are similar to basal leaves and lack stems towards the top

Inflorescence: Stem branches like an umbel; each branch teminates in a second-order umbel, the last protruding above the lateral ones; second-order umbels surrounded by numerous bracts which are lanceolate, up to 25 mm long, and white or reddish

Flowers: Male flowers have long stems and are numerous in each second-order umbel; hermaphrodite flowers have short stems; 5 sepals, narrowing to fine points, free-standing and up to 2.5 mm long; 5 petals, whitish to reddish in colour; 5 stamens; ovary inferior

Fruits: Schizocarps up to 7 mm long

Habitat: In mountain and subalpine zones in meadows, at the edge of woodland, and in woods on well-drained soil rich in nutrients

257 **Little masterwort – A. minor**
Plants 15–40 cm high, with upright stems and flowering from June to August

Astrántia minor L.
Little masterwort

Apiáceae (Umbellíferae) – Carrot family

Leaves: Basal leaves with long stalks and dark green; leaves proper 5–7-sided, up to 20 cm wide and divided into 5 or 7 lobes almost to the base; leaflets lanceolate, pointed, and sharply toothed; stem leaves similar to basal leaves but with much shorter stems

Inflorescence: Stem branches like an umbel; each branch ends in a second-order umbel, the last protruding above the lateral ones; second-order umbels surrounded by numerous bracts which are lanceolate , up to 10 mm long, and whitish green

Flowers: Long-stemmed and numerous in every umbel; 5 sepals. narrow oval, free-standing, and up to 1 mm long; 5 petals, longer than the sepals; 5 stamens; inferior ovary

Fruits: Schizocarps 3–4 mm long; 5 mericarps, upward-pointing teeth

Habitat: In mountain and subalpine zones in grassland, scree, and thickets, on well-drained to dry, generally lime-free, particularly acid soil; found most often in the central and southern Alps

258 **Masterwort – P. ostrúthium**
Plants 25–100 cm high, perennials with knobbly rhizomes; flowering from June to August

Peucédanum ostrúthium (L.) Koch
Masterwort

Apiáceae (Umbellíferae) – Carrot family

Leaves: Lowest are ternate; leaflets on stalks, oval or round in outline, up to 15 cm in diameter and divided very deeply into 2 or 3 lobes; sections are narrow oval, generally widest towards the ends, coming to a blunt point, with irregular fine teeth; points awn-like

Inflorescence: Umbels with up to 50 second-order umbels

Flowers: 5 petals, white or reddish and up to 1.5 mm long; 5 stamens; inferior ovary of 2 fused carpels

Fruits: Schizocarps, round, up to 5 mm in diameter; marginal ridges form broad wings as wide as the fruit itself

Habitat: In subalpine and alpine zones by running water, camp sites, in shrubby undergrowth and green alder thickets, on well-drained to moist, deep, loose clay and loamy soil rich in nutrients and bases; plants grow in light and half-shade; widespread and frequently found in the Alps

259 **Angelica – A. silvéstris** Plants 50 cm high, covered with grey hairs under the umbel branches and flowering from July till September

Angélica silvéstris L.
Angelica

Apiáceae (Umbellíferae) – Carrot family

Leaves: With stalks, up to 60 cm long and generally doubly pinnate; second-order leaflets lanceolate to oval,10–14 cm long, at the base often asymmetric, divided into 2 or 3 parts and with a double row of fine teeth; at its base the main stem has a large swollen sheath

Inflorescence: Umbels with 20–40 flowers, somewhat domed

Flowers: 5 petals, 5–15 mm long, white, greenish, or rarely red; 5 stamens; inferior ovary of 2 fused carpels; fertilisation by insects

Fruits: Oval schizocarps, up to 4 mm long, 2 –4 mm wide and with bristly marginal ridges

Habitat: From hill to subalpine zones in marshes, wet meadows, undergrowth, river forest and on river banks, on wet or periodically flooded gravel, sandy, loamy, or clay soil rich in nutrients, generally alkaline, and containing humus; very deep-rooted; used in earlier times as a medicinal plant; an indicator of very fertile soil

260 **Hogweed – H. sphondylium**
Plants 50–150 cm high, with thick bristly stems; flowering from July to October

Herácleum sphondylium L.
Hogweed

Apiáceae (Umbellíferae) – Carrot family

Leaves: Basal leaves simply pinnate with stalks almost completely enclosed in a light green sheath; 3 or 5 leaflets, generally with stalks, irregularly lobed or pinnate, with deeply toothed edges and covered with bristles beneath (rough to the touch!)

Inflorescence: Umbels with 15–30 second-order umbels

Flowers: Mostly hermaphrodite; 5 white, pink, or greenish petals, those on the outside zygomorphic and up to 1 cm long; 5 stamens; inferior ovary of 2 fused carpels

Fruits: Oval or round schizocarps, up to 10 mm long and with marginal ridges

Habitat: From hill to subalpine zones in rich meadows, shrubby undergrowth, woodland, at forest margins and also in well-manured places round houses and barns, on well-drained to wet, deep, loose clay and loamy soil rich in bases, nutrients, and humus; an indicator of fertility and excessive manuring

261 **Giant hogweed – H. mantegazziánum**
Plants 1–3.5 m high, sometimes cultivated as ornamental plants to attract bees; flowering from July/Sept.

Herácleum mantegazziánum Sommier
Giant hogweed

Apiáceae (Umbellíferae) – Carrot family

Leaves: The lowest, without the stalk, is up to 1 m long, in 3 or 5 sections and with deep pinnate incisions; these are also toothed; diameter of stalk at ground level up to 10 cm

Inflorescence: Umbel with many second-order umbels, the whole up to 50 cm in diameter

Flowers: Mostly hermaphrodite; 5 petals, usually white, those on the outside zygomorphic and up to 1.5 cm long; 5 stamens; inferior ovary of 2 fused carpels

Fruits: Schizocarps, oval or round, 9–15 mm long, up to 8 mm wide and with bristly marginal ridges

Habitat: In hill zone (also planted in gardens in higher regions) in well-manured fields, parks, gardens, and by rubbish dumps, on well-drained to moderately dry, deep, loose clay and loamy soil rich in nutrients; this Caucasian plant was introduced to Europe at the end of the 19th century as a garden plant; nowadays it is naturalised and sometimes a troublesome weed

262 **Sermountain – L. síler**
Plants 40–150 cm high, with finely grooved stems 2 cm thick; flowering from June till August

Laserpitium síler L.
Sermountain

Apiáceae (Umbellíferae) – Carrot family

Leaves: On stalk up to 1 m long; leaf proper triangular, trebly or quadruply pinnate and blue-green; last-order leaflets lanceolate to narrow oval, 2–6 cm long, usually narrowing to a wedge shape at the base, entire, smooth, with branching veins

Inflorescence: Umbels up to 25 cm in diameter, with 20–40 second-order umbels; bracts oval, with smooth translucent skin

Flowers: Usually all the same size and hermaphrodite; 5 sepals, small and lanceolate; 5 petals, white to reddish; 5 stamens; ovary of 2 fused carpels

Fruits: Schizocarps, smooth and up to 12 mm long

Habitat: In mountain and subalpine zones in scree, sparse woodland (e.g. pine woods), scrub, and at the edge of sunny thickets, on moderately dry, fairly flat, loamy soil rich in lime, stony and poor in nutrients; plants grow in light and half-shade; they are early coloisers of freshly dug soil

263 **Water parsnip – S. latifólium**
Plants 50–150 cm high, without spreading roots, smooth-leaved; flowering in July and August

Sium latifólium L.
Water parsnip

Apiáceae (Umbellíferae) – Carrot family

Leaves: Some up to 40 cm long, light to dark green, unequally pinnate and with 4–10 pairs of leaflets, which are lanceolate to oval, sessile, up to 6 cm long and with regular fine teeth on the edges; the lowest pair is not far from the rest; the end leaflet is the same shape as the others; where submerged leaves are present they are finely divided into linear points

Inflorescence: Umbels with up to 25 second-order umbels, at the end of the stem; bracts of first-order umbels usually entire

Flowers: 5 petals, white; 5 stamens; ovary of 2 fused carpels

Fruits: Schizocarps, oval to round, up to 3 mm long with large, blunt ridges

Habitat: In hill zone, found either in reed beds or in slowly flowing water rich in nutrients, with a variable water level and a muddy bottom containing humus; depth of water up to 60 cm

264 **Hairy chervil – C. cicutária**
Plants 30–100 cm high, perennials with thick stems; flowering from May to August

Chaerophyllum cicutaria Vill.
C. hirsutum ssp. cicutaria (Vill.) Briq.
Hairy chervil

Apiáceae (Umbellíferae) – Carrot family

Leaves: Trebly or quadruply pinnate and on stalks; sheathes of top leaves 1–6 cm long; last leaflets only slightly divided or irregularly toothed; in the lowest leaves the last leaflets are almost as large as the rest of the leaf put together

Inflorescence: Umbels, with 10–20 second-order umbels

Flowers: 5 petals, white to pink; 5 stamens; ovary of 2 fused carpels; fertilised by insects

Fruits: Schizocarps 8–12 mm long, broad rounded ridges, no beaks

Habitat: From mountain to alpine zones, frequently found in mountain forests and spinneys, bushy undergrowth, by streams and springs, and in moutain meadows, on moist to wet clay soil rich in nutrients, usually rich in lime and containing humus, in humid places; plants grow in light and half-shade; deep-rooted

265 **Rough chervil – C. témulum**
Plants 25–100 cm high, with sweeping, branched stems; flowering from May to July

Chaeróphyllum témulum L.
Rough chervil

Apiáceae (Umbellíferae) – Carrot family

Leaves: Doubly or trebly pinnate, on stalks; last leaflets only slightly cut; leaf points and teeth aristate

Inflorescence: Umbels with 6–13 second-order umbels on stems with bristles pointing forwards

Flowers: 5 petals, without hairs and white, rarely pink or yellowish; 5 stamens; ovary of 2 fused carpels; fertilised by insects

Fruits: Schizocarps up to 8 mm long, with thread-like longitudinal ridges, smooth and without beaks; axis on ripe fruit often vertical or inclined

Habitat: In hill and mountain zones in thickets, forest clearings, hedgerows, gardens, close to rubbish dumps, by the edge of woods, close to houses and by the roadside, in shady places on well-drained soil rich in nutrients, dry and containing humus; plants grow in half- shade; indicator of nitrogen in the soil; northern range extends to Denmark and southern Sweden

266 **Golden chervil – C. aúreum**
Plants 25–120 cm high, with backward-pointing hairs on stems; flowering from June to August

Chaeróphyllum aúreum L.
Golden chervil

Apiáceae (Umbellíferae) – Carrot family

Leaves: Trebly or quadruply pinnate, on stalks; last leaflets pinnately cut, with stems thickly covered in hairs beneath; teeth and leaf ends come to fine points

Inflorescence: Umbels with 7–20 second-order umbels with smooth stems and lanceolate bracts coming gradually to long points

Flowers: 5 petals, without hairs, smooth and white; 5 stamens; ovary of 2 fused carpels; fertilised by insects

Fruits: Schizocarps 7–12 mm long, with clearly visible thread-like longitudinal ridges, smooth and without beaks; axis of ripe fruits bent backwards or upright

Habitat: In mountain and subalpine (rarely hill) zones in weed beds near villages, over-manured fields, by ditches, hedgerows, river banks, rubbish dumps, by the roadside and at the edges of fields, on moist to dry, stony and loose clay and loamy soil rich in bases and nutrients, containing lime and humus; indicator of nitrogen in the soil; plants grow in light and half-shade

267 **Cow parsley – A. silvéstris**
Plants 20–150 cm high, biennials or perennials; flowering from April to August

Anthríscus silvéstris (L.) Hoffm.
Chaeróphyllum silvéstre L.

Apiáceae (Umbellíferae) – Carrot family

Leaves: Doubly or trebly pinnate; the two lowest leaflets are smaller than the rest of the leaf; without spots on the stem

Inflorescence: Umbels with 6–18 second-order umbels with smooth stems; bracts wide lanceolate, crinkly at the edges and drawing to long points

Flowers: 5 petals, smooth and white; 5 stamens; ovary of 2 fused carpels; fertilised by insects, especially beetles and flies

Fruits: Schizocarps 5–10 mm long, without distinct ridges, brown, glossy, and with a ring of long bristles at the base; axis on ripe fruits only slightly inclined

Habitat: From hill to subalpine zones in meadow land, in hedgerows, and by the wayside on well-drained, loose, deep, clay and loam rich in nutrients and humus and often manured; nitrogen indicator; plants grow in light and half-shade; only of limited value as fodder

268 **Moon carrot – S. libanótis** Plants 30–120 cm high, perennials with rhizomes crowned with clumps of fibres; flowering from July/Sept.

Séseli libanótis (L.) Koch
Holy-root

Apiáceae (Umbellíferae) – Carrot family

Leaves: Basal, doubly or trebly pinnate, with stalks and blue-green, especially beneath; segments mostly lanceolate, pointed and smooth; bottom pair of leaflets does not point backwards; leaves on stem, which often branches, are absent or reduced

Inflorescence: Umbels with 20–40 second-order umbels, hairy on top; bracts very often longer than the flower stems

Flowers: Sepals up to 1 mm long; 5 petals, wide oval, about 1 mm long, and white, yellowish or reddish; 5 stamens; ovary of 2 fused carpels; fertilised by insects (mainly flies and beetles)

Fruits: Schizocarps 3–5 mm long, narrow yellow longitudinal ridges

Habitat: From hill to subalpine zones in sparse, dry woodland and scrub, at the edge of woods and by roadsides, on rocks and in scree, on moderately dry thin soil which is warm in summer, infertile, and often contains lime and humus

269 **Caraway – C. cárvi** Plants 25–80 cm high, biennials with sparse branches and thick rhizomes; flowering from May to July

Cárum cárvi L.
Caraway

Apiáceae (Umbellíferae) – Carrot family

Leaves: Doubly or trebly pinnate, with stalks and dark green or blue-green; leaf tips and teeth come to fine points; the bottom pair of leaflets points backwards; stem has several leaves

Inflorescence: Umbels with 6–18 second-order umbels; no bracts on first-order umbels

Flowers: 5 petals, up to 5 mm long, white, pink, or, especially in mountain regions, vivid red; 5 stamens; ovary of 2 fused carpels; fertilised by insects (especially flies and beetles); hermaphrodite and male flowers present

Fruits: Schizocarps 3 –4 mm long, only slightly flattened, with distinctly protruding ridges and axis bent back; stems clearly of different lengths

Habitat: From hill to subalpine zones in fields and meadows, at the edge of woodland and by roadsides, on well-drained clay and loam rich in bases and nutrients, at least fairly deep and containing humus, in cool and humid places; indicator of high nutrients content; used today as flavouring (Kümmel) or medicinal herb

270 **Milk parsley – P. palústre**
Plants 50–150 cm high, without clumps of fibres at the base of the stem; flowering in July and August

Peucédanum palústre (L.) Moench
Milk parsley

Apiáceae (Umbellíferae) – Carrot family

Leaves: Basal leaves doubly or multiply pinnate; last leaflets singly or doubly pinnately cut, with narrow pointed tips, smooth or rough edges, and a clearly visible network of veins beneath

Inflorescence: Umbels with 15–30 second-order umbels; bracts of first-order umbels present in great numbers and have rough transparent edges; stems of second-order umbels up to 3 cm long when fruit is ripe

Flowers: Mostly hermaphrodite; 5 petals, up to 3 mm long, deeply notched and white, pink, yellowish, or greenish; 5 stamens; ovary of 2 fused carpels; fertilised by insects

Fruits: Schizocarps oval, flat, up to 6 mm long and with clearly visible ridges; axis up to 1 mm long and usually bent back

Habitat: In hill and mountain zones in reedy meadows, alder woods, and on river banks, on wet, sometimes flooded peaty or marshy soil which rich in bases and nutrients; plants grow in light or half-shade

271 **Wild carrot – D. caróta**
Plants 20–80 cm high, dying off after fruiting for the first time and flowering from June to August

Daúcus caróta L.
Wild carrot

Apiáceae (Umbellíferae) – Carrot family

Leaves: Usually doubly pinnate; last leaflets pinnately cut; tips narrow and sharply or bluntly toothed

Inflorescence: Umbels with numerous second-order umbels; first-order umbels domed or flat but concave during flowering; bracts at edges transparent, light in colour, and crinkled

Flowers: 5 petals, white or pink; 5 stamens; ovary of 2 fused carpels; fertilised by insects (especially flies and beetles)

Fruits: Oval schizocarps, round in cross-section and with axes not fused together at the base

Habitat: In hill and mountain zones in infertile and fertile fields, quarries, thickets, ditches and by the roadside, on moderately dry to well-drained clay or loamy soil more or less rich in bases and nutrients and with or without humus; an early coloniser of newly dug soil; today distributed worldwide except for the arctic and tropical regions

272 **Hogweed – L. hálleri**
Plants 10–60 cm in height, with large clumps of fibres on the bases of the stems; flowering from June to Aug.

Laserpítium hálleri Crantz
Hogweed

Apiáceae (Umbellíferae) – Carrot family

Leaves: Main stem divided near the ground, with many fine grooves; leaf stalks only a few centimetres long; leaves proper doubly or trebly pinnate, 20–50 cm long, rhombic or triangular in outline, and dark green or blue-green; last leaflets doubly pinnately cut; tips are pointed

Inflorescence: Umbels with 10–40 second-order umbels; bracts lanceolate, pointed, with white or transparent margins and also hairs sticking out

Flowers: 5 petals, white and hairy on the outside; 5 stamens; ovary of 2 fused carpels

Fruits: Schizocarps 4–9 mm long, oval, with ridges protuding only slightly, somewhat winged and with bristles; axis generally upright

Habitat: In subalpine and alpine zones on sunny, dry, stony soil which is acid to neutral and poor in nutrients; fairly frequently found in the Alps on bare rock; also found in the Cottian Alps and the Dolomites

273 **Hog fennel – P. oreoselínum**
Plants 20–100 cm hegh, perennials flowering from July to September

Peucédanum oreoselínum (L.) Moench
Hog fennel

Apiáceae (Umbellíferae) – Carrot family

Leaves: Finely formed, doubly or trebly pinnate, long, with thin stems, and blue-green; first leaflets usually at a right angle and with long stems; last ones wide lanceolate in outline, oval, triangular or rhombic, deeply toothed or in 3 parts which are also toothed

Inflorescence: Umbel with 10–20 second-order umbels; bracts directed backwards

Flowers: 5 petals, white or vivid pink; 5 stamens; ovary of 2 fused carpels; fertilised by insects

Fruits: Schizocarps, round, up to 8 mm wide, with wide ridges on edges

Habitat: In hill and mountain (rarely subalpine) zones at the edge of sunny thickets, sparse oak and pine woods, shrubberies, and on roadside banks, on moderately dry to dry, loose stony or sandy soils which are warm in summer, base-rich, often lime-poor and neutral and containing humus; plants grow in half-shade or light

274 **Candy carrot – A. creténsis** Plants 10–30 cm high, perennials, often looking grey-green due to hairy covering; flowering from May to July

Athamánta creténsis L.
Candy carrot

Apiáceae (Umbellíferae) – Carrot family

Leaves: Trebly pinnate, on stalks and with a broad sheath at base; third-order leaflets pinnately cut with pointed leaflets up to 3–5 mm long

Inflorescence: Umbels with 4–15 second-order umbels and 1–4 bracts, each with a green stripe down the middle; second-order umbels with many longish transparent bracts

Flowers: Usually hermaphrodite; 5 petals, somewhat hairy on the outside, white; 5 stamens; ovary of 2 fused carpels

Fruits: Schizocarps, narrow oval and widest at the bottom, thickly set with long hairs; compartments have 5 clearly visible main ridges

Habitat: From mountain to alpine zones in limestone scree and broken rock, on moist to dry rocky soils rich in lime, humus and trace elements, in sunny places; plants grow in light; appears early on disturbed rock; found as far south as the Spanish mountains

275 **White waterlily – N. álba**
Plants up to 3 m high, with branching rhizomes up to 10 cm thick; flowering from June to August

Nymphaéa álba L.
White waterlily

Nymphaeáceae – Waterlily family

Leaves: Wide oval or round in outline, 10–30 cm in diameter, with a deep basal sinus, more or less entire, with a stalk up to 3 m long, and floating

Inflorescence: Solitary flowers

Flowers: Very large, up to 9 cm in diameter; generally 4 sepals, not overlapping at the edges; 14–33 petals, narrow oval, generally pointed, and white; up to 125 stamens; anthers not or only slightly widening at the middle; ovary superior, more or less spherical, with up to 25 compartments; up to 25 yellow stigmatic rays; fertilised by insects or self-fertile

Fruits: Usually spherical capsules with numerous seeds in each compartment

Habitat: In hill (rarely mountain) zones in water which is still or slowly moving, e.g. in ponds, stagnant water, or undisturbed branches of lakes, with a depth of up to 3 m on a muddy bottom containing humus; the pink form is cultivated as an ornamental plant; range extends into Scandinavia up to 61 degrees

276 **Wild violet – V. arvénsis**
Plants 10–20 cm high, annuals or biennials, with solitary hairs; flowering from April to September

Víola arvénsis Murray
Wild violet

Violáceae – Violet family

Leaves: Lower leaves are on stalks and have oval, deeply serrated or notched sections; leaves in the upper part of the plant are narrow oval to lanceolate and also serrated; bracts are large and often pinnately cut

Inflorescence: Flowers spring singly or in pairs from the axes of the leaves

Flowers: Zygomorphic; 5 sepals, free-standing, generally light green and with appendages 6 – 12 mm long; 5 petals, free-standing; the top two are white to bluish, the side pair are white to pale yellow, the bottom one is widened, with a yellow spot and with a spur up to 16 mm long; 5 stamens with short stamens thickened at the base and each with a a club-shaped continuation; ovary superior, consisting of 3 carpels

Fruits: 3 rattle-shaped capsules which burst open to disperse seed

Habitat: From hill to subalpine zone in gardens, vineyards, fields, and by the wayside on well-drained to moderately dry loamy and clay soil which is base-rich, rich in nutriment and usually poor in humus

277 **Flax dodder – C. epilínum**
Parasitic plants flowering in June and July

Cúscuta epilínum Weihe
Flax dodder

Cuscutáceae – Dodder family

Leaves: Plants have no roots, their thread-like branched stems winding round the host plant; leaves only resemble scales, yellowish like all other parts of the plant because of reduced chlorophyll content

Inflorescence: Balls containing many flowers

Flowers: 5 (or 4) sepals, very wide and shortly before the end narrowing into little points; 5 (or 4) petals, fused together for much of their length, bell-shaped; throat has a ring of divided or fringed scales; 5 stamens (or 4); superior ovary of 2 touching carpels; 2 styles

Fruits: Capsules, flattened and more or less divided into 2 cells

Habitat: In hill and mountain zones in places with hot summers, where it parasitises flax; nettles and hops can also be hosts; this European plant is seldom found now in the region, as the flax and linen industry has declined since the Second World War

278 **Black nightshade – S. nígrum**
Plants 10–70 cm high, smooth or covered thickly with stiff bristles; flowering from June to September

Solánum nígrum L.
Black nightshade

Solanáceae – Nightshade family

Leaves: Broad oval to triangular, with short stalks; the leaves proper come to fine points and are sometimes entire, but more often coarsely toothed

Inflorescence: Cymes on long stalks springing from leaf nodes

Flowers: Stems up to 25 mm long; 5 sepals, blunt or pointed and with backward pointing teeth when fruit is ripe; 5 petals (rarely 6), narrow triangular, with short hairs on the outside, 3–6 mm long and white; 5 stamens, with anthers cohering in a tube; ovary superior and 2-celled; fertilised by insects

Fruits: Black berries; dispersal by birds

Habitat: In hill and mountain zones on rubbish dumps, in fields, gardens, weed patches, by the roadside and on river banks on moist, loose, loam and clay soil rich in nutrients, generally neutral, and containing humus; an indicator of nitrogen; binds soil; nowadays distributed worldwide

279 **Potato – S. tuberósum**
Plants 30–60 cm high, with underground tubers; flowering from June to August

Solánum tuberósum L.
Potato

Solanáceae – Nightshade family

Leaves: Unequally and irregularly pinnate on branching stems, with stalks and hairy, particularly on the underside; leaflets broad lanceolate to oval, entire or wavy in places, and of different sizes; the end segment often comes to a sharp point

Inflorescence: Cymes on stalks

Flowers: 5 sepals; 5 petals, fused together, 2–3 cm in diameter, usually white, with the edge bending upwards; 5 stamens, with anthers cohering in a tube; ovary superior and 2-celled

Fruits: Spherical, juicy berries up to 2 cm in diameter

Habitat: In hill and mountain (rarely subalpine) zones on rubbish dumps and in fields, on well-drained loamy soil rich in nutrients and humus, in cool atmosphere; various varieties are cultivated; has been known since the 16th century in Europe; first began to be widely grown in the 18th century; now one of the most important cultivated plants

280 **Bindweed – C. arvénsis**
Plants 20–70 cm long, with spreading rhizomes; flowering from June to September

Convólvulus arvénsis L.
Bindweed

Convolvuláceae – Bindweed family

Leaves: Alternate, with stalks; arrow- or spear-shaped, entire and up to 4 cm long; the stem grows along the ground

Inflorescence: Flowers solitary or in threes (the 2 others are smaller!) in the axes of the leaves

Flowers: 5 sepals, up to 5 mm long, 3 longer than the other 2; 5 petals, fused and forming a funnel, white or pink and up to 25 mm long; 5 stamens with capitate glandular hairs towards the bottom on the inner side; ovary superior, with 2 threadlike stigmas on the pistil; fertilised by insects, especially flies and bees

Fruits: Egg-shaped capsules 4–8 mm long; dispersal by means of spreading shoots

Habitat: In hill and mountain (rarely subalpine) zones by roadsides, rubbish dumps, in fields and vineyards on well-drained to moderately dry loamy and clay soil rich in bases and nutrients and generally poor in humus; roots go up to 2m deep

281 **Bindweed – C. sépium**
Plants 1–3 m long, perennials with climbing stems, rarely recumbent, and flowering from June to Sept.

Calystégia sépium (L.) Br.
Convólvulus sépium L.
Bindweed

Convolvuláceae – Bindweed family

Leaves: Alternate with stalks; heart- or spear-shaped, usually entire and 5–15 cm long, and blunt or more often pointed; the stalk is shorter than the leaf proper

Inflorescence: Solitary flowers growing from the leaf axes

Flowers: With long stems (longer than the flowers!) and sepals usually covered by 2 bracteoles; 5 sepals, up to 10 mm long and pointed; 5 stamens with glandular hairs towards the bottom on the inner side; ovary superior, with 2 oval stigmas on the pistil; fertilised by insects, especially bees

Fruits: Egg-shaped capsules 6–12 mm long

Habitat: In hill and mountain zones in hedges, gardens, thickets, by fences and river banks, and on the edge of woodland on moist to well-drained loamy soil which rich in bases and nutrients and moderately acid; plants grow in light and twine anticlockwise

282 **Horned rampion – P. spicátum**
Plants 20–80 cm high, perennials with roots thickened like turnips; flowering in May and June

Phyteúma spicátum L.
Horned rampion

Campanuláceae – Bellflower family

Leaves: Basal leaves with stalks, cordate, smooth, and singly or doubly toothed; stem-leaves with very short stalks or almost none, long lanceolate, round at the base, with long points and toothed

Inflorescence: Spikes, initially egg-shaped, later cylindrical, with narrow lanceolate bracts

Flowers: 5 sepals, narrow lanceolate; 5 petals, linear, up to 15 mm long, white or yellowish, and on first flowering fused at the base and the tip; 5 stamens, free-standing, with hairy filaments, widening at the base to enclose the pistil; ovary inferior, with 2 stigmata; fertilised by insects

Fruits: Capsules opening by lateral pores; dispersal by wind

Habitat: In hill and mountain (rarely subalpine) zones in deciduous mixed woodland with good ground cover, in thickets and fields on damp to well-drained loose alkaline soils rich in nutrients and humus

283 **Stemless Carline thistle – C. acaúlis**
Plants 2–10 cm high, perennials with thick, woody tap-roots; flowering from July to September

Carlína acaúlis L.
Stemless Carline thistle

Asteráceae (Compósitae) – Sunflower family

Leaves: Usually arranged in a rosette, short, with thick grooved stalks, narrow oval, often somewhat curved, very deeply pinnately cut with 6–12 spiny segments on each side

Inflorescence: Composite flowers 5–12 cm in diameter, with silvery bracts shining and spreading in dry weather

Flowers: Pappus (= reduced calyx) 10–15 mm long, white, somewhat yellow at the base, in 1 row of 9–11 bristles; petals 9–12 mm long, fused into a tube (= tubular florets) and white or reddish; 5 stamens; inferior ovary of 2 carpels, with a pistil and 2 stigmata; fertilised by bees and bumble-bees

Fruits: Achenes 4–6 mm long, the seed husk fused with the wall of the fruit; dispersal by wind and by birds

Habitat: From hill to subalpine zones in sparse woodland, on infertile meadow-land, on moderately dry, neutral, loamy and clay soil which is warm in summer and has medium to deep topsoil; plant has deep roots

284 **Globe thistle – E. sphaerocéphalus**
Plants 40–150 cm high, perennial, with angular stems; flowering from July to September

Echinops sphaerocéphalus L.
Globe thistle

Asteráceae (Compósitae) – Sunflower family

Leaves: Lanceolate to oval in outline, green above, covered with fine white hairs beneath, and pinnately cut on both sides almost to the middle vein; segments triangular, slightly toothed, with spiny bristles; stem leaves lanceolate and sessile

Inflorescence: Spherical and 2–6 cm in diameter

Flowers: Bracts 1–2 cm long, lanceolate, with short hairs; pappus up to 1 mm long, consisting of hairy scales; 5 petals, fused into florets, up to 1 cm long and light blue; 5 stamens with blue-green anthers; inferior ovary

Fruits: Cylindrical achenes, usually heptagonal, with partly joined hairs, grey and 6–8 mm long

Habitat: Usually in the hill zone by roadsides, railway embankments, and in shrubby undergrowth, on loam and clay soil which is dry in summer, rich in bases and nutrients, neutral, and contains humus; also used as an ornamental plant in gardens; attracts bees

285 **Cut-leaf teasel – D. laciniátus**
Plants 40–200 cm high, biennials, branching out towards the top, spiny; flowering in July and August

Dípsacus laciniátus L.
Cut-leaf teasel

Dipsacáceae – Teasel family

Leaves: Stem leaves opposite, irregularly pinnately cut almost to the middle vein, with pinnately cut or toothed leaflets edged with bristles

Inflorescence: Egg-shaped to cylindrical flower heads, erect and 3–8 cm long, with protruding bracts; these are shorter than the flower heads and covered with bristles on the edges

Flowers: Exterior calyx invisible; calyx is 4-angled, bent upwards, entire, with many teeth; 4 petals fused into a tube, white, 8–14 mm long and with 4 points at the end; 4 stamens, fused to the tubular corolla; anthers are free-standing; inferior ovary

Fruits: Achene-like, up to 5 mm long and crowned by the persistent calyx with spiny scale leaves

Habitat: In hill zone by ditches, at roadsides, on the edge of woodland, on river banks, and in weed beds, on moist to well-drained loam and clay soil rich in bases, nutrients, and humus, and contains lime; plant likes warmth; mainly found in southern Europe; also grown in gardens

286 **Michel aster – B. michélii**
Plants 5–20 cm high, perennial, with branched rhizomes; flowering from May to July

Bellidiástrum michélii Cass.
Aster bellidiástrum (L.) Scop.
Michel aster

Asteráceae (Compósitae) – Sunflower family

Leaves: Forming a basal rosette, with long stem which may have narrow wings; leaves proper lanceolate oval or obovate, with wavy or bluntly toothed edges and hairs, especially underneath

Inflorescence: Stem without leaves and bears a composite flower 2–4 cm across without scales; bracts greenish to reddish, lanceolate, arranged in 2 rows with sparse hairs

Flower: Ray-florets form 1 row, white, often reddish on the outside; tubular florets light to dark yellow

Fruits: Achenes 2–3 mm long, with a pappus of rough bristles

Habitat: From mountain to alpine zones on landslides, by springs, woody slopes, moors, and scree on moist, stony, loamy, or marshy soils which are base-rich and contain lime and humus, in cool, humid places; plants grow in light or half-shade

287 **Daisy – B. perénnis**
Plants 3–15 cm high, perennials with stout fibrous roots; flowering from February to November

Béllis perénnis L.
Daisy

Asteráceae (Compósitae) – Sunflower family

Leaves: All in a basal rosette, oval or spatulate, on long stems or narrowing into a winged stem, entire, with a wavy or toothed edge and smooth or rather hairy

Inflorescence: Stem without leaves, with a composite flower 1–3 cm across at the end; bracts 3–6 mm long, smooth or with scanty hairs at the edges, blunt or pointed, and green to dark green

Flowers: Ray-florets generally in 1 row, white or slightly pink above and white to deep red beneath; tubular florets hermaphrodite and yellow; fertilised by insects

Fruits: Achenes up to 1 mm long, oval, rather hairy, without pappus; dispersal by wind

Habitat: In hill and mountain (rarely subalpine) zones in meadows, parks, and garden lawns on well-drained, heavy, sandy or pure loam and clay soil which contains nutrients and humus; definitely an indication of fertility; heliotropic

288 **Karvinski's fleabane – E. karvinskiánus**
Plants 5–20 cm high, perennials with erect stems; flowering from April till November

Erígeron karvinskiánus DC
Karvinski's fleabane

Asterácreae (Compósitae) – Sunflower family

Leaves: Lanceolate to oval, the lower ones deeply toothed or with 3 lobes, the upper ones sessile, with narrow bases and fine points; leaves sparsely hairy and green or reddish

Inflorescence: Composite flowers at end of stalk, up to 15 mm across; bracts greenish and 3–5 mm long

Flowers: Ray-florets widening, white inside and often pink outside; tubular florets yellowish

Fruits: Achenes up to 1 mm long, with a white pappus 2–3 mm long; in ray-florets this is short and in 1 row; in tubular florets in 2 rows

Habitat: In the hill zone on walls, rocky slopes, and moist scree, on well-drained and generally rather stony soil; found particularly in southern areas; comes from Central and South America; widespread in the south of Switzerland, also on rocky slopes by Lake Geneva

289 **Ox-eye daisy – C. leucánthemum**
Plants 20 – 70 cm in height, with branched rhizome and upright stem; blooms from May to October

Chrysánthemum leucánthemum L.
Leucánthemum vulgare Lam.
Ox-eye daisy, marguerite

Asteráceae (Compósitae) – Sunflower family

Leaves: Lower leaves longish, obovate or spathulate, up to 3 cm long, notched, toothed, or pinnately cut, and gradually narrowing towards the stem; upper leaves sessile, with deep teeth

Inflorescence: Composite flower up to 5 cm across at the end of the stem; bracts lanceolate to narrow lanceolate, generally smooth with dark brown edges

Flowers: Ray-florets white, up to 6 mm wide; tubular florets yellowish; fertilised by insects, mainly beetles, bees, and butterflies

Fruits: Achenes up to 4 mm long, generally with 10 ridges and black resin glands in between; dispersal by wind and by passing through animals

Habitat: From hill to subalpine zones in meadow land, fields, rubbish dumps, and by the roadside on well-drained, loose, generally loamy soil of any kind, rich in bases and nutrients; deep-rooted

290 **Corn chamomile – A. arvénsis**
Plants 10–40 cm high, annuals, slightly scented, and flowering from May to October

Anthemis arvénsis L.
Corn chamomile

Asteráceae (Compósitae) – Sunflower family

Leaves: Doubly or trebly pinnate, without stalks, usually alternate and smooth or sparsely haired; last leaflets narrow lanceolate, often rather toothed and pointed with spines

Inflorescence: Composite flowers 2–3 cm across at the end of the stem; bracts light green and downy; receptacle spherical; scales lanceolate

Flowers: Ray-florets white, up to 1.3 cm long, standing in several rows and with a distinct notch at the end; tubular florets yellow; fertilised by insects (flies and wasps)

Fruits: Achenes up to 3 mm long, smooth, with 5–10 ridges or furrows and often with a toothed edge at the end

Habitat: In hill and mountain (rarely subalpine) zones on rubbish dumps, fields, by the roadside, in cereal fields and vineyards, on damp, loamy, or clay soil rich in bases and nutrients and generally lime-poor; an indicator of acidification; roots up to 30 cm deep; accompanies cultivation

291 **Scentless mayweed – T. inodórum**
Plants 15–50 cm high, with recumbent or upright stems; flowering in June and July

Tripleurospérmum inodórum (L.) Sch.- Bip.
Chrysánthemum inodórum L.
Scentless mayweed

Asteráceae (Compósitae) – Sunflower family

Leaves: Generally smooth, doubly or trebly pinnate, with narrow linear or thread-like leaves, segments of the last leaflets being somewhat spiny

Inflorescence: A composite flower 5 cm across at the end of each stem; bracts smooth, often light green and with transparent edges

Flowers: Ray-florets white, up to 2 cm long; tubular florets yellowish; fertilised by insects

Fruits: Achenes up to 2 mm long, the areas between the ridges with sparse warts, and often with 2 round black oil glands at the top of the outer face

Habitat: From hill to subalpine zones on rubbish dumps, by the roadside, in cereal and other fields, frequently found in weed beds; favoured soil is well-drained to moderately dry, neutral and sandy or pure clay and loam rich in nutrients, generally lime-poor and containing humus; an early coloniser of newly turned soil; accompanies cultivation; root goes down more than 1 metre

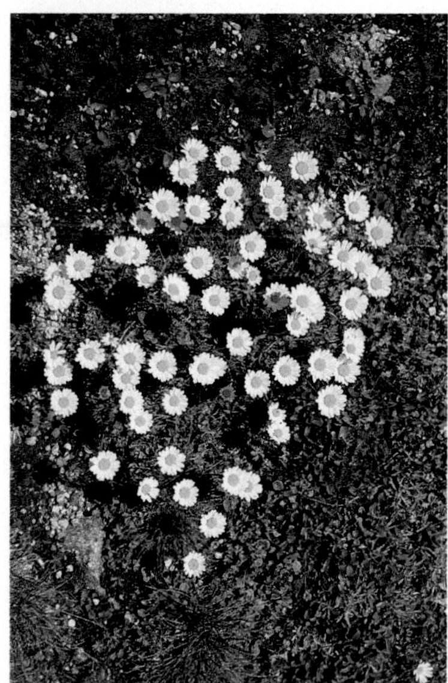

292 **Alpine moon daisy – C. alpínum**
Plants 5–15 cm high, perennials, often forming clumps; flowering from July to August

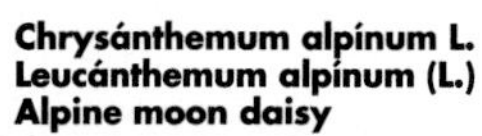

Chrysánthemum alpínum L.
Leucánthemum alpínum (L.)
Alpine moon daisy

Asteráceae (Compósitae) – Sunflower family

Leaves: Basal leaves with stems, pinnately cut in the form of a comb, with the segments clearly separated from each other and 4–8 times as long as wide; the middle part of the leaf does not taper towards the end

Inflorescence: Composite flower 3 cm across at the end of each stem; bracts narrow triangles, smooth or rather hairy, with dark brown edges

Flowers: Ray-florets white, up to 15 mm long, and turning rather pink when the flower opens; tubular florets yellowish; fertilised by insects

Fruits: Achenes up to 3 mm long, with 5 ridges and a toothed edge on top

Habitat: In alpine (rarely subalpine) zone in grassland, scree, snow valleys, and on ground with long periods of snow cover, on stony or pure clay or loam soil saturated with snow, lime-poor or with the lime leached out, moderately acid, with humus based on rotted matter; here and there carried down into the valleys

293 **Wild chamomile – M. chamomílla**
Plants 15–40 cm high, sweetly scented, with upright stems; flowering from May to September

Matricária chamomílla L.
Wild chamomile

Asteráceae (Compósitae) – Sunflower family

Leaves: Alternate, smooth, doubly or trebly pinnate with narrow lanceolate to thread-like last leaflets

Inflorescence: Plants branch in the upper region; at the end of each branch is a composite flower 15–25 mm across; bracts greenish and smooth

Flowers: Ray-florets white, standing in several rows, narrow oval, with distinctly clipped ends and bending back towards the stem after flowering; tubular rosettes yellowish; receptacle is always conical, hollow, and without scales

Fruits: Achenes up to 1.5 mm long, often with glands and a narrow toothed margin

Habitat: From hill to subalpine zones in rubbish dumps, wasteland, fields, vineyards, by walls and roadsides, on well-drained, generally lime-poor, sandy or pure loam and clay soil rich in nutrients and containing at least some humus; found in warm places; known for a long time to accompany cultivation

294 **Alpine fleabane – E. alpínus**
Plants 3–25 cm high, with gnarled rhizomes and reddish stems; flowering from June till August

Erígeron alpínus L.
Alpine fleabane

Asteráceae (Compósitae) – Sunflower family

Leaves: Basal leaves lanceolate to narrow spathulate, entire, gradually narrowing towards the stem and with hairs on both sides; leaves on the stem lanceolate, towards the top of the stem sessile on a restricted base, hairy on both sides and entire; main vein often reddish

Inflorescence: Generally 1–5 composite flowers on each stem, 15–30 mm across; bracts form narrow triangles, come to a sharp reddish point, and are hairy

Flowers: Ray-florets pink or mauve, erect before flowering and longer than the bracts; tubular florets yellowish or reddish; between tubular and ray-florets stand very slender female florets; fertilised by insects

Fruits: Achenes 2–3 mm long with pappus 3–5 mm long and whitish or reddish; dispersal by wind

Habitat: In subalpine and alpine zones in meadows and scree on well-drained, often lime-poor, loose and stony loam or clay; range extends throughout the Alps

295 **Annual fleabane – E. ánnuus**
Plants 20–140 cm in high, with woody roots, one stem often at an angle; flowering from June till Oct.

Erígeron ánnuus (L.) Pers.
Annual fleabane

Asteráceae (Compósitae) – Sunflower family

Leaves: Alternate, light green, with sparse hairs on both sides; lower leaves oval or obovate or elliptical, narrowing sharply to a long stem, the leaf proper more than 3 cm long, deeply toothed; leaves on middle and upper part of the stem long lanceolate, entire or with slight, fine teeth

Inflorescence: Umbel-like panicle of composite flowers 1–2 cm across and nodding before flowering; bracts light green, with sparse hairs or smooth, and up to 5 mm long

Flowers: Ray-florets narrow lanceolate, numerous, spreading, and lilac or white; tubular florets yellowish

Fruits: Achenes up to 1 mm long with a pappus 2 mm long

Habitat: In hill (rarely mountain) zone by river banks, ditches, roadsides, on rubbish dumps, in clearings of river woodland and near thickets, on wet to moist sandy or stony loam rich in nutrients; an early coloniser of newly turned soil

296 **Quickweed – G. ciliáta**
Plants 10–50 cm high, with tap-roots; flowering from July to October

Galinsóga ciliáta (Rafin.) Blake
Fringed quickweed

Asteráceae (Compósitae) – Sunflower family

Leaves: Alternate, with scattered hairs on both sides, oval to lanceolate, pointed, with short stalks and irregular deep teeth

Inflorescence: Composite flowers, soliary or in small umbel-like racemes and panicles; flowers 3–6 mm across; yellow bracts narrow oval, 2–4 mm long, with scattered hairs

Flowers: 4 or 5 ray-florets, female and white or purple; tubular florets numerous, hermaphrodite, hairy outside, and yellow

Fruits: Achenes about 1 mm long, dark-coloured, hairy and with a short pappus; dispersal by wind or burrs

Habitat: In hill (rarely mountain) zone in weed-beds, mown fields, gardens, vineyards, and by roadsides on well-drained to moderately dry, loose, sandy or pure loam and clay rich in nutrients, preferably lime-poor, particularly rich in nitrogen, neutral and containing more or less humus

297 **Edelweiss – L. alpínum**
Plants 5–20 cm high, perennials with white thread-like hairs, without runners; flowering in July and August

Leontopódium alpínum Cass.
Edelweiss

Asteráceae (Compósitae) – Sunflower family

Leaves: Narrow lanceolate, alternate, gradually tapering towards the base, 2–5 cm long, pointed, blue-green above and grey-green beneath because of the thread-like hairs

Inflorescence: Composite flowers in umbel-like bunches of between 1 and 10, 4–7 mm across and surrounded by numerous transparent brown-edged bracts; these are covered with crinkly white hairs and are narrow lanceolate

Flowers: Tubular and thread-like in form; the outer flowers of the whole inflorescence and the outer florets of the inner flowers are thread-like and female; the inner florets of the inner flowers are apparently hermaphrodite, but have only fertile stamens; fertilised by insects

Fruits: Achenes about 1 mm long, with scattered short hairs

Habitat: In alpine (rarely subalpine) zone on fairly well-drained, loose, dry soil which is base-rich, neutral, generally containing lime and humus

298 **Yarrow – A. millefólium**
Plants 15–40 cm high, with scattered hairs and branching upper parts; flowering from June to Oct.

Achilléa millefólium L.
Yarrow, milfoil

Asteráceae (Compósitae) – Sunflower family

Leaves: Lanceolate to linear-lanceolate and doubly or trebly pinnate as far as middle vein, with up to 50 leaflets on each side, the last ones narrow lanceolate; there is a sterile basal rosette

Inflorescence: Composite flowers in terminal corymbs; flowers with numerous brown, blunt bracts up to 6 mm long, overlapping like tiles

Flowers: Ray-florets white or pink, female and only slightly drawn back after flowering; tubular florets whitish, hermaphrodite and without spores; fertilised by insects or self-fertile

Fruits: Achenes up to 2 mm long, flat oval, without a toothed margin or pappus bristles; dispersal by wind and insects

Habitat: From hill to subalpine zones in fields, meadows, ravines, woodland, by the roadside and on exposed sand, on well-drained to moderately dry sandy, stony, or loamy soil rich in nutrients

299 **Fragrant yarrow – A. nóbilis**
Plants 10–50 cm high, perennials with short, branching rhizomes; flowering from June to Sept.

Achilléa nóbilis L.
Fragrant yarrow

Asteráceae (Compósitae) – Sunflower family

Leaves: Broad oval in outline and doubly or trebly pinnate nearly up to the middle vein, which is usually winged; between the main leaflets solitary teeth are found on the middle vein; 4–12 pinnate, hairy sections present on each side of the middle vein

Inflorescence: Composite flowers in terminal corymbs; bracts covered with scattered hairs and light brown scales

Flowers: Ray-florets white or yellowish white, round, incised at the end, 4–6 per flower; tubular florets whitish, hermaphrodite but without spores; fertilised by insects

Fruits: Achenes up to 1 mm long, flat oval, without a toothed margin or pappus bristles; dispersal by insects

Habitat: In hill and mountain zones on rocks, moderately dry to dry grassland, by thickets and roadsides, and as an accompaniment of cultivation on dry, base-rich, loose rock or sandy loess usually containing lime but little humus; an early coloniser of newly turned soil; likes light and warmth

300 **Black yarrow – A. atráta**
Plants 5–25 cm high, perennials with much-branched rhizomes; flowering in July and August

Achilléa atráta L.
Black yarrow

Asteráceae (Compósitae) – Sunflower family

Leaves: Lanceolate or narrowly rectangular, pinnately cut almost to the middle vein and with scattered hairs; the winged middle vein is at most 2 mm wide; 6–12 5-toothed leaflets on either side

Inflorescence: Corymbs with few composite flowers; these are 5 –8 mm long, with bracts having scattered hairs and a dark brown to black border

Flowers: Ray-florets white, female, with 2 distinct notches at the end; 6–12 in each flower; tubular florets whitish to brownish, hermaphrodite but without spores; fertilised by insects

Fruits: Achenes up to 2 mm long, flat oval, without a toothed margin or pappus bristles; dispersal by insects

Habitat: In alpine (rarely subalpine) zone in slate and limestone scree on moist to wet, stony, lime-rich soil; widespread and frequently found in the Alps; these plants of the eastern Alps spread readily on scree

301 **Musk milfoil – A. moscháta**
Plants 5–20 cm high, perennials with branching rhizomes; sweet-smelling and flowering from June/Aug.

Achilla moschaáta Wulf.
Musk milfoil

Asteráceae (Compósitae) – Sunflower family

Leaves: Elongated, pinnately cut almost to the middle vein and smooth or rather hairy; the winged middle vein is at most 2 mm wide; 5–12 entire or more rarely somewhat toothed leaflets on either side

Inflorescence: Corymbs few composite flowers; these have light brown bracts 3–5 mm long, with scattered, but a dark brown border

Flowers: Ray-florets white, female, with notches at the end, and 5–8 in each flower; tubular florets whitish, hermaphrodite but without spores; fertilised by insects or self-fertile

Fruits: Achenes up to 2 mm long, flat oval, without a toothed margin or pappus bristles; dispersal by insects

Habitat: In the alpine (rarely subalpine) zone on rocks, scree, grassy slopes, and in stony grassland on wet, lime-poor, disturbed or undisturbed, loamy, scree soil; an early coloniser of newly turned soil; spreads readily on scree

302 **Claven yarrow – A. clavénnae**
Plants 10–25 cm high, with branching rhizomes and sterile rosettes of leaves; flowering in July and Aug.

Achilléa clavénnae L.
Claven yarrow

Asteráceae (Compósitae) – Sunflower family

Leaves: Oval, irregularly pinnate almost to middle vein and with close silky hairs; winged middle vein 2–5 mm wide with close silky hairs on both sides; 3–8 5-pointed segments on either side

Inflorescence: Corymbs of few composite flowers; these are up to 1.5 cm across, with light grey bracts 3–5 mm long, with hairs and a black border; 3–20 per stem

Flowers: Ray-florets white, with wavy ends, 5–9 in each flower; tubular florets white, hermaphrodite but without spores; fertilised by insects

Fruits: Achenes up to 2 mm long, flat oval, without toothed margins or pappus bristles

Habitat: In the alpine (rarely subalpine) zone on scree, stony meadows, and sunny rock faces, on well-drained, loose, stony loam which is lime-rich, neutra, and contains humus; mainly in the eastern Alps, but found westwards to Bavaria and Lake Lugano

303 **Dwarf milfoil – A. nána**
Plants 5–15 cm high, perennials with much-branched rhizomes; flowering in July and August

Achilléa nána L.
Dwarf milfoil

Asteráceae (Compósitae) – Sunflower family

Leaves: Narrow lanceolate, pinnately cut almost to middle vein, covered with fine woolly hairs, with 4–12 entire or 2–5-toothed segments on either side; middle vein covered with thick woolly hairs

Inflorescence: Corymb with 5–20 composite flowers; these are up to 1 cm across, with bracts up to 5 mm long, covered with fine woolly hair and with black margins

Flowers: Ray-florets white, narrow oval, with few teeth at the ends, female, and 5–8 in each flower; tubular florets hermaphrodite and light yellow to light brown; fertilised by insects

Fruits: Achenes up to 2 mm long, flat oval, without toothed margins or pappus bristles

Habitat: In alpine zone on moraines and scree on base-rich, loose, moderately acidic soil which is wet with snow; this western alpine plant is particularly widespread in the central and southern Alps; it is seldom found in the northern Alps

304 **Cat's foot – A. dioica**
Plants 5–20 cm high, perennials with runners above ground; flowering from May to July

Antennária dioica Gaertner
Cat's foot

Asteráceae (Compósitae) – Sunflower family

Leaves: Lower leaves are narrow oval, narrowing towards the base or spathulate, coming to short points and with white thread-like hairs beneath; upper leaves narrow lanceolate, pointed, and often reddish; rosettes of leaves

Inflorescence: Several composite flowers up to 8 mm across in a terminal umbel; bracts are oval, overlapping like tiles, with gossamer-like hairs, transparent in their upper part; red (rarely white) on female flowers and mostly white on male flowers

Flowers: Plants are dioecious; in female flowers all florets are female with thread-like corollas; in male plants florets appear hermaphrodite (but only the pollen is fertile) and have tubular corollas; fertilised by insects (particularly butterflies)

Fruits: Achenes up to 1 mm long, cylindrical or egg-shaped and with pappus bristles 6–10 mm long

Habitat: In subalpine and alpine zones in infertile meadows, sparse woodland, and heathland, on well-drained to dry soil

305 **Norwegian cudweed – G. norvégicum**
Plants 10–40 cm high, with thin rhizomes sending up several shoots; flowering in July and August

Gnaphalium norvegicum Gunnerus
Norwegian cudweed

Asteráceae (Compósitae) – Sunflower family

Leaves: Lanceolate, narrowing towards the base, up to 15 cm long and 2 cm wide, pointed, with 3 veins (a distinct main vein and 2 further longitudinal veins) and felted hairs on both sides

Inflorescence: Composite flowers, up to 7 mm long, in loose spikes; bracts generally blunt, entire, with wide dark brown or black margins; the lowest bract is longer than the florets

Flowers: Outer flowers are female and thread-like; inner ones are tubular, hermaphrodite, and fertile; anthers lie in the lower region of the floret and have elongated points; fertilised by insects or self-fertile

Fruits: Achenes up to 3 mm long, pappus of 1 row of fine bristles; dispersal by wind

Habitat: In subalpine zone in sparse woodland, scrub, meadow-land, grassy slopes, and by the roadside, on damp to well-drained loam which is lime-poor but contains humus; grows well in grassy clearings

306 **Canadian fleabane – E. canadénsis**
Plants 20–100 cm high, with tap-roots, and glossy or grooved stems; flowering from June to September

Erígeron canadénsis L.
Conyza canadénsis (L.) Cronq.
Canadian fleabane

Asteráceae (Compósitae) – Sunflower family

Leaves: Lanceolate, entire or with a few small teeth, and close hairs; lower leaves taper to a winged stem, while upper ones are sessile, with narrow bases

Inflorescence: Panicles of several composite flowers; these are 3–5 mm long, with bracts up to 4 mm long, green and almost smooth

Flowers: Ray-florets inconspicuous, upright and whitish; no thread-like florets present; tubular florets yellow-white; self-fertile

Fruits: Achenes up to 1.5 mm long, with scattered hairs, yellow, and with pappus 2–3 mm long; dispersal by wind

Habitat: In hill and mountain zones in warmer places on scree, fields, ditches, margins of woods, and in forest clearings, on well-drained to moderately dry clay or loamy soil, rich in nutrients but with little humus; these plants, originally from North America, have now spread worldwide

307 **White butterbur – P. álbus**
Plants up to 30 cm high in flower and up to 70 cm high when fruiting; perennials flowering in May

Petasítes álbus (L.) Gaertner
White butterbur

Asteráceae (Compósitae)- Sunflower family

Leaves: Basal leaves first appear after flowering; they are long-stalked, round to heart-shaped, up to 40 cm wide; margins have alternating large and small teeth; grey hairs developing beneath with age, hairy above at first, becoming smooth later; flowering stems covered with pale green lanceolate to oval scales

Inflorescence: Composite flower heads in short thick racemes; individual flower heads on stems, up to 1 cm across; bracts thickly covered with glandular hairs

Flowers: All florets tubular, with yellow-white corollas; fertilised by insects

Fruits: Achenes 2–3 mm long, with white pappus up to 12 mm long; dispersal by wind

Habitat: In mountain (rarely hill and subalpine) zones in beech, oak, and spruce woods with good ground cover, woods in gorges, thickets, by the roadside and in hill scrub, on wet clay soil containing lime, fine earth and nutrients but little humus; an indicator of ground water; a plant of shade or half-shade; an early coloniser of newly turned land

308 **Cabbage thistle – C. oleráceum**
Plants 30–140 cm high, perennials with thick, gnarled rhizomes; flowering from June to September

Círsium oleráceum (L.) Scop.
Cabbage thistle

Asteráceae (Compósitae) – Sunflower family

Leaves: Entire or pinnately cut down to the middle, amplexicaul, with toothed edges and soft-stemmed; end leaflet wider than lateral ones; all leaflets pointed

Inflorescence: Several composite flower heads in short racemes, surrounded by pale, wide oval sheaths; bracts finely pointed, greenish, usually with soft spines

Flowers: All florets tubular, with thread-like tips, up to 2 cm long and pale yellow; fertilised by insects

Fruits: Achenes 3–5 mm long, yellowish-brown, with pappus 15 mm long; dispersal by wind

Habitat: In hill and mountain (rarely subalpine) zones in wet meadows, marshes, riverside woodland, ditches, shrubby undergrowth, on banks of streams and springs, on wet, loose, sandy or pure clay soil which is base-rich, rich in nutrients, and contains a moderate amount of humus; plants grow in light or half-shade; deep roots; plants attract bees; an indicator of well manured soil

309 **Spiniest thistle – C. spinosíssumum**
Plants 20–50 cm high, perennials covered with leaves, with rhizomes; flowering in July and August

Círsium spinosíssimum Scop.
Spiniest thistle

Asteráceae (Compósitae) – Sunflower family

Leaves: Lanceolate, entire, pointed, pinnately cut towards the middle vein, with spiny leaflets, stiff to the touch; upper leaves amplexiaul; main and lateral veins pale yellow and raised distinctly above the surface of the leaf

Inflorescence: Composite flower heads grouped in knots at the end of the stem, surrounded by numerous pale green, spiny, lanceolate, and pinnate leaves; bracts have pile-like resin glands and (particularly the outer ones) spines

Flowers: All florets tubular, 1–2 cm long and pale yellow; fertilised by insects

Fruits: Achenes 3–5 mm long, yellowish-brown and with pappus 15 mm long; dispersal by wind and by means of burrs

Habitat: In subalpine and alpine zones in meadows, by camp sites, scree, and thickets, on moist to well-drained stony loam or clay soil containing nutrients and humus; an indicator of nitrogen

310 **Hyssop – H. officinális** Sub-shrubby plants (the lower part is woody) 20–50 cm high; perennials flowering from July to September

Hyssópus officinális L.
Hyssop

Lamiáceae (Labiátae) – Mint family

Leaves: Linear to lancolate, narrowing gradually towards the base, coming to blunt points at the tips, almost sessile, entire, up to 25 mm long and with margins slightly bent down

Inflorescence: Spike-like, 3–10 cm long, with several whorls of 3–7 flowers in leaf axils at the ends of the stems

Flowers: Short-stemmed; 5 sepals, fused, with 5 identical teeth; 4–8 mm long, with about 15 clearly protruding veins, green; 5 petals, generally blue, rarely white, 8–12 mm long, with a double-lipped edge; the upper lip is flat, the lower lip in 3 parts; 4 stamens, emerging from the corolla; superior ovary of 2 merged carpels

Fruits: 4 ovoid-triangular nutlets, each 2 mm long

Habitat: In hill (rarely mountain) zone in dry grassland, rocky waste, and sunny slopes, on dry, stony, flat, lime-poor soil

311 **White dead-nettle – L. álbum** Plants 15–45 cm high, with rectangular stems; flowering from May to August

Lámium álbum L.
White dead-nettle

Lamiáceae (Labiátae) – Mint family

Leaves: Opposite, oval, coming to long points, with only short stems or sessile, with deep, irregular teeth, usually blunt; covered in scattered hairs

Inflorescence: Flowers in dense axillary, sessile, whorls towards the ends of the stems

Flowers: 5 sepals, fused, 6–11 mm long, generally with violet spots at the base, covered in scattered hairs, and with a long point; 5 petals, fused, generally white, up to 25 mm long and with 2 lips, the upper one shaped like a helmet, the lower one in 3 parts; 4 stamens, protruding from below the upper lip, with black anthers having white hairs; pollen pale yellow; superior ovary of 2 carpels placed together

Fruits: 4 ovoid-triangular nutlets, smooth and each up to 3 mm long

Habitat: From hill to subalpine zones by walls, camp sites and the wayside, in hedgerows, thickets and ditches on well drained loose loam or clay soil containing nutriment and humus

312 **Downy hemp-nettle – G. ségetum** Plants 10–30 cm high, annuals or biennials, with rectangular, hairy stems; flowering in July and Aug.

Galeópsis ségetum Necker
Downy hemp-nettle

Lamiáceae (Labiátae) – Mint family

Leaves: Lancolate to oval, on short stems, with 3–9 teeth on each side and velvety hairs, especially beneath

Inflorescence: Flowers in axillary, sessile, whorls towards the ends of the stems

Flowers: 5 sepals, light yellow, with green stripes, 6 –10 mm long, with erect hairs; sepal teeth are usually dark green; 5 petals, up to 35 mm long, whitish to pale yellow, with 2 lips, the upper one helmet-shaped, erect, entire, and hairy, the lower one in 3 parts, with a very large middle section bearing yellow or purple marks; 4 stamens, protruding under the upper lip and having white filaments; ovary superior; fertilised by insects

Fruits: 4 egg-shaped, slightly triangular nutlets, smooth, and each up to 3 mm long

Habitat: In hill and mountain zones in scree, gravel beds, woodland clearings, fields, and by the roadside on fairly well-drained, stony soil more or less rich in nutrients and poor in lime; one of the first plants to appear on newly turned soil

313 **Catnip – N. catária** Plants 40–130 cm high, perennials with tap-roots; flowering from July to September

Népeta catária L.
Catnip

Lamiáceae (Labiátae) – Mint family

Leaves: Narrow oval to cordate, on stalks, pointed, coarsely toothed, and grey-green beneath; points directed forwards

Inflorescence: Short-stemmed flowers in axillary whorls towards the ends of the stems; whorls have distinct stems lower down

Flowers: 5 sepals, light to dark green, 5–8 mm long, thickly covered in hairs; teeth of calyx coming to long narrow points; 5 petals, fused into a long tube widening towards the front and whitish to yellowish or reddish; upper lip flat or slightly swollen; lower lip in 3 parts; 4 stamens; ovary superior; fertilised by insects

Fruits: Ovoid-triangular nutlets, smooth and up to 2 mm long

Habitat: In hill and mountain zones by railway lines,walls, rubbish dumps, by the roadside, and in thickets, on moderately dry, generally sandy or stony loam in warmer places rich in nutrients; used as a medicinal plant in earlier times

314 **Common eyebright – E. rostkoviána** Plants 5–25 cm high, annuals covered in frizzy hairs, with reddish stems; flowering from June to Oct.

Euphrásia rostkoviána Hayne
Common eyebright

Scrophulariáceae – Figwort family

Leaves: Oval to broad oval, with a short, wide stalk, 3–7 pointed teeth on each side, a blunt or rounded tip, and covered with glandular hairs

Inflorescence: Solitary flowers or small whorls growing in the axils of upper leaves on the stem

Flowers: Bell-shaped calyx, in 4 parts, 4–6 mm long, usually covered in glandular hairs and with 4 pointed teeth; 5 petals, fused into a tube, up to 14 mm long and with 2 lips; upper lip white, usually with violet veins, swollen and notched at the end; lower lip longer than the top one, wide and flat, in 3 parts, and white, with violet veins and a yellow spot; corolla tube also yellow inside; 4 stamens, enclosed by the upper lip and with hairy anthers in the lower part; ovary superior

Fruits: Capsules bursting in 2 parts, with many seeds

Habitat: In all zones on meadow land and moors on moist to dry soil

315 **Pretty eyebright – E. pulchélla** Plants 2–10 cm high, with dark red to brown stems; flowering in July and August

Euphrásia pulchélla Kerner
Pretty eyebright

Scrophulariáceae – Figwort family

Leaves: Oval to round, with 1–3 teeth, blunt or with short points, on each side, and distinctly hairy; the tip is blunt or rounded

Inflorescence: Whorls in leaf axils of upper stem leaves; flowers in lateral whorls have stems

Flowers: Bell-shaped calyx, in 4 parts, 2–4 mm long, covered in hairs and with 4 pointed teeth; 5 petals, fused into a tube, 6–8 mm long and with 2 lips, the upper one whitish to pale red, usually with dark red veins and slightly swollen; the lower lip longer than the upper one, usually wide and flat, in 3 parts, and white, with red stripes and a yellow spot in the lower part; the segments of the lower lip are clearly notched; 4 stamens, enclosed by the top lip; ovary superior

Fruits: Capsules bursting in 2 parts, with many seeds

Habitat: In subalpine and alpine zones on meadow land and at the edge of thickets, on moist to well-drained soil containing humus

316 **Canary clover – D. herbáceum**
Sub-shrubs 25–60 cm high, woody at the base, often forming mats; flowering in June and July

Dorycnium herbáceum Vill.
Canary clover

Fabáceae (Papilionáceae) – Pea family

Leaves: Trefoil leaflets lanceolate, pointed, and covered with erect hairs

Inflorescence: Spherical umbels bearing 5–12 flowers with long stems

Flowers: 5 fused sepals, up to 2 mm long, with points; 5 petals, white and 3–5 mm long; the keel has a dark purple tip and points slightly upwards; the standard is fiddle-shaped; the 2 wings are fused at the point; 10 stamens; the anthers enclose the superior ovary of 2 carpels

Fruits: Pods springing open in 2 parts

Habitat: In hill (rarely mountain) zone in sparse woodland, on rocky hillsides, on rough grassland, and in shrubbyy places on dry, loamy, stony soil containing lime, in warm places; a south-east European plant with a range extending westwards to the Rhône and northwards to the Danube

317 **Alpine clover – T. nivále**
Plants 5 –20 cm high, perennials with recumbent or erect stems; flowering till October

Trifólium nivále Sieber
Alpine red clover

Fabáceae (Papilionáceae) – Pea family

Leaves: Trifoliate, on long stems; leaflets narrow oval to oval, entire, up to 3 cm long, pointed, rounded, or somewhat incised and rather hairy

Inflorescence: Racemose heads 20–35 mm across with many flowers

Flowers: 5 fused sepals, covered thickly with hairs, lightly coloured in the lower part and mid to dark green at the long teeth, often with a reddish overlay; 5 petals, white, yellowish, or reddish, 15–20 mm long, fused to form 2 narrow wings, a keel, and a standard swelling outwards a little; 10 stamens; superior ovary of 2 carpels

Fruits: Pods springing open in 2 parts

Habitat: In subalpine and alpine zones on meadow land and near the edge of woodland on well-drained loamy soil fairly rich in nutrients; range extends south to the Pyrenees and the northern Spanish mountains

318 **White clover – T. répens**
Plants 5–20 cm high, perennials with creeping stems rooting at the nodes; flowering till September

Trifólium répens L.
White clover, Dutch clover
Fabáceae (Papilionáceae) – Pea family

Leaves: Trifoliate and long-stalked; leaflets narrow oval to oval, very finely toothed, up to 4 cm long, rounded or somewhat incised, smooth and sometimes with a white mark in the middle

Inflorescence: Racemose heads 15–20 mm across with many flowers

Flowers: With distinct stems; 5 fused sepals, covered thickly with hairs, lightly coloured in the lower part and mid to dark green at the long teeth, often with a reddish overlay; calyx tube has green veins; 5 petals, white to pale reddish, 6 –12 mm long, fused to form 2 narrow wings, a keel, and a standard swelling outwards a little; 10 stamens; superior ovary 2 carpels

Fruits: Pods

Habitat: From hill to subalpine zones on lawns, arable fields, playing fields, meadow land, etc., on moist to well-drained heavy loam or clay soil rich in nutrients

319 **White melilot – M. álbus**
Plants 30–140 cm high, annuals or perennials, sweet-smelling in dry atmosphere; flowering June to Aug.

Melilótus álbus Desr.
White melilot

Fabáceae (Papilionáceae) – Pea family

Leaves: Trifoliate and short-stalked; leaflets of the lower leaves are long oval, those of the upper ones almost linear; all are entire or finely toothed, and pointed

Inflorescence: Racemes on stems, with many flowers; lengthening when seeds are ripe

Flowers: Pendant with long stems; 5 fused sepals, light green and pointed; 5 fused white petals; standard distinctly longer than the 2 wings and the keel; 10 stamens; superior ovary of 2 carpels

Fruits: Pods 3–4 mm long, smooth, with a network of ridges and many seeds

Habitat: In hill and mountain zones on excavations, rubbish dumps, quarries, by the roadside, and in gravel pits, on moderately dry soil of any kind rich in bases and nutrients, and may be rich in humus or uncultivated; roots go up to 90 cm deep; an early coloniser of newly turned soil; attracts bees; accompanies cultivation; also used as fertiliser

320 **Alpine milk vetch – A. álpinus**
Plants 5–15 cm high, perennials with erect stems; flowering in July and August

Astrágalus álpinus L.
Alpine milk vetch

Fabáceae (Papilionáceae) – Pea family

Leaves: Unequally pinnate, with 7–12 pairs of lateral leaflets, which are narrow oval, entire, with blunt ends, and hairy or smooth on both sides

Inflorescence: Long-stemmed spherical racemes

Flowers: 5 fused sepals, greenish to reddish; teeth of sepals almost as long as the calyx tube; 5 fused petals; the keel is violet and almost as long as the standard, the white wings shorter than the keel; 10 stamens; superior ovary of 2 carpels

Fruits: Pods up to 15 mm long, up to 5 mm thick, with short dark hairs

Habitat: In subalpine and alpine zones im meadow land and scree on well-drained to moderately dry stone/clay soil, which is neutral to moderately acid and contains humus; these arctic-alpine plants are found in arctic regions of Europe, Asia, and North America; their southward range extends to the Pyrenees

321 **Wood vetch – V. silvática**
Plants 40–140 cm high, perennials with recumbent or climbing stems; and flowering till August

Vícia silvática L.
Wood vetch

Fabáceae (Papilionáceae) – Pea family

Leaves: Unequally pinnate, with 6–12 pairs of lateral leaflets, which are long to narrow oval, 6–18 mm long, with many oblique side veins, alternate and smooth; the end leaflet is formed into branching tendrils

Inflorescence: Nodding racemes on stems, formed of 10–12 flowers

Flowers: 5 fused sepals, generally light green, pointed and of unequal length; 5 fused petals up to 18 mm long; standard and wings white with blue or violet veins; keel often has a violet tip; 10 stamens, superior ovary

Fruits: Pods 2–3 cm long, flat, smooth and containing 4–8 seeds

Habitat: In mountain and subalpine (rarely hill) zones in sparse woodland, forest clearings, near thickets, and by woodland paths on well-drained to moderately dry, stony or sandy loam or clay soil more or less rich in nutrients and containing humus; plants grow in light and half- shade

322 **Lady's slipper – C. calcéolus**
Plants 20–50 cm high, with creeping rhizomes; not often found; flowering in May and June

Cypripédium calcéolus L.
Lady's slipper

Orchidáceae – Orchid family

Leaves: 5 –13 cm long, oval, widest in the middle, amplexicaul in the lower part of the stem, with parallel veins, 2–4 on each plant, light to mid green, often with hairs on veins and at the edges

Inflorescence: 1–3 flowers

Flowers: 6 petals; 4 wide lanceolate, fused from the middle to the tip, up to 5 cm long, erect and brownish-red; 2 fused along four-fifths of their length to form a yellow shoe or slipper 3–4 cm long; 2 stamens; ovary inferior; stigma, in 3 parts, covered by an erect sterile stamen (= staminode) in the opening of the lip

Fruits: Capsules

Habitat: In hill and mountain zones in fairly sparse deciduous or coniferous forests with good grass and ground cover and in hedgerows, on intermittently watered loam and clay soil which is dry in summer, base-rich, and generally contains lime; prefers sheltered places with mild weather; plants grow in half-shade and attract bees

323 **Small white orchid – L. álbida**
Plants 10–25 cm high, with deeply palmate tubers; flowering in June and July

Leucórchis álbida E. Meyer
Pseudórchis álbida Seg.
Gymnadénia álbida C. Rich
Small white orchid

Orchidáceae – Orchid family

Leaves: 5–15 cm long, narrow oval, widest above the middle, amlexicaul in lower part of stem, with parallel veins; light to mid green

Inflorescence: Raceme 3–6 cm long, cylindrical, with dense flowers

Flowers: Pale yellow or whitish oval petals, 2–3 cm long; lip somewhat longer than outer petals, narrowing in a wedge towards the base and with 3 deep lobes; spur cylindrical, rising upwards and as long as the ovary; 2 stamens; inferior ovary; fertilised by insects

Fruits: Capsules

Habitat: In subalpine (rarely mountain or alpine) zone in rough pasture, and sparse coniferous forest on fairly well-watered, stony or pure loam which is lime-free, rich in rotted peat and humus, rather acid and poor in nutrients; often found with mountain tobacco; common in the Jura, the Black Forest, the Vosges, and the Alps

324 **Wild tulip – T. sylvéstris**
Plants 20–50 cm high, with runners (with bulbs at the end); flowering in May and June

Tulipa silvestris L.
Wild tulip

Liliáceae – Lily family

Leaves: 10–20 cm long, up to 2 cm wide, narrow lanceolate, green to blue-green, with parallel veins, alternate, 2–3 on each plant

Inflorescence: Solitary flowers

Flowers: Generally nodding before blooming; 6 petals, not always of the same size, 4–6 cm long, yellow, pointed and with fine hairs at the tip; 6 stamens; anthers usually thickly covered with hairs at the base; superior ovary; stigma small, narrower than the ovary and sitting directly on it

Fruits: Capsules

Habitat: In hill zone in orchards, thickets, fields, and vineyards on warm, fairly well-watered, deep loose loam or chalk which is base-rich and rich in nutrients; also found in wet scrub; plants attract bees; often found with tassel hyacinth and star of Bethlehem; an ornamental plant from southern Europe, naturalised on the northern slopes of the Alps; found especially in climatic conditions favouring vineyards, otherwise seldom found

325 **Wild daffodil – N. pseudonarcíssus**
Plants 10–40 cm high, flowering in March and April; a garden plant since the Middle Ages

Narcíssus pseudonarcíssus L.
Wild daffodil

Amaryllidáceae – Daffodil family

Leaves: All basal, 10–30 cm long, linear, rather fleshy, blunt at the ends, and blue-green

Inflorescence: Generally solitary flowers, rarely multiple ones, and sometimes with a translucent bract; the specimen shown here is a garden variety

Flowers: More or less upright, on stems, and 5–10 cm in diameter; 6 light yellow petals; tube 5–15 mm long, the segments usually oval ; it is as long as the surrounding petals, spreading at the mouth, and deep yellow; 6 stamens, visible in the tube and not always of the same height; inferior ovary; 1 pistil with 3 stigmata

Fruits: Capsules with 3 compartments, many seeds in each

Habitat: From hill to subalpine zone in gardens (cultivated varieties) or wild in meadow land and sparse woodland, on loamy soil poor in lime but rich in bases and nutrients

326 **Yellow flag – I. pseudacórus**
Plants up to 1 m high, with thick rhizomes; flowering in May and June

Iris pseudacórus L.
Yellow flag

Iridáceae – Iris family

Leaves: 30–110 cm long, 1–3 cm wide, gradually pointed towards the end, basal and blue-green

Inflorescence: Generally with several flowers on each stem, sometimes enclosed by a bract or spathe

Flowers: 6 petals, yellow and fused at the base into a short tube; 3 outer petals 3–8 cm long, without erect hairs and bent back or erect; 3 inner petals generally upright, rather narrower and not longer than the tip of the pistil; 3 stamens; anthers as long as filaments; inferior ovary; both branches of the styles have fine points and are toothed

Fruits: Capsules of 3 compartments, with many seeds; capsules float

Habitat: In hill (rarely mountain) zone in marshland in woods and fields, in ditches and on the banks of streams, in the shallows of lakes (a typical shallow-water plant) on marshy soil containing humus which is periodically or generally flooded; plants grow in light and half- shade and tend to like warmth

327 **Scotch asphodel – T. calyculáta**
Perennial plants 10–35 cm high, with short rhizomes; flowering from June till September

Tofiéldia calyculáta (L.) Wahl.
Scotch asphodel

Liliáceae – Lily family

Leaves: Linear, 2–5 mm wide, gradually coming to a point; basal leaves arranged in 2 rows; up to 4 stem leaves, decreasing in size or absent in higher parts of the stem

Inflorescence: 2–7 cm long cylindrical raceme

Flowers: On stems up to 1 mm long, standing loosely in the lower part of the raceme; 6 petals, lanceolate, 3–4 mm long, slightly swollen, yellow-green or whitish, often with red tips; below the flower head there is sometimes a small, curved bract, often in 3 segments; 6 stamens; superior ovary

Fruits: Capsules containing many seeds

Habitat: From mountain to alpine (rarely hill) zones on moorland meadows, flat and marshy moorlands, moist meadow land, stony alpine slopes, and shady thickets, on intermittently or permanently moist, waterlogged, muddy loam or clay which is lime-rich and moderately rich in nutrients; an indicator of lime and marshy soil

328 **Blue spurge – E. myrsinites**
Plants 10–20 cm high, smooth-leaved perennials with recumbent stems; flowering from April to July

Euphórbia myrsinítes L.
Blue spurge

Euphorbiáceae – Spurge family

Leaves: Generally sessile, 2–3 cm long, rectangular or inversely egg-shaped, coming to a short point, arranged spirally and overlapping, fleshy, blue-green, rather frosty in appearance; green in winter

Inflorescence: Umbels with 7–10 branches

Flowers: Yellow bracts enclosing a female flower, generally hanging on a long stem, and several male flowers of only one stamen; ovary in 3 sections, with 3 styles fused at the base and each in 2 parts; there is no perianth

Fruits: Fruits fall apart into 3 sections

Habitat: In hill zone on dry, sunny slopes on loose, dry, porous, well-drained soil, more or less base-rich and rich in nutrients; a decorative and conspicuous spurge which may be planted in rockeries and dry walls; it originated in the Mediterranean region and has been developed as a garden plant

329 **Golden saxifrage – C. oppositifólium**
Plants 5–20 cm high, with thread-like runners; flowering from April to July

Chrysosplénium oppositifólium L. / Golden saxifrage

Saxifragáceae – Saxifrage family

Leaves: Basal leaves are broad oval (wider than long) to round, 10–20 mm in diameter, narrowing at the base to a wide wedge or cut short, crenate or notched on the edge and usually hairy on top; stem leaves appear in 1 to 3 pairs, opposite and stemmed or sessile

Inflorescence: Umbels with many flowers

Flowers: Usually 4 sepals, up to 2 mm long and yellow-green to yellow; no petals; 8 stamens; inferior ovary of 2 carpels; fertilised by insects or sel-fertile

Fruits: Capsules holding red-brown seeds with a silky gloss

Habitat: In mountain (rarely hill or subalpine) zone by shady springs or streams, on wet rocks, in scree and ash woods on cool, wet, lime-poor, stony or sandy soil, moderately base rich and containing nutrients and humus; in earlier times used as a medicinal plant against diseases of the spleen

330 **Evening primrose – O. biénnis**
Plants 50–120 cm high, with upper part of stem bending over; flowering until October

Oenóthera biénnis L.
Evening primrose

Onagáceae – Evening primrose family

Leaves: Alternate, lanceolate, 5–15 cm long, pointed, entire or slightly toothed, sessile or narrowing towards the reddish stem

Inflorescence: Solitary flowers growing from the leaf nodes in panicles at the end of the stem

Flowers: 4 sepals, fused before flowering, often reddish, coming to a fine point and later lying back along the stem; 4 petals, 15–30 mm long,, pale yellow, rounded at the end, with edges overlapping, and longer than the stamens; 8 stamens; inferior ovary in 4 sections; fertilised by insects or self-fertile

Habitat: In hill zone by rubbish dumps, railway embankments, thickets, in weed-beds, docks and quarries, on wet to dry, generally stony, gravelly or sandy loam more or less rich in nutrients; very deep-rooted; an early coloniser of newly turned soil

331 Wood – I. tinctória
Plants 25–100 cm high, biennials or perennials, with thick tap-roots; flowering from April to June

Isatis tinctória L.
Wood

Brassicáceae (Crucíferae) – Mustard family

Leaves: On the stems, longish arrowheads in shape, in the lower part of the stem usually grey-green and amplexicaul with pointed tips at the base; lower leaves also toothed

Inflorescence: A hemispherical, much-branched, corymbose panicle

Flowers: On stems; 4 sepals, up to 2 mm long and yellowish green; 4 petals, on short claws, rounded at the end and yellow; 6 stamens (the outer 2 short, the inner 4 longer); superior ovary of 2 ovules; fertilised by insects

Fruits: Longish pods, wedge-shaped, up to 20 mm long, black when ripe

Habitat: In hill and mountain zones by the roadside, in ditches, on rubbish dumps, fences, and in quarries, on dry soil warm in summer, more or less rich in bases and nutrients, and usually containing lime; used in earlier times as a dye plant, the indigo content being released during fermentation of the plant

332 **Rape – B. nápus**
Plants 50–120 cm high, annuals or perennials with thick tap-roots; flowering from April to May

Brássica nápus L.
Rape

Brassicáceae (Crucíferae) – Mustard family

Leaves: Upper stem leaves narrow arrowheads, wavy in outline or entire, usually blunt-ended, with no or a few hairs, and usually with a broadened base which is mostly amplexicaul; lower stem leaves are often deeply notched

Inflorescence: Loose racemes without leaves

Flowers: On stems; 4 sepals, blue-green, upright, and up to 8 mm long; 4 petals, with long claws, up to 15 mm long and yellow; 6 stamens; superior ovary of 2 ovules

Fruits: Pods 5–10 mm long containing up to 40 seeds

Habitat: In hill and mountain zones in fields and occasionally wild on rubbish dumps on well-drained, deep, sandy or pure loam soil rich in bases and nutrients; used as an oil-bearing plant and also as fodder and as a vegetable; there are various subspecies, e.g. nápus (oilseed rape) and rapifera (swede – a fodder plant)

333 **Bladderwort – A. utriculáta**
Plants 20–50 cm high, with woody, branching rhizomes; flowering in April

Alyssoídes utriculáta (L.) Medicus
Bladderwort

Brassicáceae (Crucíferae) – Mustard family

Leaves: Stem leaves lanceolate, pointed, 2–4 cm long, sessile or narrowing to thin stalks, entire, numerous, and dark green; basal leaves narrow oval to lanceolate and narrowing to short stalks

Inflorescence: Raceme with many flowers, often rather compact

Flowers: On stems; 4 sepals, upright, smooth or with isolated fine hairs, up to 12 mm long and light green; inner ones pouched at the base; 4 petals, oval, 15–20 mm long and yellow; 6 stamens; superior ovary of 2 ovules

Fruits: Pods up to 12 mm long and 10 mm wide, inflated and containing many seeds

Habitat: In hill and mountain zones in scree and rock debris on more or less dry and generally stony soil; this western Alpine plant may be found in the Apennines, the Rhône valley, and Val d'Aosta

334 **Buckler mustard – B. laevigáta**
Plants 10–35 cm high, with stems up to 25 mm thick; flowering from May to July

Biscutélla laevigáta L.
Buckler mustard

Brassicáceae (Cruciferae) – Mustard family

Leaves: Basal, longish, up to 13 cm long, maximum width 2 cm, arranged in a rosette, entire or somewhat toothed, with plentiful hairs

Inflorescence: Loose racemes of several flower heads on stems

Flowers: On long stems; 4 sepals, light yellow, pouched at the base; 4 petals, round at the end, on claws, and yellow; 6 stamens, 4 longer and 2 shorter; filaments without teeth at the base; superior ovary of 2 ovules; fertilised by flies

Fruits: Pods shaped like spectacles, with beaks 3–5 mm long; stem as long as fruit

Habitat: In subalpine and alpine (rarely hill and mountain) zones on stony meadows, scree, alluvial soil, among rock debris, on well-drained to moderately dry stony soil which is warm in summer, neutral, and usually contains lime and humus; also found on wet moorland; plant grows in light

335 **Yellow whitlow grass – D. aizoides**
Plants 2–10 cm high, perennials with branching rhizomes; flowering from May to July

Drába aizoídes L.
Yellow whitlow grass

Brassicáceae (Cruciferae) – Mustard family

Leaves: In a thick basal rosette; single leaves up to 2 cm long, narrow lanceolate, stiff, leathery, and fringed with stiff bristles; there are no stem leaves

Inflorescence: Racemes with 3–20 flowers

Flowers: On stems; 4 sepals, 2–4 mm long, somewhat bent and yellow-green; 4 petals, yellow at start of flowering, later turning whitish; 6 stamens, 2 longer and 2 shorter, as long as the petals; superior ovary superior of 2 ovules with a long pistil; fertilised by insects

Fruits: Pods 5–15 mm long, with a smooth stem up to 20 mm long and a beak 1–3 mm long; seeds up to 1.5 mm long

Habitat: In subalpine and alpine (rarely mountain) zones on stable scree, open spaces, alpine rock plateaux and debris, and on sunny boulders, mostly on dry, stony soil which is lime-rich but lacking good topsoil; species shows great variety

336 **Charlock – S. arvénsis**
Plants 15–60 cm high, with long tap-roots; flowering from May to October

Sinápis arvénsis L.
Charlock, wild mustard

Brassicáceae (Cruciferae) – Mustard family

Leaves: Up to 15 cm long; basal and lower stem leaves lyre-shaped, on stalks, irregularly notched or pinnate, with a large terminal lobe, and rough-haired veins beneath; upper stem leaves sessile, entire and irregularly notched

Inflorescence: Racemes

Flowers: On stems; 4 sepals, generally smooth, 4–6 mm long, yellow-green; 4 petals, with long claws, sulphur yellow; 6 stamens, 2 longer and 2 shorter; superior ovary of 2 ovules

Fruits: Smooth pods with upward-pointing hairs

Habitat: In hill and mountain (rarely subalpine) zones by the roadside, rubbish dumps, beside paths, a weed of arable land, on moderately dry to well-drained sandy soil or pure loam rich in nutrients and bases and usually neutral; also found on lime-rich soil; roots often more than 1 m deep; an indicator of lime; attracts bees

337 **Yellow rocket – B. intermédia**
Plants 10–70 cm high, biennials, smooth, with thin tap-roots; flowering from April to June

Barbárea intermédia Boreae
Yellow rocket

Brassicáceae (Cruciferae) – Mustard family

Leaves: All pinnately cut; basal leaves have 1–6 lateral lobes on each side; upper stem leaves also pinnate to the middle lobe, partially amplexicaul and with winged stalks; the terminal lobe is very large

Inflorescence: Racemes

Flowers: On stems; 4 sepals, lanceolate to narrow oval, 2–4 mm long, usually yellow or yellowish-green; 4 petals, 4–6 mm long, yellow; 6 stamens, 2 longer and 2 shorter; superior ovary superior of 2 ovules

Fruits: Pods 1.4–4 mm long, in great numbers

Habitat: From hill to subalpine zones in warmer places on riverbanks, rubbish dumps, by the roadside, in fields and gardens, on moist to well-drained soils of all kinds which are rich in nutrients and bases and may or may not contain humus; an early coloniser of newly turned soil; a western Mediterranean plant that has spread widely

338 **Greater celandine – C. május**
Plants 20–80 cm high, with orange-yellow, milky sap; flowering from April to September

Chelidónium május L.
Greater celandine

Papaveráceae – Poppy family

Leaves: Irregularly pinnately cut or pinnate, on stalks, usually blue-green with scattered hairs beneath; leaflets oval, with irregular double notches or lobes, with rounded segments

Inflorescence: 2 sepals, with scattered hairs, and yellow-green; 4 petals, golden yellow and up to 15 mm long; numerous stamens with club-like anthers; superior ovary of 2 carpels forming 1 compartment

Fruits: Pod-like capsules up to 5 cm long, linear, opening from below by 2 valves

Habitat: In hill and mountain zones in weed beds, rubbish dumps, by car parks and walls, by the roadside, in thickets and in the neighbourhood of houses, on well-drained, loose, stony, sandy ,or clay soils rich in nutrients; an indicator of nitrogen content; accompanies cultivation; used in earlier times to remove warts

339 **Tormentil – P. erécta**
Plant stem 10–50 cm long, recumbent or erect; flowering from June to September

Potentilla erecta (L.) Räusch.
Tormentil

Rosáceae – Rose family

Leaves: Basal leaves ternate, sessile, or only on short stems, and green on both sides; leaflets oval, narrowing in a wedge shape towards the base, up to 2 cm long, with greatest width towards the tip, covered with scattered silky hairs and deeply toothed, with protruding terminal teeth; stem leaves usually of 5 leaflets and larger than basal leaves; bracts have sparse hairs and are palmate

Inflorescence: Solitary flowers at the end of noticeably thin stems

Flowers: Generally with all parts in fours; up to 1 cm in diameter; outer sepals narrow to broad lanceolate, inner ones somewhat broader; usually 4 petals, heart-shaped and yellow; numerous stamens; numerous carpels on a rather domed receptacle; fertilised by insects

Habitat: From hill to subalpine zones in sparse woodland, meadows, rough meadow land, and moorland on well-drained clay and loam which contains humus and is often also acidic

340 **Lady's bedstraw – G. vérum**
Plants 10–80 cm high, with 4-angled stems; flowering till September

Gálium vérum L.
Lady's bedstraw

Rubiáceae – Bedstraw family

Leaves: 15–25 mm long, narrow linear (needle-shaped), 1–2 mm wide, with sharp points and edges rolled downwards; usually dark green above, light green beneath, often thickly covered with short hairs; 6–12 to a whorl

Inflorescence: Thick panicles at end of stem, with many flowers

Flowers: Smelling of honey; stems 1–3 mm long, usually with bracts; 4 sepals; 4 petals, golden yellow, coming to sharp points, and spreading out more or less flat; 4 stamens; inferior ovary; fertilised by insects

Fruits: Fissured fruits splitting into 2 parts

Habitat: In hill and mountain (rarely subalpine) zones on the margins of woods and thickets and by the roadside, on embankments, in pine woods, reedy and dry meadows, on moderately dry loose loam and loess rich in bases and nutrients and generally containing lime and humus; also found on sandy soils; the plant has spreading roots and grows in light; only of limited value for fodder

341 **Yellow fumitory – C. lútea**
Plants 10–30 cm high, with branching rhizomes; flowering from March to September

Corydalis lútea (L.) DC
Yellow fumitory

Fumariáceae – Fumitory family

Leaves: Long stalks, blue-green and multiply ternate; lobes entire or unequally notched

Inflorescence: Racemes at end of stem

Flowers: Situated in axils of small, narrow lanceolate, finely toothed bracts, and up to 2 cm long; 2 sepals, toothed and up to 2 cm long; 4 petals, dark yellow towards the tip and pale yellow towards the stem; the upper of the 2 outer petals has a backward-pointing spur, wider towards the front and bent upwards, while the bottom one is bent downwards and not notched at the base; the 2 inner petals are identical; 2 stamens, tripartite, the central branch bearing a 2-celled anther, the lateral branches each bearing a 1-celled anther; superior ovary

Fruits: Pod-like capsules

Habitat: In hill and mountain zones on rocks and walls, in rock debris and shingle by streams, on fairly damp, stony, lime-bearing soil

342 **Yellow alpine poppy – P. alpínum**
Plants 5–20 cm high, perennials with branching rhizomes; flowering in July

Papáver alpínum L. ssp. kérneri (Hayek) Fedde
Papaver kérneri Hayek
Yellow alpine poppy

Papaveráceae – Poppy family

Leaves: Stem without leaves; those in basal rosette singly or doubly pinnate to the middle vein, blue-green, with 2–6 lanceolate, hairy segments, the ends blunt or coming to obtuse points

Inflorescence: Solitary flowers

Flowers: 2 sepals, enclosing the flower completely and covered thickly in brown hairs; 4 petals, up to 4 cm long, with wavy edges at the ends; golden yellow; numerous stamens; filaments thread-like; ovary superior with 5–7 stigma rays

Fruits: Capsules opening by pore-like valves below stigmatic disk

Habitat: In alpine zone: if found lower, it has been washed down; found on rubbish dumps, moraines, alluvial soil, and in mobile scree with prolonged snow covering, on limestone and older rock in the central and southern Alps; roots spread out in scree; a central and southern European mountain plant; found in the mountains of the Balkan peninsula, the eastern Pyrenees, and the southern Alps

343 **Fringed waterlily – N. peltáta**
Plants grow together in clusters of floating leaves; flowering from July to September

Nymphoídes peltáta O. Kuntze
Fringed waterlily

Menyanthácae – Bogbean family

Leaves: On long stems, almost circular, similar to the leaves of true waterlilies, with a heart-like incision where the stalk joins the leaf, large, the underside with scattered glands and grey-green or reddish-violet, dark green above

Inflorescence: Axillary panicles

Flowers: 5 sepals, lanceolate to narrow oval and greenish to reddish; 5 petals, up to 3 cm long, golden yellow, broad oval, with fringed, toothed edges, and bearded at the throat; 5 stamens; arrow-shaped anthers; superior ovary of 2 ovules; fertilised by insects, mainly bumble-bees

Fruits: Capsules containing many seeds; dispersal by water by means of the floating fruit

Habitat: In hill (rarely mountain) zone in masses of floating leaves in tranquil water or in slowly flowing water which is warm in summer, rich in nutrients, base-rich and eutrophic, over a muddy base containing humus; also found in stagnent water; an early coloniser of shallow waters

344 **Yellow waterlily – N. lútea**
Plants rooting up to 6 m deep, with rhizomes 3–8 cm thick; flowering in July and August

Núphar lútea (L.) S.
Yellow water-lily

Nymphaeáceae – Waterlily family

Leaves: 10–30 cm long, floating, broad oval in outline, entire, though often with somewhat wavy edges, with a heart-shaped incision at the stalk; the lateral veins are triply forked and radiate throughout the leaf (no cross veins are present)

Inflorescence: Solitary flowers

Flowers: Up to 5 cm in diameter and with a simple perianth; 5 petals, golden-yellow and 2 to 3 cm long; numerous honey glands and stamens; superior ovary, many-celled, summit more or less convex with a central depression and 15 –20 stigmatic rays; fertilised by insects or self-fertile

Fruits: Flask-shaped capsules 2–4 cm long

Habitat: In hill and mountain zones in still or slow-flowing water on a base of sand or gravel containing humus; optimal depth of water 80–200 cm; range extends northwards to Scotland and in Scandinavia to 67 degrees

345 **Yellow pimpernel – L. nemórum** Plants without underground runners, upright stems; root in the lowest region at leaf nodes; flower May/July

Lysimáchia nemórum L.
Yellow pimpernel

Primuláceae – Primrose family

Leaves: 1–3 cm long, opposite, broad oval to oval, coming to blunt points, entire, smooth, dotted with transparent glands

Inflorescence: Long-stemmed and solitary, arising in the axils of the upper leaves

Flowers: 5 sepals, narrow lanceolate, without red dots, smooth and up to 5 mm long; 5 petals, yellow, 4–8 mm long and oval or lanceolate; 5 stamens; superior ovary

Fruits: Capsules, 3–5 mm long, not pitted

Habitat: In hill and mountain (rarely subalpine) zones in wet beech woods or woodland in gorges with ground cover, in alder woods, by the roadside or woodland springs, on wet, sandy, stony or pure loam and clay rich in nutrients, lime-poor, neutral to moderately acid and containing humus, in humid climatic conditions; plant grows in shade; range extends northwards to 62 degrees

346 **White mullein – V. lychnítis**
Plants up to 1.5 m in height, with clearly angled upper stems; flowering from June to September

Verbáscum lychnítis L.
White mullein

Scrophulariaceae – Figwort family

Leaves: 10–30 cm long, up to 15 cm wide, finely notched, with scattered hairs above, and dense grey or white powdery filaments beneath; upper leaves sessile with rounded bases, lower leaves have short stalks

Inflorescence: 2–8 flowers in axils of upper leaves

Flowers: Stems up to 10 mm long; 5 sepals, each 2–4 mm long; 5 petals, spread flat or funnel-shaped, up to 2 cm in diameter, yellow (rarely white) with hairs on the outside; 5 stamens, with kidney-shaped anthers at right angles to each other; filaments covered with whitish-yellow woolly hairs; superior ovary; style with simple stigma; fertilised by insects

Fruit: Capsules

Habitat: From hill to subalpine zones in forest clearings, dry grassland, hedges, oak woods, by the roadside, and in thickets, in warmer places on dry, loose, stony soils more or less base-rich and rich in nutrients, and containing lime and humus; deep-rooted

347 **Yellow loosestrife – L. vulgáris**
Plants 20–130 cm high, without branches or with branches in upper part only; flowering June to Aug.

Lysimáchia vulgáris L.
Yellow loosestrife

Primulaceae – Primrose family

Leaves: Up to 15 cm long, opposite or in whorls of 3 to 4, long oval, with very short stems or sessile, with narrowing bases; edges slightly rolled, pointed, dotted with red glands, and covered with hairs – scattered above and thick beneath

Inflorescence: Terminal panicles springing from the axils of the top stem leaves

Flowers: Usually 5 sepals, lanceolate, covered with glands, with hairs at the base and with a red edge; 5 petals, up to 14 mm long, oval, with blunt points, yellow, with short glandular hairs on the inner side; 5 stamens; superior ovary; fertilised by insects or self-fertile

Fruits: Capsules 4–5 mm long

Habitat: In hill and mountain zones in woodland on moors and by river banks, ditches, moorland grass and stubble, by banks of streams and in thickets, on wet loam and clay soils which are more or less base-rich, with deep topsoil; acts as manure for soil; deep-rooted

348 **Tufted loosestrife – L. thyrsiflóra**
Plants 30–70 cm high, with long underground runners; flowering from May to July

Lysimáchia thyrsiflóra L.
Tufted loosestrife

Primuláceae – Primrose family

Leaves: 5–12 cm long, opposite, with succeeding pairs at right angles, narrow oval to lanceolate, pointed, with wavy edges slightly rolled upwards, narrowing toward the base, sessile, generally dotted with red glands and with hairs beneath on the middle vein

Inflorescence: Racemes up to 3 cm long springing from the axils of the middle stem leaves; bracts up to 4 mm long, lanceolate and smooth

Flowers: 5 or 6 sepals, narrow lanceolate and smooth; generally 5 or 6 petals, up to 6 mm long, narrow lanceolate, yellow, with red glands towards the end; 5 stamens, usually protruding beyond the petals; superior ovary

Fruits: Capsules 2–4 mm long, dotted with red glands

Habitat: In hill and mountain zones in ditches, bogs, by ponds or slowly flowing water, on wet, periodically flooded, peaty, soil usually rich in bases and nutrient;s plants grow in half- shade; germinates in frosty weather

349 **Yellow mountain saxifrage – S. aizoídes**
Plants 3–25 cm high, with upright stems, forming mats; flowering from June to August

Saxifrága aizoídes L.
Yellow mountain saxifrage

Saxifragáceae – Saxifrage family

Leaves: 1–3 cm long, linear or linear-lanceolate, fleshy, pointed, semicircular in cross-section, covered at the base and at the edge with stiff hairs

Inflorescence: Racemes or panicles containing 3–10 flowers and covered with glandular hairs

Flowers: On long stems; 5 sepals, smooth, 4–6 mm long, oval and lying close to the petals; 5 petals, narrow oval, yellow, longer than the sepals, often with dark spots; 10 stamens; superior ovary; fertilised by insects, mostly flies

Habitat: From mountain to alpine zones on banks of streams, by springs, on moist scree, on wet or flooded rocks, on wet, base-rich, generally lime-containing, stony or pure clay and marl soils, but also on rock without fine soil; also reproduces vegetatively; the range of this alpine/arctic species extends to Iceland and Spitzbergen

350 **Yellow mountain saxifrage – S. aizoídes**
Plants 3–25 cm high, with upright stems, forming mats; flowering from June to August

Saxifrága aizoádes L.
Yellow mountain saxifrage
(a variety with yellow and red flowers)

Saxifragáceae – Saxifrage family

Leaves: 1–3 cm long, linear, or linear-lanceolate, fleshy, pointed, semicircular in cross-section, covered at the base and at the edge with stiff hairs

Inflorescence: Racemes or panicles containing 3–10 flowers and covered with glandular hairs

Flowers: On long stems; 5 sepals, smooth, 4–6 mm long, oval and lying close to the petals; 5 petals, narrow oval, reddish yellow, longer than the sepals, often with dark spots; 10 stamens; superior ovary; fertilised by insects, mostly flies

Habitat: From mountain to alpine zones on banks of streams, by springs, on moist scree, on wet or flooded rocks, on wet, base-rich, generally lime-containing clay and marl soils, but also on rock without fine soil; also reproduces vegetatively; the range of this alpine/arctic species extends to Iceland and Spitzbergen

351 **St John's wort – H. perforátum**
Plants up to 100 cm in height; stems woody at base, with 2 raised lines; flowering from June to Sept.

Hypéricum perforátum L.
St John's wort

Hypericáceae – St John's wort family

Leaves: 1–3 cm long, oval to long oval or narrow linear, abundantly covered with translucent glandular dots; entire, without stems, and opposite

Inflorescence: Panicles with flowers on long stems

Flowers: 5 sepals, lanceolate, 3–5 mm long and ending in fine points; 5 petals, much longer than the sepals, yellow and toothed on one side; numerous stamens, combined in bundles; superior ovary, usually with 3 styles; fetilised by insects or self-fertile

Fruits: Capsules opening with flaps

Habitat: In hill and mountain (rarely subalpine) zones in thickets, dry grassland, forest clearings, heath land, and colonising newly turned soil, on fairly dry, moderately acid to neutral soils of all kinds which contain humus and have deep topsoil; roots up to 50 cm deep; an indicator of exhausted soil; formerly used medicinally; grows in half-shade

352 **Rock-rose – H. nummulárium**
Plants 10–40 cm high, perennials, lower part woody; flowering from May to October

Heliánthemum nummulárium (L.) Mill.
Rock-rose

Cistáceae – Rock-rose family

Leaves: 1–5 cm long, up to 1 cm wide, narrow oval to linear, entire, with edges somewhat turned back, opposite, with lanceolate stipules longer than the leaf stalk, leathery, with bristles above and a thick web of fine hairs beneath

Inflorescence: Lax raceme-like cymes

Flowers: On stems; 5 sepals of unequal size, the 3 inner ones 4–7 mm long, with bristles on the veins and soft hairs between them, larger than the 2 outer sepals; petals 5–12 mm long, golden-yellow; many stamens; superior ovary 3 cells and with a pistil; fertilised by insects or self-fertile

Habitat: In hill and mountain (rarely subalpine) zones on dry grassland, in thickets, ridges of fields, in pine woods and dry woodland, in warmer places, on base-rich loess and loam which is warm in summer; roots go deep

353 **Great spearwort – R. língua**
Plants 40–130 cm in height, perennials, branching freely; flowering from June till August

Ranúnculus língua L.
Great spearwort

Ranunculáceae – Buttercup family

Leaves: When plants are not in bloom the basal leaves are oval and often heart-shaped at the base of the leaf proper; when the shoots are submerged the leaves are oval; otherwise basal and stem leaves up to 25 cm long, lanceolate, pointed, and usually entire

Inflorescence: Solitary flowers

Flowers: 3–4 cm in diameter (the largest local buttercup); 5 sepals, yellow and lying close to the petals; 5 petals, yellow, widest towards the end of the petal; many stamens; numerous carpels on a domed receptacle; fertilised by insects

Fruits: Achenes

Habitat: In hill (rarely mountain) zone in ponds, ditches, and marshes, on flat muddy ground covered with standing water or alternately flooded and dry, base-rich, lime-poor, and containing humus; flowers attract flies; plant tends to prefer warmth

354 **Grass-leaved buttercup – R. gramíneus**
Plants 10–40 cm high, smooth-leaved perennials without rhizomes; flowering in April and May

Ranúnculus gramíneus L.
Grass-leaved buttercup

Ranunculáceae – Buttercup family

Leaves: Basal leaves grass-like, narrow linear to lanceolate, 10–20 cm long, up to 1 cm wide, coming gradually to a fine point; stem leaves similar in form, but a maximum of 3 cm long; at the base of the plant there is a thick bunch of fibres consisting of the remains of dead leaves

Inflorescence: Solitary flowers

Flowers: 1.5–2 cm in diameter; 5 sepals, narrow oval, greenish, often with yellow edges, coming to sharp points and lying close to the corolla; 5 petals, yellow, widest towards the end of the petal; many stamens; numerous carpels on a domed receptacle; fertilised by insects

Fruits: Achenes up to 2.5 mm long, containing a single seed and beaked; ripe

fruits form a cylindrical or more or less egg-shaped spike

Habitat: In hill and mountain zones in dry, poor meadows and grass on deep, lime-rich soil

355 **Lesser spearwort – R. flámmula**
Plants 15–65 cm high, usually with curved stems, branching freely; flowering from June till August

Ranúnculus flámmula L.
Lesser spearwort

Ranunculáceae – Buttercup family

Leaves: All entire or faintly toothed; basal leaves narrow oval to lanceolate with long stalks; stem leaves narrow lanceolate, with short stems or sessile

Inflorescence: Solitary flowers

Flowers: 0.5–1.5 cm in diameter; 5 sepals, yellow and lying close to the petals; 5 petals, yellow, broadest towards the end of the petal; many stamens; numerous carpels on a domed receptacle; fertilised by insects

Fruits: Round achenes, up to 1.5 mm in diameter, with smooth surfaces and short,straight beaks

Habitat: In hill and mountain (rarely subalpine) zones in ditches, marshes, moorland, by springs, on the banks of streams, on wet, often more or less acid, sandy or muddy soil containing humus; an early coloniser of newly turned soil; spreads by rooting at nodes of runners; dispersed by waterfowl; poisonous; found in almost the whole of Europe except Iceland

356 **Wall pepper – S. ácre** Plants 3–15 cm high, perennials with numerous creeping stems forming mats, blooming in June and July

Sédum ácre L.
Wall pepper

Crassuláceae – Stonecrop family

Leaves: 2–5 cm long, up to 3 mm wide, narrow oval, rounded at the end, fleshy, flat above, swollen beneath, without stems; alternate, and with a hot, acrid taste

Inflorescence: 2 or 3 main branches, each with 2–4 flowers at the fork

Flowers: 5 sepals, oval, up to 3 mm long and greenish; 5 petals, narrow oval to lanceolate, coming to fine points, standing out almost horizontally and golden yellow; 10 stamens in 2 circles; 5 superior, free-standing carpels

Fruits: Follicles 3–5 mm long

Habitat: From hill to subalpine zones in dry sunny places like rocks, dry grassland, dunes, shingle , walls, and roof-tops, on warm, dry, loose, sandy and rocky ground more or less base-rich and rich in nutrients but lacking fine topsoil; roots shallowly

357 **Soft stonecrop – S. sexanguláre**
Plants 5–10 cm in height, erect or climbing, without glands; flowering in June and July

Sédum sexangulare L.
Sédum mite Gilib.
Soft stonecrop

Crassuláceae – Stonecrop family

Leaves: 3–6 mm long, up to 1.5 mm wide, narrow oval, blunt at the ends, fleshy, alternate, and with an appendage about 0.3 mm long at the base

Inflorescence: With 2 or more main branches, each with several flowers at the fork

Flowers: 5 sepals, oval, 2–3 mm long and greenish; 5 petals, 4–8 mm long, narrow oval to lanceolate, coming to fine points, standing out almost horizontally and lemon yellow; 10 stamens in 2 circles; 5 superior, free-standing carpels

Fruits: Follicles 2–6 mm long

Habitat: From hill to subalpine zones, like wall pepper, in dry, sunny places like rocks, dry grassland, dunes, shingle, walls and roof-tops, on warm, dry, loose, sandy and rocky ground, more or less base-rich and rich in nutrients but lacking fine topsoil; roots shallowly

358 **Alpine wall pepper – S. alpéstre**
Plants 2–9 cm high, recumbent or with curved stems; flowering in July and August

Sédum alpéstre Vill.
Alpine wall pepper

Crassuláceae – Stonecrop family

Leaves: 2–6 mm long, narrow oval, wide and rounded at the end, fleshy, alternate, widely spaced in lower part of stem, closely spaced near the flowers, and without a spur at the leaf base

Inflorescence: Spirals containing a few flowers

Flowers: 5 sepals, rarely 6, oval, 2–3 mm long and often yellow-green; 5 petals, 4–6 mm long, narrow oval to oval, blunt-ended or coming to short points, and yellow; 10 stamens in 2 circles; 5 superior, free-standing carpels

Fruits: Follicles 1–4 mm long

Habitat: In alpine (rarely subalpine) zone in grassland, scree, and stony fissures on well-drained, base-rich, lime-free, loamy soils low in humus and weakly acidic, having long periods of snow cover and permanently leached of nutrients; found mainly in masses in snow valleys; also an early coloniser of bare rock

359 **Kingcup – C. palústris**
Plants up to 50 cm high, with curved or prostrate aerial stems, which are hollow; flowering in July

Cáltha palústris L.
Kingcup, marsh marigold

Ranunculáceae – Buttercup family

Leaves: On long stalks, dark green, alternate, and glossy; leaf proper is roundish, up to 15 cm in diameter, heart-shaped or kidney-shaped at the base, and with a notched or toothed edge; upper stem leaves sessile

Inflorescence: Groups of flowers in upper parts of stem branches

Flowers: usually 5 petals (no sepals), broad oval, 1–2 cm long, bright golden-yellow, glossy and dropping off immediately after flowering; up to 40 stamens; no nectaries; 3–9 superior ovaries containing many seeds on a domed receptacle; fertilised by insects

Fruits: Flat follicles up to 2.5 cm long, often somewhat bent and forming a group spread out like a star

Habitat: Found in all zones in marshes, fens, ditches, and wet woods on moderately acid marshy soil or on loam or clay containing humus; range extends north into the Arctic up to 76 degrees, southwards to Greece

360 **Upright wood sorrel – O. fontána**
Plants 10–30 cm high, with tuberous underground runners, flowering from June to September

Oxalis fontána Bunge
Oxalis europaéa Jordan
Upright wood sorrel

Oxalidáceae – Wood sorrel family

Leaves: Trefoil leaves with heart-shaped leaflets pointed at base; leaflets 5–20 mm long, up to 25 mm wide, meeting in a flexible joint at the stalk, entire and rounded at the ends; leaves opposite or in whorls

Inflorescence: Umbels containing few flowers; the stem is knotty at the base of the inflorescence

Flowers: On stems and nodding before flowering; flower stem upright after flowering starts; 5 sepals, pointed and green; 5 petals, round at the end, 4–8 mm long and mid to deep yellow; 10 stamens, the outer 5 being rather shorter; superior ovary with 5 free-standing styles

Fruits: Capsules dispersing seeds by ejection

Habitat: In hill zone in warmer locations in waste places and as a garden or arable weed, on well-drained, loose sand or loam rich in nutrientsand generally lime-poor and containing humus; accompanies cultivation

361 **Large-flowered cinquefoil – P grandiflóra**
Plants 20–40 cm high, perennials with hairy stems; flowering in July and August

Potentilla grandiflóra L.
Large-flowered cinquefoil

Rosáceae – Rose family

Leaves: Basal leaves have erect hairy stalks 4–8 cm long, and generally have 3 radial leaflets; these are oval or roundish, 1–3 cm long, widest towards the end, slightly hairy, narrowing in a wedge shape towards the base and with 4–9 blunt teeth on each side; upper stem leaves often sessile

Inflorescence: Stem branches frequently, bearing several or many flowers

Flowers: 10 sepals, hairy and light to dark green, 5 longer and 5 shorter; 5 petals, golden- yellow, 1–2 times as long as sepals and not fused together; many stamens with smooth filaments and outwardly pointing open anthers; numerous carpels; fertilised by insects

Fruits: Numerous achenes with one seed each

Habitat: In subalpine and alpine zones on rough pasture, sunny, often rocky slopes, or in grassland, on base-rich, generally lime-poor, stony loam

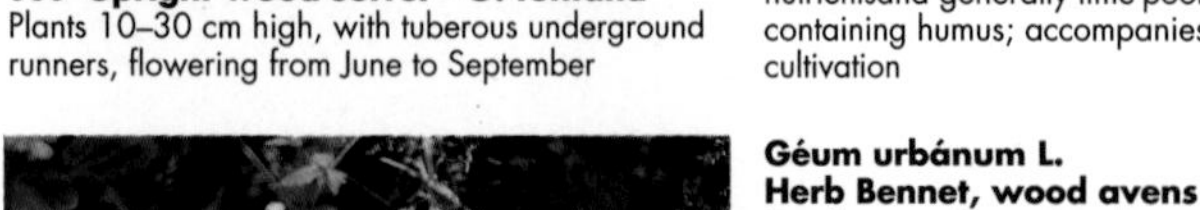

362 **Herb Bennet – G. urbánum**
Plants 25–85 cm high, with thick rhizomes; flowering from May to August

Géum urbánum L.
Herb Bennet, wood avens
Rosáceae – Rose family
Leaves: Basal leaves lobed or pinnate with 2–3 pairs of unequal lateral leaflets 5–10 mm long and a large terminal leaflet 5–8 cm long; stem leaves large, the lower ones like the basal leaves or ternate or deeply trilobed, the upper usually simple; all leaves crenate or deeply toothed; stipules 1–3 cm long, ranging from linear to leaf-like with large teeth or lobes
Inflorescence: Stem carries one or more flowers
Flowers: Double calyx; sepals narrow or broad lanceolate to triangular, hairy, green, up to 8 mm long and bent back after flowering; generally 5 petals, wide oval to round, up to 8 mm long and golden-yellow; many stamens; numerous superior carpels; style is hook-shaped in upper part, and covered in bristles; fertilised by insects
Fruits: Groups of achenes
Habitat: In hill and mountain (rarely subalpine) zones in woods, scrub, hedge banks, rubbish dumps, by the roadside and in shady places on well-drained loam or clay soils containing nutrients and humus

363 **Woolly buttercup – R. lanuginósus**
Plants 20–100 cm high, perennials with stems thickly covered in hairs; flowering from May to August

Ranúnculus lanuginósus L.
Woolly buttercup

Ranunculáceae – Buttercup family

Leaves: Basal leaves have 5–7 lobes or are almost trifoliate; topmost stem leaves often divided into 3 to the base; segments or leaflets deeply toothed or 3- lobed and covered in fine woolly hairs

Inflorescence: Solitary flowers on round, forking stems

Flowers: 1.5–3 cm in diameter; 5 sepals, greenish-yellow and lying close to the petals; 5 petals, deep yellow amd widest towards the end; many stamens; numerous superior carpels on a smooth, slightly domed receptacle; fertilised by insects

Fruits: Round or oval achenes up to 4 mm long, with a hook-shaped or rolled-up beak

Habitat: In mountain and subalpine zones in shrubby undergrowth, in mountain and wet woodlands on wet, loose, stony, or pure clay or loam rich in lime and nutrients, and containing humus; typically found on rubbish tips

364 **Mountain buttercup – R. montánus**
Plants 20–50 cm high, with smooth rhizome s; flowering from May to August

Ranúnculus montánus Willd. s.l.
R. grenieranus Willd.
Mountain buttercup

Ranunculáceae – Buttercup family

Leaves: Basal leaves rectangular oval, roundish, 5-sided, or triangular, the lateral segments incised as far as the base; all segments toothed; upper stem leaves with 3 or 5 segments, widest in the middle

Inflorescence: Stem bearing one or more flowers

Flowers: 2–3 cm in diameter; 5 sepals, yellowish, graduating to greenish towards the end, hairy and lying close to the petals; 5 petals, yellow and rounded at the ends; many stamens with smooth filaments; numerous carpels on a hairy receptacle; fertilised by insects

Fruits: Roundish, smooth achenes with hook-shaped beaks

Habitat: In subalpine and alpine (rarely mountain) zones on rubbish dumps, wet grassland, sparse woodland, and moorland on lime-poor soil containing humus

365 **Golden cinquefoil – P. aúrea**
Plants 5–20 cm high, with curving or upright stems; flowering from June to August

Potentílla aúrea L.
Golden cinquefoil

Rosáceae – Rose family

Leaves: Basal leaves have stalks 2–5 cm long with close hairs and 5 radial leaflets; these are oval, up to 2.5 cm long, widest towards the end, narrowing to a wedge at the base, smooth above, with scattered hairs beneath only on the veins, and with close hairs at the edge; each leaflet has 2–5 teeth pointing towards the end; the end tooth is very small; topmost stem leaves often lanceolate

Inflorescence: Stem bears 1–5 flowers

Fruits: Smooth achenes

Flowers: 5 each of inner and outer sepals, oval and of equal size; 5 petals, wide heart-shaped, generally with a dark gold spot at the base, 5–10 mm long and golden-yellow; numerous stamens; numerous carpels; mainly fertilised by flies

Habitat: In subalpine and alpine zones on meadows and grassland, in debris left by snow, and in scrub, on moderately dry to intermittently wet soils which are base-rich, lime-poor, acidic, and contain humus

366 **Spring cinquefoil – P. praecox**
Plants 10–30 cm high, with scattered, erect stems; flowering from April to July

Potentilla praecox F.Schultz
Spring cinquefoil

Rosáceae – Rose family

Leaves: With 5 leaflets; these are longish, narrowing to a wedge at the base, with 3–7 blunt teeth on each side, with shaggy hairs beneath; the end tooth is generally of the same size as the lateral teeth; upper stem leaves are also trifoliate; there are numerous leaf rosettes which do not carry flowers

Inflorescence: Stem bears 1 or more flowers

Flowers: 10–15 mm in diameter; 5 each of inner and outer sepals, greenish, narrow oval; 5 petals, wide heart-shaped, round at the end and golden-yellow; numerous stamens; numerous carpels; fertilised by insects

Fruits: Achenes

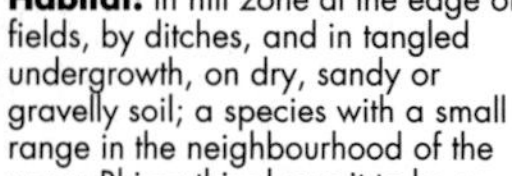

Habitat: In hill zone at the edge of fields, by ditches, and in tangled undergrowth, on dry, sandy or gravelly soil; a species with a small range in the neighbourhood of the upper Rhine; this shows it to be a plant emerging after the Ice Age; now found in Alsace and round Schaffhausen, but not frequently

367 **Upright cinquefoil – P. récta**
Plants 25–70 cm high, rigidly upright, branching in the upper part of the stem; flowering in June and july

Potentílla récta L.
Upright cinquefoil

Rosáceae – Rose family

Leaves: Basal leaves have stalks up to 20 cm long and 5 or 7 radial leaflets; these are narrow oval, often thickly covered in hair, with 7–18 deep, blunt teeth on each side; topmost leaves on stem are trifoliate

Inflorescence: Stem branches plentifully, bearing many flowers in the form of an umbel

Flowers: 2–2.5 cm in diameter, the stem having clearly visible bristles; outer 5 sepals narrow lanceolate, inner 5 broad lanceolate; 5 petals, heart-shaped, generally longer than the sepals and light to golden-yellow; many stamens with smooth filaments; numerous superior carpels; fertilised by insects

Fruits: Smooth achenes

Habitat: In hill (rarely moutain) zones on warm, dry hillsides, by ditches, sunny slopes, in gravel pits, and warm waste ground, on sandy or gravel soil which is generally dry in summer, more or less base-rich and rich in nutrients, and lime-poor

368 **Dwarf cinquefoil – P. pusilla**
Plants 5–15 cm high, perennials with numerous bushy bristles; flowering in April and May

Potentilla pusilla Host.
Potentílla pubérula Kras.
Dwarf cinquefoil

Rosáceae – Rose family

Leaves: Basal leaves have 5 or 7 leaflets; these are obovate, wedge-shaped, covered with fine hairs on both sides and with 4–8 blunt teeth on each side; leaf stalks are hairy; all leaves more or less grey-green

Inflorescence: Stems,usually upright, bear 3–8 flowers,

Flowers: 1–2 cm in diameter; outer 5 sepals narrow oval, inner sepals somewhat broader and usually pointed; 5 petals, heart-shaped, usually somewhat longer than the petals, and light to golden-yellow; many stamen; numerous superior carpels

Fruits: Smooth achenes

Habitat: From hill to subalpine zones on grassland and rough pasture, and on warm dry hillsides, on warm, dry, base-rich, lime-containing, stony loam in sunny places; also found in dry places; a mid and south European mountain plant; found throughout the Alps in warmer regions

369 **Hoary cinquefoil – P. argéntea**
Plants 10–40 cm high, with stems covered in white down; flowering from June to August

Potentílla argéntea L.
Hoary cinquefoil

Rosáceae – Rose family

Leaves: Formed of 5 leaflets, with long, hairy stalks; leaflets narrow oval, narrowing in a wedge to the base, with 2–5 blunt teeth on either side, dark green above, covered with white down beneath and more or less leathery

Inflorescence: Stems bear 2 or more flowers and are also upright when fruits are ripe

Flowers: 1–2 cm in diameter; outer 5 sepals lanceolate, inner 5 wider and roughly of equal length; 5 petals, broad oval, usually somewhat trimmed and yellow; many stamens with smooth filaments; numerous superior carpels; fertilised mostly by bees and bumble-bees

Fruits: Smooth achenes; seeds dispersed by shaking

Habitat: From hill to subalpine zones on top of boulders, on walls, by the roadside, in rocky areas and dry sandy grassland with sparse plant cover, on dry, loose, sandy or stony soils which are warm in summer, usually rich in bases and nutrients, lime-poor, and rather acidic; an indicator of sand; deep roots

370 **Staghorn cinquefoil – P. multifida**
Plants 4–30 cm hight with recumbent or erect stems; flowering in June and July

Potentílla multifída L.
Staghorn cinquefoil

Rosáceae – Rose family

Leaves: Basal leaves oval or round, up to 3 cm long and palmate; first-order leaflets 1–2 cm long, usually incised to the middle vein; second-order leaflets entire with blunt ends, with edges turned down and a thick covering of white hairs beneath; upper stem leaves also deeply incised

Inflorescence: Many-flowered cyme, hairy stem

Flowers: 7–15 mm in diameter; outer 5 sepals narrow lanceolate or narrow wedge-shaped and pointed; inner 5 sepals broader and rather longer; all sepals visibly covered with hairs; 5 petals, with flat ends, as big as or somewhat longer than sepals and pale yellow; many stamens with smooth filaments; many carpels

Fruits: Smooth achenes

Habitat: In alpine zone in places frequented by sheep and wild animals, in meadow land and on rocks, on soils rich or poor in lime but rich in nitrogen; range extends to Spitzbergen

371 **Globe flower – T. europaéus**
Plants 20–60 cm in high, with horizontal rhizomes; flowering in May and June

Tróllius europaéus L.
Globe flower

Ranunculáceae – Buttercup family

Leaves: Basal leaves have long stalks and 5 leaflets; these are oval, narrowing to a wedge shape at the base, and unequally toothed, particularly in the upper part; lower stem leaves similar, still with stalks; upper stem leaves sessile and simpler in form

Inflorescence: 1–3 flowers at end of stem

Flowers: No sepals; 5–15 petals, oval, up to 2.5 cm long, grouped in a sphere; numerous nectaries up to 7 mm long, rounded at the top and narrowing towards the base; many stamens; numerous carpels with beak-like styles; fertilised by insects

Fruits: Follicles up to 1 cm long, containing many seeds

Habitat: Generally in mountain and subalpine zones on moorland, wet pastures, scrub and woodland, on wet loam and clay rich in bases and nutrients but poor in lime

372 **Mountain buttercup – R. montánus**
Plants 20–50 cm high, erect and blooming until August. (A shorter form is found in the alpine zone)

Ranúnculus montánus Willd.
Mountain buttercup

Ranunculáceae – Buttercup family

Leaves: Basal leaves rectangular oval, round or pentagonal in outline, lateral segments incised almost to the base; all segments toothed; upper stem leaves have 3, 5, or 7 segments, broadest in the middle

Inflorescence: 1 or more flowers at end of stem

Flowers: 2–3 cm in diameter; 5 sepals, yellowish, graduating to greenish at the ends and lying close to the petals; 5 petals, yellow and rounded at the ends; many stamens, smooth at the base; numerous carpels on a hairy receptacle; fertilised by insects

Fruits: Round, smooth achenes with hook-shaped beaks

Habitat: In subalpine and alpine (rarely mountain) zones in rubbish dumps, wet grassland, meadow land, sparse woodland and moorland, on neutral soils rich in nutrients and containing lime and humus

373 **Meadow buttercup – R. ácris**
Plants 15–90 cm high, perennials with short rhizomes; flowering from April to September

Ranúnculus ácris L. s.l.
Meadow buttercup

Ranunculáceae – Buttercup family

Leaves: Basal leaves on stalks with 3 or 5 leaflets, deeply incised, narrowing in a wedge towards the base; segments also incised and often overlap; upper stem leaves usually without stalks and with narrow lanceolate segments

Inflorescence: Solitary flowers on long, round stems without branches

Flowers: 2–3 cm in diameter; 5 sepals, narrow oval, yellow and greenish and lying close to the petals; 5 petals, golden-yellow and rounded at the ends or slightly incised; many stamens; numerous carpels; fertilised by flies and bees

Fruits: Smooth achenes with straight or slightly bent beaks

Habitat: From hill to subalpine zones on meadow land and by the roadside, and in the alpine zone close to cattle-sheds, on wet to dry, loose, stony, or unmixed clay or loam rich in nutrients, lime, and humus; an indicator of fertile ground; poisonous when fresh

374 **Yellow wood anemone – A. ranunculoídes**
Plants 10–25 cm high, with horizontal creeping rhizomes; flowering in March and April

Anémone ranunculoídes L.
Yellow wood anemone

Ranunculáceae – Buttercup family

Leaves: Hardly any basal leaves appear before or during flowering; there are 3 stem leaves arranged in a whorl on the upper part of the stem, with short stalks or sessile, 3–8 cm long, rhomboidal or pentagonal in outline and with 3 leaflets; these are coarsely toothed and often deeply incised

Inflorescence: 1 or 2 flowers at end of stem

Flowers: 2–3 cm in diameter, with a hairy stem; generally 5 (but sometimes up to 9!) petals, oval, hairy outside and yellow; many stamens; numerous carpels; fertilised by insects

Fruits: Flat achenes thickly covered with short hairs; dispersed by ants

Habitat: In hill and mountain (rarely subalpine) zones in thickets, well-watered meadows, orchards; abundantly gregarious in deciduous woodland on all but the most base-deficient or waterlogged soils; grows well on wasteland; poisonous; found in Scandinavia up to 69 degrees

375 **Creeping buttercup – R. répens**
Plants 15–50 cm high, recumbent or inclined; flowering from May to September

Ranúnculus répens L.
Creeping buttercup

Ranunculáceae – Buttercup family

Leaves: Basal leaves smooth or hairy in many shapes; usually singly or doubly pinnate; the lower first-order leaflets usually on stalks; second-order segments irregularly toothed; lower stem leaves similar to basal leaves; upper stem leaves simpler in form, with narrower segments

Inflorescence: Flowers on long, branching stems

Flowers: 2–3 cm in diameter; 5 sepals, linear, hairy, often greenish at the ends and lying close to the petals; 5 petals, broad oval, yellow, rounded at the ends and narrowing to a wedge shape at the base; many stamens; numerous carpels on a hairy receptacle

Fruits: Smooth, oval achenes

Habitat: From hill to subalpine zones in wet meadows, pastures, and woodlands, on dunes and gravel pits, etc., and as a weed especially on heavy soils; grows on well-watered loam and clay soils rich in nutrients and humus;an indicator of heavy soil

376 **Bulbous buttercup – R. bulbósus**
Plants 15–50 cm high, with enlarged stem-bases; flowering from May to July

Ranúnculus bulbósus L.
Bulbous buttercup

Ranunculáceae – Buttercup family

Leaves: Basal leaves either hairy or smooth, on stalks and ternate; leaflets deeply toothed and often deeply incised; middle leaflet often itself on a stalk; lower stem leaves similar to basal ones but often sessile; upper leaves have narrow, deeply incised leaflets

Inflorescence: Solitary flowers on erect or recumbent, hairy, forked stems

Flowers: 2–3 cm in diameter; 5 sepals, wedge-shaped, yellowish and bent downwards; 5 petals, golden-yellow, broad oval, narrowing in a wedge shape towards the base; many stamens; numerous carpels

Fruits: Smooth, flat, oval achenes up to 4 mm long, with recessed edges and bent beaks

Habitat: From hill to subalpine zones in dry meadows and pastures, on grassy slopes and stable dunes, and in waste places, especially on calcareous substrates, on moderately dry, loose and loamy soils fairly rich in nutrients and bases and containing humus

377 **Goldilocks – R. auricómus**
Plants 15–30 cm high, with short rhizomes; flowering from April to June

Ranúnculus auricómus L.
Goldilocks

Ranunculáceae – Buttercup family

Leaves: Basal leaves variable, round or kidney-shaped in outline, toothed or divided into leaflets; stem leaves deeply lobed, generally with linear, entire, or slightly toothed segments

Inflorescence: Stem roundish, ending in solitary flowers

Flowers: 1–3 cm in diameter; 5 sepals; 5 petals, yellow and often partly atrophied; many stamens; numerous carpels on a slightly domed receptacle; fertilised by insects

Fruits: Achenes up to 4 mm long, somewhat swollen and generally with bent beaks; dispersal by ants

Habitat: In hill (rarely mountain) zone in wet woodland with good ground cover, in wet grassland and moorland, on wet to well-drained clay or loam soil rich in bases and nutrients, and containing lime and humus; plants grow in light or half-shade; found on waste ground

378 **Corn crowfoot – R. arvénsis**
Plants 15–60 cm high, without rhizomes, growing erect; flowering from May to July

Ranúnculus arvénsis L.
Corn crowfoot

Ranunculáceae – Buttercup family

Leaves: Lowest basal leaves spathulate, deeply toothed and on stalks; succeeding leaves are ternate, stalked, and with narrow, deeply toothed or incised leaflets; top stem leaves sessile and ternate with deeply incised leaflets

Inflorescence: Numerous flowers on much-branching stems

Flowers: 0.5–1.5 cm in diameter; 5 sepals, narrow oval, greenish-yellow and lying close to the petals; 5 petals, golden-yellow and narrowing in a wedge shape towards the base; many stamens; numerous carpels on a receptacle covered in scattered hairs

Fruits: 4–8 achenes up to 7 mm long, oval, flat and with distinct borders; on the surface inside this border are several hook-shaped beaks

Habitat: In hill and mountain zones on wasteland and arable land on moderately dry clay and loam rich in bases, nutrients, and humus; an indicator of loamy soil; visited by flies; range extends north to Scotland, south to North Africa

379 **Silverweed – P. anserína**
Plants up to 1 m in height, recumbent perennials, rooting at the nodes; flowering from May to Sept.

Potentílla anserína L.
Silverweed

Rosáceae – Rose family

Leaves: Basal leaves up to 20 cm long, with stalks 2–5 cm long, alternate and pinnate; leaflets narrow oval, 2–4 cm long, with numerous narrow teeth, dark green above, with silvery silky hairs beneath; leaflets decrease in size towards the base of the leaf

Inflorescence: Solitary flower at end of upright stem 4–20 cm long

Flowers: 2–3 cm in diameter; outer 5 sepals usually with 3 teeth; inner 5 sepals broad lanceolate; 5 petals, oval, generally rounded at the ends (rarely flat) and twice as long as the sepals; many stamens with smooth filaments; many carpels

Fruits: Achenes, occasionally with hairs when unripe, smooth when ripe; dispersal by animals and ants

Habitat: In hill and mountain zones in ditches and waste places, roadsides, damp pastures and dunes, on wet to dry heavy loam or clay soil rich in bases and nutrients; colonises newly turned soil by means of runners; can withstand salt; range almost worldwide

380 **Common agrimony – A. eupatória**
Plants 50–140 cm high, with erect, hairy stems; flowering from June to September

Agrimónia eupatória L.
Common agrimony

Rosáceae – Rose family

Leaves: All stem leaves, lower ones up to 20 cm long, with 5–8 pairs of simple leaflets; these are oval, up to 6 cm long, with broad teeth, dark green and hairy above and completely covered wholly beneath in short grey hairs; topmost leaves have fewer leaflets

Inflorescence: Several racemes 10–40 cm long, individual flowers having short stems

Flowers: 5 sepals, edged with hook-shaped bristles on the outside, greenish and bending together after flowering; calyx shows distinct grooves; 5 petals, oval, longer than the sepals and yellow; many (10–20) stamens; superior ovary of 2 ovules

Fruits: Enclosed by a hard, grooved calyx tube

Habitat: In hill and mountain (rarely subalpine) zones in hedge-banks, roadsides, dry and rough pasture, sparse woodlands, on moderately dry soils which are more or less rich in bases and nutrients; plant likes warmth

381 **Sickle hare's-ear – B. falcátum**
Plants 30–80 cm high, with branching stems, found in scattered locations, flowering from July to Sept.

Bupleúrum falcátum L.
Sickle hare's-ear

Apiáceae (Umbellíferae) – Carrot family

Leaves: Basal leaves on long stalks, narrow oval or spathulate, entire, with 5–7 veins and widest towards the end; upper stem leaves linear or linear-lanceolate, often slightly bent and half-amplexicaul

Inflorescence: Umbels on terminal and lateral shoots

Flowers: 3–12 flowers in each umbel; 5 sepals, greenish and narrow oval; 5 petals, not fused, yellow; 5 stamens lying between the petals; inferior ovary of 2 ovules

Fruits: Schizocarps dividing into 2 carpels with 1 seed each; these have 5 clearly developed ridges

Habitat: In hill and mountain (rarely subalpine) zones on dry slopes, banks, dry meadows, and sparse woodland (e.g. oak and pine woods) on infertile, generally dry soil rich in lime and containing humus; roots up to 120 cm deep; range extends north to northern Germany, south to Spain

382 **Buttercup hare's-ear – B. ranunculoídes**
Plants up to 50 cm in height, almost without dead leaf sheaths; flowering in July and August

Bupleúrum ranunculoídes L.
Buttercup hare's-ear

Apiáceae (Umbellíferae) – Carrot family

Leaves: Basal leaves lanceolate or linear, 5–10 cm long, narrowing towards the base, entire, with 5–20 lateral veins, more or less stalked and partly amplexicaul; stem leaves linear, generally without stalks and smaller than basal leaves; 2–5 first-order bracts (at the base of the umbel), similar to upper stem leaves; second-order bracts generally 5 in number, broad oval or roundish, with short points, 5–7 longitudinal veins, and pale yellow-green

Inflorescence: Primary and secondary umbels

Flowers: Up to 20 flowers in each second-order umbel, with 5 small sepals and 5 small petals; these are yellow-green to yellow; 5 stamens; inferior ovary

Fruits: Schizocarps dividing into 2 carpels 4 mm long, dark brown and with 5 narrow ridges

Habitat: In mountain and subalpine zones on rocks and subalpine scree on sunny, well-drained to dry, more or less stony soils containing lime

383 **Wild celery – A. graveólens**
Plants 30–90 cm high, the cultivated form with knotted rhizomes; flowering from June to October

Apium graveólens L.
Wild celery

Apiáceae (Umbellíferae) – Carrot family

Leaves: Simply pinnate (often doubly pinnate in cultivated varieties), dark green amd glossy; leaflets wedge-shaped to rhomboidal, distinctly toothed and often deeply incised

Inflorescence: Primary and secondary umbels without bracts

Flowers: 5 sepals, small; 5 petals, yellowish, whitish or greenish, small and round; 5 stamens; superior ovary of 2 ovules; fertilised by insects or self-fertile

Fruits: Schizocarps; 2 carpels with 5 inconspicuous, angular, yellow ridges

Habitat: In hill and mountain zones; cultivated forms very rarely found on rubbish dumps; true wild species grows by the coast, in salt marshes, and damp places by rivers and ditches, on moist to wet muddy soil rich in nutrients and containing lime or salt; the cultivated variety came from southern to central Europe in the 8th century; plants like warmth

384 **Dill – A. graveólens**
Plants 40–120 cm high, smooth-leaved, bluish; flowering from June to October

Anéthum graveólens L.
Dill

Apiáceae (Umbellíferae) – Carrot family

Leaves: Trebly or quadruply pinnate; thread-like leaflets 2–7 cm long, and generally less than 1 mm wide; leaf sheaths 1–2 cm long, eared on both sides at the points

Inflorescence: Umbels, very large and with many rays; secondary umbels also have many rays

Flowers: On long stems; 5 petals, small and yellow; 5 stamens between the petals; ovary of 2 ovules; pistil shorter than the base and bent back; self-fertile

Fruits: Schizocarps: 2 carpels up to 4 mm long, lenticular, with 5 broad ridges

Habitat: Cultivated in the hill zone and found wild here and there in stony places, rubbish dumps, and vineyards; likes warmth; apparently used as a vegetable and a medicinal herb from Roman times; originally from the eastern Mediterranean and south-west Asia, this plant has now spread worldwide; it does not have a bulb

385 **Yellow gentian – G. lútea**
Plants 50–130 cm high, with thick rhizomes; flowering from June to August

Gentiána lútea L.
Yellow gentian

Gentianáceae – Gentian family

Leaves: Opposite and blue-green, 6–20 cm long, oval to broad oval, with 5–7 parallel veins

Inflorescence: Whorls of 3–10 flowers in the axils of upper leaves on the stem

Flowers: Stem half to whole length of flowers; calyx translucent, with 2–6 teeth and split on one side like a sheath; 5 or 6 petals, narrow oval, pointed, 2–3.5 cm long and yellow; stamens free-standing, 5, or as many as the sepals; superior ovary with 2 cells containing many seeds; fertilised by flies and bumble-bees

Fruits: Capsules formed from the cells of the ovary

Habitat: In mountain and subalpine (rarely hill) zones in meadow land, shrubby undergrowth, wet meadows, sparse woodland amd moorland, on intermediate damp to fairly damp clay rich in bases and nutrients and usually containing lime; roots deeply; plants are in retreat in some places (used to flavour spirits!)

386 **Mountain avens – G. montánum**
Plants 10–40 cm high, prerennials without runners, with thick rhizomes; flowering from May to August

Géum montánum L.
Mountain avens

Rosaceae – Rose family

Leaves: Pinnate and up to 8 cm long; lateral leaflets small, with 3–5 big teeth; end one very large, somewhat lobed, with wide, round teeth, often heart-shaped at base; basal leaves form a rosette

Inflorescence: Solitary flowers

Flowers: Parts in fives or sixes, twice as many sepals as petals, hairy and green; outer and inner petals of different sizes; 5–6 petals, roundish, up to 2 cm long and yellow; many stamens; numerous carpels; fertilised by insects

Fruits: Single-seeded achenes with conspicuously elongated beaks with feathery hairs, remaining after flowering; dispersal by wind

Habitat: In subalpine and alpine (rarely mountain) zones in meadows, rough pasture and scrub, on moderately dry to well-watered or moist stony soils which are lime-free and more or less base-poor and so somewhat acid

387 **Winter aconite – E. hiemális**
Plants 5–20 cm high, smooth, with knotted rhizomes; flowering from January to March

Eránthis hiemális (L.) Salisb.
Winter aconite

Ranunculáceae – Buttercup family

Leaves: Basal leaves roundish, with 3–7 radially arranged leaflets, on long stems, usually appearing after the flowers

Inflorescence: Solitary flowers; a whorl of 3 palmate stem leaves is found below the flower

Flowers: Generally 6 (also 5–8) petals, oval, up to 2 cm long, widest towards the end, golden-yellow; nectaries on stems, cup-shaped and half length of stamens; many stamens; 4 –8 carpels, each with several seeds and beaked; mainly pollinated by insects

Fruits: Follicles containing many seeds

Habitat: Naturalised in hill zone in vineyards, orchards, and parks; frequently cultivated as a garden plant; found on well-drained or fairly dry loose loam rich in nutrients and generally neutral, containing humus and deep top-soil; plants grow in light or half-shade; this southern European plant is rarely found in the wild; it was introduced later to west and central Europe

388 **Yellow alpine pasque flower – Pulsatilla sulphúrea** Plants 15–25 cm high, with erect stems covered in upright hairs; flowering May to October

Pulsatílla sulphúrea (L.) DT. et S. / Pulsatílla apiifólia (Scop.) Schult.
Yellow alpine pasque flower

Ranunculáceae – Buttercup family

Leaves: Generally only one basal leaf, long-stalked when fruit is ripe, triangular in outline, ternate and up to 25 cm wide; leaflets also ternate, their segments deeply toothed or incised and hairy beneath; 3 stem leaves (or bracts) not fused at the base and of the same form as the basal leaf but rather smaller

Inflorescence: Solitary flowers on hairy stems

Flowers: 2–6 cm in diameter; usually 6 radial petals, smooth inside, usually hairy outside and sulphur-yellow on both sides; many stamens; no nectaries; numerous carpels

Fruits: Achenes; when these are ripe the hairy beaks are 3–5 cm long

Habitat: Generally in subalpine zone, in scrub and meadows on moderately well-watered, lime-poor, infertile, rather acid soils with rotted humus; in the Alps found mainly near primitive rocks

389 **Tomato – S. lycopérsicum**
Plants 30–160 cm high, annuals, branching freely; flowering from July to October

Solánum lycopérsicum L.
Tomato

Solanáceae – Nightshade family

Leaves: Intermittently and irregularly pinnate; leaflets stalked, oval to narrow oval, covered in glandular hairs, toothed, and often themselves pinnate

Inflorescence: Numerous flowers in several panicle-like heads

Flowers: Stems and stamens covered in glandular hairs; 5 or 6 sepals, narrow wedge-shape, spread flat, fused together at the base, up to 1 cm long, hairy outside and coloured yellow; 5 stamens, with filaments and anthers fused in a tube; superior ovary in 2 compartments

Fruits: Juicy berries, 2–10 cm in diameter, oval, egg-shaped, or spherical, orange, yellow, or bright red

Habitat: In hill (rarely mountain) zone, mainly as many cultivated varieties in gardens; needs moderately moist soils rich in nutrient sand containing lime; occasionally found wild on rubbish dumps; the tomatoes shown here are growing with haricot beans on a balcony

390 **Wild mignonette – R. lútea**
Plants 20–55 cm high,biennials or perennials with tap-roots; flowering from June to September

Réseda lútea L.
Wild mignonette

Resedáceae – Mignonette family

Leaves: Alternate, with narrowly winged stalk, divided into 3 or more leaflets; these in turn have 1, 2,or 3 narrow lanceolate to narrow oval segments; edges slightly toothed

Inflorescence: Racemes of many flowers; bracts 2–3 mm long

Flowers: Stem 3–5 mm long; 6 sepals, up to 3 mm long and greenish; 6 petals, yellowish; the 2 top ones up to 5 mm long, in 3 parts, with a short middle segment; 4 lower petals not divided or have only small lateral points; many stamens; ovary superior, with 3 carpels and 3 styles; fertilised by insects or self-fertile

Fruits: Capsules containing many seeds

Habitat: In hill (rarely mountain) zone on waste places, disturbed ground, and arable land, in vineyards and docks, on warm, dry, generally sandy, stony or loamy soil rich in bases and nutrients but containing little humus; an early coloniser of newly turned soil; roots up to 75 cm deep; originally a Mediterranean plant

391 **Lesser celandine – R. ficária** Plants 5–25cm high, perennials without rhizomes but with club-like roots; often roots at the nodes; flowering March/April

Ranúnculus ficária L.
Ficária vérna Hudson
Lesser celandine

Ranunculáceae – Buttercup family

Leaves: Basal leaves heart- or kidney-shaped, long-stalked, smooth, glossy, with flat, broad notches or entire, and fleshy; stem leaves similarly shaped, often with axial buds; leaves sometimes have black spots

Inflorescence: Solitary flowers on long stems

Flowers: 2–3 cm in diameter; 3–5 sepals, rarely up to 7, oval and with sack-shaped spurs;

8–12 petals, narrow oval, blunt and mid to deep yellow; many stamens; numerous carpels

Fruits: Achenes, usually spherical, on stems and somewhat hairy

Habitat: In hill and mountain zones in deciduous woods, meadows, hedgerows, grassy banks and riversides, on moist loam and clay; range extends over whole of Europe except for the arctic zone

392 **Pheasant's eye – A. vernális**
Plants 10–30 cm in height, perennials with branching stems; flowering in April and May

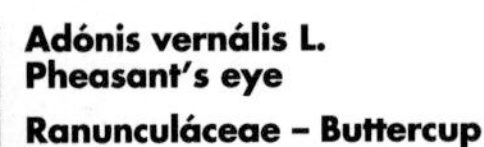

Adónis vernális L.
Pheasant's eye

Ranunculáceae – Buttercup family

Leaves: Usually sessile, closely set, doubly or trebly pinnate; leaflets long and very narrow linear

Inflorescence: Soliary flowers at end of stems

Flowers: 3–7 cm in diameter; 5 sepals, green with recumbent hairs; 10–20 petals, narrow oval, entire or slightly toothed, golden-yellow, often with reddish-brown spots towards the tip on the underside; many stamens; numerous carpels on cylindrical receptacle

Fruits: Egg-shaped achenes, up to 5 mm long, with scattered white hairs; dispersal by ants

Habitat: In hill and mountain zones on rough pasture, in dry meadows, sparse pine woods, and on warm, dry, scrubby hillsides, on dry, base-rich, lime-rich, neutral, loose loam and loess containing humus; deep-rooted; poisonous, but used medicinally; often used as an ornamental plant in gardens

393 **Yellow wood violet – V. biflóra**
Plant 5–15 cm high, perennials with thin stems; flowering from May till August

Víola biflóra L.
Yellow wood violet L.

Violáceae – Violet family

Leaves: 3–4 cm in diameter, kidney-shaped, with shallow teeth, slightly hairy, alternate anbd dark green; only 2–4 leaves per stem; bracts are small, lanceolate, smooth or with small hairs at the edge, and entire

Inflorescence: Flowers are solitary, on long stems springing from leaf axils

Flowers: 5 sepals, narrow wedge-shaped, 4–6 mm long and light green; 5 petals, not fused and yellow; lateral petals have dark streaks, are without hairs and point upwards; the bottom petal has a straight spur up to 15 mm long; 5 stamens; superior ovary of 3 placentae with a double style

Fruits: Capsule with bursting flaps

Habitat: From mountain to alpine zones in alder thickets, scrubby undergrowth ,and rock debris and on banks of streams, on wet to moist soils rich in bases and nutrients, neutral, and generally containing lime; plants prefer shade and humid atmosphere

394 **Wild pansy – V. trícolor**
Plants 5–30 cm high, branching in the lower parts; flowering from March to September

Víola trícolor L.
Wild pansy, heartsease

Violáceae – Violet family

Leaves: Oval to lanceolate, with 2–5 teeth on each side, rarely entire, stalked and 1–3 cm long; stipules deeply incised, often as long as leaves proper and with end sections with blunt teeth

Inflorescence: Solitary flowers on long stems arising from leaf axils

Flowers: 5 petals, not fused, lanceolate, pointed, with appendages up to 16 mm long; 5 petals, free-standing, roundish, violet, yellow, or white, or any combination of these; lowest petal has spur 10–25 mm long; 5 stamens; superior ovary with 3 placentae and a visible style protruding beyond the elongated bottom lip

Fruits: Capsule with 3 bursting flaps

Habitat: In mountain and subalpine (rarely hill) zones on grassland, manured fields, dunes, and grassy slopes, and by the roadside on well-drained, sandy loam rich in nutrients and containing lime and humus; fairly common

395 **Subalpine violet – V. trícolor ssp. subalpína** Plants 10–30 cm high, sweet-scented perennials; flowering from May to September

Víola trícolor ssp. subalpína Gaudin
Subalpine violet

Violáceae – Violet family

Leaves: Lower leaves roundish, with heart-shaped bases and notched edges; upper leaves lanceolate, narrowing in a wedge shape at the bases; stipules deeply incised, often as long as leaves proper and having end sections with blunt teeth

Inflorescence: Solitary flowers on long stems arising from leaf axils

Flowers: 5 petals, not fused, lanceolate, pointed, with appendages up to 20 mm long; 5 petals, free-standing, oval to roundish, much longer than the calyx and yellow to blue; lowest 3 petals marked with dark streaks; lowest petal has a marked spot; 5 stamens; superior ovary with 3 placentae and a yellowish style clearly protruding beyond the elongated bottom lip

Fruits: Capsule with 3 bursting flaps

Habitat: In mountain and subalpine zones on grassland and fields, on well-drained loam and peat which is generally base-rich, lime-poor, rich in nutrients and contains humus; native to almost all Europe

396 **Oxlip – P. elátior**
Plants 10–30 cm high, with hairy flower stems; blooming from March to August

Prímula elatior (L.) Hill
Oxlip

Primuláceae – Primrose family

Leaves: 10–20 cm long, up to 6 cm wide, long oval, narrowing to winged stalks which are broader above and very narrow beneath, with edges rolled downwards when young, with irregular fine teeth, hairy (especially on the veins) and light to dark green in colour

Inflorescence: Umbel with many flowers, pointing to one side

Flowers: 5 fused sepals, 6–15 mm long, with sharp angles and pale green; calyx teeth up to 6 mm long and narrow triangular; 5 petals, fused, pale yellow, without a golden-orange spot, slightly scented and with 5 lobes spread out in a cone and shallowly notched at the ends; 5 stamens fused to the corolla; superior ovary, single-celled, with a pistil having a stigma shaped like a pinhead

Fruits: Capsules 10–15 mm long

Habitat: From hill to subalpine zones in shady grassland, thickets, and woodland on wet, neutral soils rich inbases and nutrients

397 **Cowslip – P. véris**
Plants 10–30 cm high, with velvety hairs; flowering from April to August

Prímula véris L.
Cowslip

Primuláceae – Primrose family

Leaves: 5–15 cm long, up to 4 cm wide, long oval, often narrowing abruptly at the bases into the winged stems, and at the base of the leaf proper often flat or heart-shaped, somewhat wrinkled, when young rolled down at the edge, with irregular fine (sometimes coarser) teeth, but without sharp points, lightly covered with hairs and light to dark green

Inflorescence: Umbel with many flowers, pointing to one side

Flowers: Sweetly scented; 5 fused sepals, up to 15 mm long, with sharp angles, hairy, somewhat swollen and pale green; calyx teeth 3–5 mm long and lanceolate; 5 petals, fused, dark yellow, and in the case of pin-eyed plants often with a golden-orange spot; 5 lobes spread out in a cone and slightly notched at the ends; 5 stamens; superior ovary, single-celled

Fruits: Capsules 10–15 mm long

Habitat: From hill to subalpine zones in rough grassland, at the edge of arable fields and woodlands, in sparse oak woods, hedgerows, and stubble, on moderately dry to well-watered soils which are neutral or contain lime

398 **Auricula – P. aurícula**
Plants 10–25 cm high, with few glandular hairs on the stems; flowering in May

Prímula aurícula L.
Auricula

Primuláceae – Primrose family

Leaves: 6–12 cm long, up to 6 cm wide, oval, gradually narrowing to short, winged stems, fleshy, entire, or (especially in the upper part) with shallow teeth, with more or less mealy specks, and with glandular hairs, particularly at the edges

Inflorescence: Umbels on stems; bracts are oval

Flowers: 5 fused sepals, 6–15 mm long, without angles and with a mealy appearance; calyx teeth up to 2 mm long and generally pointed; 5 petals, fused, shiny yellow, with a mealy ring at the entrance to the throat, and scented; lobes spread out in a cone, up to 10 mm long and flat at the ends; 5 stamens; superior ovary; fertilised by insects

Fruits: Capsules 6–15 mm long

Habitat: In subalpine and alpine zones on rocks and stony expanses, and in moorlands of Alpine foothills, on wet stony soils usually containing lime but neutral and lacking humus; plant has hair roots

399 **Primrose – P. vulgáris**
Plants 5–10 cm high, apparently with basal flowers flowering in March and April

Prímula vulgáris Hudson
Primrose

Primuláceae – Primrose family

Leaves: 10–15 cm long, up to 6 cm wide, narrow oval, narrowing to winged stalks, somewhat wrinkled, with edges slightly rolled down, with irregular deep, finely pointed teeth; pale to dark green

Inflorescence: Stem is reduced, so that flowers look solitary

Flowers: Stems 4–10 cm long and more or less thickly covered in hairs; 5 fused sepals, hairy, up to 15 mm long and with sharp angles; points of calyx up to 10 mm long and narrow triangular; 5 fused petals, oval, notched at the ends, pale yellow with a dark yellow or orange spot at the throat; lobes spread out flat; 5 stamens; superior ovary

Fruits: Capsules 8–15 mm long

Habitat: In hill and mountain zones in hedgerows, rough pasture, sparse woodland, orchards, and thickets, on wet, stony loam rich in bases and nutrients, usually lime-free and containing humus; plants prefer mild winters

400 **Vitaliana – A. vitaliána**
Plants 4–10 cm high, forming numerous rosettes of leaves; blooms in June and July

Andrósace vitaliána (L.) Lapeyr.
Douglásia vitaliána (L.) Bentham et Hooker
Vitaliana

Primuláceae – Primrose family

Leaves: 4–12 mm long, up to 2 mm wide, narrow lanceolate, thickest towards the base, with hairs on margins, light green, coated, white and hairy

Inflorescence: Flowers with hairy stems 2–5 mm long arising from topmost leaf axils

Flowers: 5 fused sepals, 3–5 mm long, covered with hair and with teeth 2–4 mm long; 5 fused petals, pale to dark yellow, with lobes 4–8 mm long, broad oval and rounded at the tips; 5 stamens fused in a corolla tube up to 15 mm long; superior ovary; fertilised by insects

Fruits: Capsules up to 5 mm long, opening to the middle with 5 teeth

Habitat: In alpine zone, up to over 3000 m; at the foot of rocks, in stable rock debris and scree, on fairly moist ground, generally containing lime and rich in topsoil; many dead leaves are found beneath the leaf rosette

401 **Greater honeywort – C. major**
Plants 25–60 cm high, smooth-leaved, with thick rhizomes; flowering from June to August

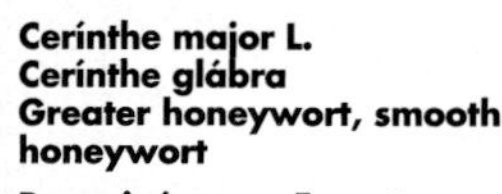

Cerínthe major L.
Cerínthe glábra
Greater honeywort, smooth honeywort

Boragináceae – Forget-me-not family

Leaves: Those of the sterile leaf rosette are oval, up to 30 cm long, 2 to 4 cm wide, contracting into the winged stems; stem leaves oval, up to 15 cm long, with round tips and blunt basal points

Inflorescence: Racemes

Flowers: 5 sepals, fused at the base only, narrow oval, 8–12 mm wide and generally blunt; 5 fused petals, 8–12 mm long, yellowish and with the small points of the corolla bent outwards; 5 stamens, fused to the corolla; anthers have thread-like appendages at the base; superior ovary of 2 cells; these are divided by false septa so that the ovary appears 4-celled

Fruits: 4 nutlets

Habitat: In subalpine zone in scrubby undergrowth, in alder thickets, on rocky slopes, and near cattle sheds, on well-watered soils rich in nutrients and containing lime

402 **Gromwell – L. officinále**
Plants 20–90 cm high, with several stems to each rhizome; flowering from May to July

Lithospérmum officinále L.
Gromwell

Boragináceae – Forget-me-not family

Leaves: Alternate, 5–10 cm long, narrow oval to lanceolate, entire, narrowing gradually at base and tip, with prominent lateral veins

Inflorescence: Llower flowers solitary in leaf axils; terminal cymes appear among bracts

Flowers: 5 sepals, fused at base only, linear, thickly covered in hairs and blunt at the tips; 5 fused petals, yellowish, yellowish-white, or greenish and with 5 hairy folds in the 5 mm long corolla tube; 5 stamens, not protruding from the corolla tube; superior ovary; fertilised by insects

Fruits: 4 smooth nutlets

Habitat: In hill (rarely mountain) zone in sparse wet and mixed deciduous woodland, in thickets on the banks of streams, on rocky slopes and sunny hillsides, on well-drained to dry, preferably sandy soils containing humus and rich in bases and nutrients

403 **Comfrey – S. officinále**
Plants 40–140 cm high, with long rhizomes, covered in upright hairs; flowering May to August

Symphytum officinále L.
Comfrey

Boragináceae – Forget-me-not family

Leaves: 10–25 cm long, oval to narrow oval, drawing gradually to points at the tips, with the leaf proper narrowing to the winged stalk at the base, and rough-haired; wings up to 4 mm wide and sometimes run down to the point of origin of the next leaf

Inflorescence: Thick panicles containing many flowers

Flowers: 5 sepals, fused at base only, narrowly triangular, covered in hairs; 5 fused petals, up to 2 cm long, yellow, reddish, or mauve, widening in the middle and narrowing again towards the 5 small points at the tip; 5 stamens; 5 scales in the throat, without hairs and not protruding from the corolla tube; superior ovary; fertilised by insects or self-fertile

Fruits: More or less warty nutlets

Habitat: In hill and mountain zones in moist to wet meadows, moorland, ditches, and woodland on intermittently wet gravelly, sandy, or pure loam or clay rich in bases and nutrients

404 **Large yellow foxglove – D. grandiflóra**
Plants 30–80 cm high, perennials thickly covered in hair in upper parts; flowering from June to August

Digitális grandiflóra Miller
Large yellow foxglove

Scrophulariáceae – Figwort family

Leaves: Narrow to broad lanceolate, running to points at the tips, with irregular fine teeth, distinctly hairy at the edges and on the veins beneath, with stalks only in the lower part of the stem

Inflorescence: Long raceme with flowers on short stems generally pointing to one side only

Flowers: 5 sepals, lanceolate, pointed, and covered in glandular hairs; 5 petals, fused into a bell, pale yellow, 2.5–4 cm long, up to 2 cm wide at the mouth, covered with glandular hairs on the outer side and with light brown marking on the inside; 4 stamens, not protruding from the corolla tube (the 2 lower ones are longer); superior ovary; fertilised by insects or self-fertile

Fruits: Capsules

Habitat: In mountain and subalpine (rarely hill) zones in woodland clearings, sparse woodland with good ground cover, scrubby undergrowth, thickets, and sunny scree, on well-drained soils which are warm in summer, rich in nutrient sand bases and containing humus

405 **Yellow bellflower – C. thyrsoídes**
Plants 10–50 cm high, biennials with closely spaced leaves; flowering in June and July

Campánula thyrsoídes L.
Yellow bellflower

Campanuláceae – Bellflower family

Leaves: 5–15 cm long, narrow oval to linear, rough-haired, blunt at the ends, usually entire and with hardly any stalk; stem leaves semi-amplexicaul

Inflorescence: Groups of 1–3 in lower leaf axils, forming a thick terminal spike

Flowers: Without stems; sepals fused towards base; free portion of sepals lanceolate, blunt and hairy; 5 petals, fused into a bell, up to 2.5 cm long, yellowish and thickly covered in hairs; 5 stamens, widening at base and enclosing pistil; inferior ovary; fertilised by insects

Fruits: Hairy capsules, erect, and opening by 3 basal pores

Habitat: In subalpine and alpine zones in mountain meadows and sunny slopes on well-drained, frequently stony clay and loam soils which are warm in summer, base-rich, more or less rich in nutrients, and usually contain lime; deep-rooted; prefers light; a mountain plant of central and southern Europe

406 **Spotted gentian – G. punctáta**
Plants 20–60 cm high, perennials with thick rhizomes; flowering in July and August

Gentiána punctáta L.
Spotted gentian

Gentianáceae – Gentian family

Leaves: 10–20 cm long, up to 8 cm wide, broad lanceolate to oval, pointed, generally with 5 parallel main veins, entire, glossy green, lacking stalks or with only short ones

Inflorescence: Terminal bunches of several flowers or axillary groups of 1–3 above the upper leaves

Flowers: Without stems; calyx divided irregularly into 5–8 parts, with upright, lanceolate lobes; petals fused into a bell, pale yellow with black spots, with 5–8 blunt oval lobes; stamens equal in number to calyx lobes; stamens mostly stuck to each other; superior ovary of 2 cells; fertilised by insects

Fruits: Capsules springing open at the seams; dispersal by wind

Habitat: In subalpine and alpine zones on rough pasture, grassland, and scrub, on well-drained loamy or clay soil which is lime-poor, usually acid and contains humus, where the topsoil is fairly deep; found especially in the southern and central Alps

407 **Groundsel – S. vulgáris**
Plants 5–30 cm high, with thin tap-roots; flowering all the year

Senécio vulgáris L.
Groundsel

Asteráceae (Compósitae) – Sunflower family

Leaves: Lanceolate, pinnately cut on both sides with oblong, blunt, irregularly toothed lobes, eared in the upper part of the stem, alternate and with scattered gossamer hairs on the underside; all tips toothed

Inflorescence: Panicles of composite flowers

Flowers: Composite flowers 5–8 mm long; outer bracts blackish, up to 3 mm long; inner bracts greenish, generally with dark tips; sepals metamorphosed to a pappus (hairs); petals fused into a tube (tubular florets), generally with 5 points; 5 stamens; anthers fused into a tube; inferior ovary of 2 cells; self-fertile

Fruits: Achenes with fused fruit and seed wall

Habitat: In hill and mountain (rarely subalpine) zones in forest clearings, gardens, fields, by the roadside and on waste land, on well-drained to fairly moist loose clay rich in nutrients; roots up to 40 cm deep

408 **Wormwood – A. absínthium** Plants 30–100 cm high, perennials with rhizomes producing shoots at many points; flowering in July and August

Artemísia absínthium L.
Wormwood

Asteráceae (Compósitae) – Sunflower family

Leaves: 5–15 cm long, wide oval, trebly pinnate and covered in thin grey hairs, particularly beneath; lobes lanceolate and pointed; lower leaves long-stalked, upper ones sessile, simply pinnate, or undivided

Inflorescence: Composite flowers 4–8 mm long with stems covered in thin grey hairs, in much-branched panicles; inner bracts linear, yellow-brown, with fine grey hairs;

Flowers: All petals fused into tubular florets and yellow; 5 stamens; anthers stuck together; inferior ovary; receptacle flat and hairy, without fine scales

Fruits: Cylindrical or egg-shaped achenes with fine longitudinal grooves

Habitat: From hill to subalpine zones on waste land, by roadsides, ditches and walls, on rocky slopes and in thickets, on moderately dry, neutral, sandy or stony loam rich in bases and nutrients; formerly used as a medicinal and flavouring plant; prefers a climate warm in summer

409 **Tansy – C. vulgáre** Plants 30–120 cm high, perennials with rhizomes producing shoots at many points; flowering from June to September

Chrysánthemum vulgáre B.
Tanacétum vulgáre L.
Tansy

Asteráceae (Compósitae) – Sunflower family

Leaves: Up to 40 cm long and pinnate; upper leaves sessile, lower ones on long stalks; leaflets lanceolate, with coarse teeth pointing towards the tip, and mid-green

Inflorescence: Composite flowers forming thick umbel-like panicles

Flowers: All florets tubular and golden yellow; petals fused into tubular florets; 5 stamens; anthers stuck together; inferior ovary of 2 cells; receptacle lacks scales; fertilised by insects

Fruits: Achenes up to 2 mm long, dotted with glands and generally 5-sided

Habitat: In hill and mountain (rarely subalpine) zones on rubbish dumps, in woodlands, thickets, forest clearings, weedy waste land, by ditches and riverbanks, on well-drained, neutral, sandy loam or clay which is warm in summer; accompanies cultivation; formerly much used as medicine and pot-herb, as a vermifuge, and to keep away moths

410 **Goldilocks – A. linósyris**
Plants 20–55 cm high, perennials with short thick rhizomes; flowering from April to September

Aster linósyris (L.) Bernh.
Linósyris vulgáris DC
Goldilocks

Asteráceae (Compósitae) – Sunflower family

Leaves: Up to 5 cm long, narrow lanceolate, entire, with a single vein and sessile

Inflorescence: Composite flowers forming an umbel-like raceme

Flowers: All florets tubular and golden-yellow; bracts hairy on the edges and pointed; petals fused into tubular florets; 5 stamens; anthers stuck together; inferior ovary of 2 cells; fertilised by insects

Fruits: Achenes 2–3 mm long, covered with thick hairs; pappus yellowish and up to 7 mm long; dispersal by wind

Habitat: In hill (rarely mountain) zone on dry warm hillsides, rocks, in dry meadows, sparse pine and oak woods, and on the margins of woodland, on dry, base-rich, generally lime- and humus-containing loose loam or clay; roots up to 55 cm deep; gregarious; likes light; prefers warmer places; a southern European plant, with a range extending to the south of England, southern Belgium, and Gotland

411 **Mountain tobacco – A. montána**
Plants 15–60 cm high, perennials with short rhizomes; flowering from June to August

Arnica montána L.
Mountain tobacco

Asteráceae (Compósitae) – Sunflower family

Leaves: Oval to narrow oval, entire (rarely somewhat toothed), opposite and usually rather hairy on the edges; basal leaves form a rosette; stem leaves form 1–3 opposite pairs

Inflorescence: Solitary composite flowers with numerous ray- and disc-florets up to 8 cm in diameter including ray-florets; at end of stems, 1–5 per plant

Flowers: Ray-florets female only, with corolla 2–3 cm long and a hairy corolla tube; tubular disc-florets hermaphrodite, with 5 stamens with anthers stuck together; receptacle flat, without scales, and with hairs

Fruits: Cylindrical achenes, with 5–10 ridges and pappus 8 mm long

Habitat: In subalpine and alpine zones in mountain meadows, siliceous rough grassland, sparse woodland, and acid moorland, on well-drained to intermittently damp clay or loam rich in nutrients, lime-poor, acid, and containing a moderate amount of humus; used as a medicinal plant; also grows on peat

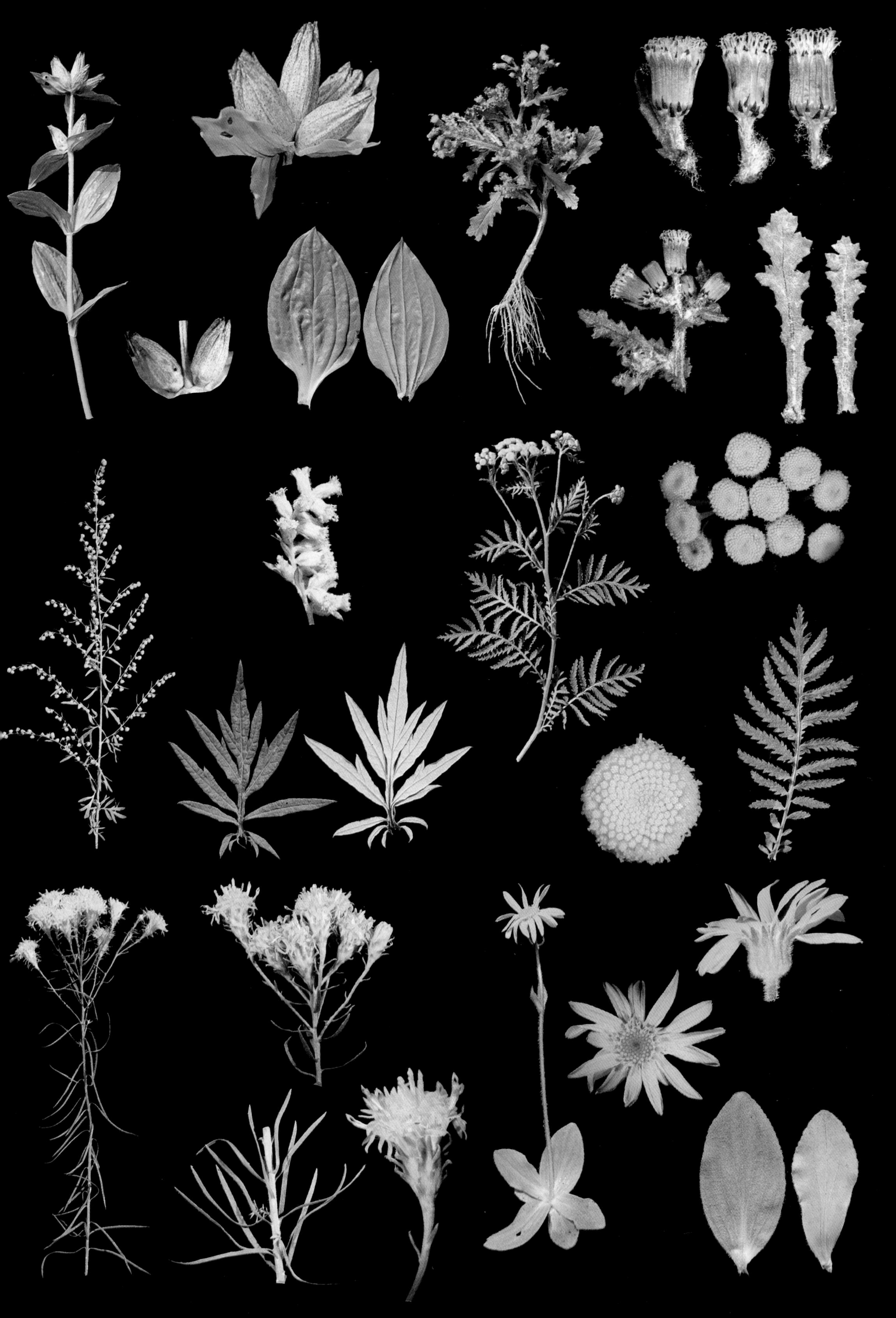

412 **Mountain tobacco – A. montána**
Plants15–60 cm high, perennials with short rhizomes; flowering from June to August

Arnica montána L.
Mountain tobacco

Asteráceae (Compósitae) – Sunflower family

Leaves: Oval to narrow oval, entire (rarely somewhat toothed), opposite, usually rather hairy on the edges; basal leaves form a rosette; stem leaves form 1–3 opposite pairs

Inflorescence: Solitary composite flowers with numerous ray- and disc-florets up to 8 cm in diameter including ray-florets; at end of stems, 1–5 on each plant

Flowers: Ray-florets are female only, with corolla 2–3 cm long and a hairy corolla tube; tubular disc-florets hermaphrodite, with 5 stamens with anthers stuck together; receptacle flat, without scales and with hairs

Fruits: Cylindrical achenes, with 5–10 ridges and pappus 8 mm long

Habitat: In subalpine and alpine zones in mountain meadows, siliceous rough grassland, in sparse woodland and on acid moorland, on well-drained to intermittently damp clay or loam rich in nutrients, lime-poor, acid, and contains a moderate amount of humus; used as a medicinal plant; illustration 412 shows a plant with several flower heads on each stem

413 **Ploughman's spikenard – I. británnica**
Plants 20–80 cm high, with short rhizomes, smelling of garlic; flowering in July and August

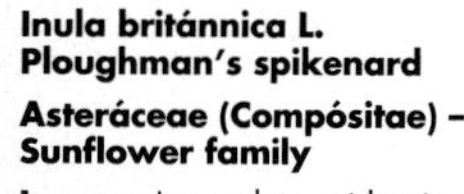

Inula británnica L.
Ploughman's spikenard

Asteráceae (Compósitae) – Sunflower family

Leaves: Lanceolate, widest in the middle, entire or with fine teeth, sessile in the upper part of the plant, narrowing to a short stalk in lower part, blunt or pointed at the end, with protruding veins; smooth or hairy

Inflorescence: Composite flowers 3–5 cm in diameter and solitary or in loose umbel-like panicles; outer bracts usually form 1 row, with hairs lying close on the outside and often with tips bent back; inner bracts smooth

Flowers: Ray-florets 15–25 mm long, female, yellow and spread out; tubular florets hermaphrodite and yellow; 5 stamens; inferior ovary

Fruits: Achenes up to 1.5 mm long, ribbed, with close hairs and with a pappus 4 – 6 mm long

Habitat: In hill and mountain zones in marshland, moist meadows, riversides, ditches, and wet woods on moist sandy or pure loam which may be dry in summer and is base-rich, rich in nutrients, and generally contains lime; range extends southwards to central Italy

414 **Yellow ox-eye – B. salicifólium**
Plants 15–60 cm high, with or without branching stems; flowering from June to September

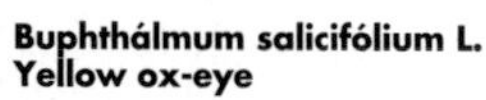

Buphthálmum salicifólium L.
Yellow ox-eye

Asteráceae (Compósitae) – Sunflower Family

Leaves: 8–15 cm long, oval to lanceolate, entire or finely toothed, blunt or with an obtuse point at tip, with sparse hair and sessile in upper part of stem; lower leaves narrow with long stalks

Inflorescence: Composite flowers 3–6 cm in diameter, solitary or in loose racemes at the end of the stem; bracts arranged like tiles, being broad lanceolate, covered with close hairs and pointed; scales lanceolate and pointed

Flowers: Ray-florets are female and yellow; inner tubular florets are hermaphrodite and yellow; 5 stamens; each half-anther is pointed underneath; inferior ovary

Fruits: Smooth achenes, 2–4 mm long, triangular with winged angles or cylindrical; pappus up to 6 mm long

Habitat: From hill to subalpine zones in warmer places on moors, dry meadows, waste land, sparse woodland, and stony slopes, on moderately dry stony or pure loam and clay with lime, which is warm in summer

415 **Fleabane – P. dysentérica**
Plants 20–60 cm high, perennials with rhizomes having short runners; flowering from June to August

Pulicária dysentérica Bernh.
Fleabane

Asteráceae (Compósitae) – Sunflower family

Leaves: 4–10 cm long, narrow oval to lanceolate, entire or crenate, pointed, rather wavy, with scattered white hairs on top and rather thicker pale grey hairs beneath; leaves in lower part sessile; those in upper part amplexicaul, with cordate bases

Inflorescence: Composite flowers 1–3 cm in diameter, forming an umbel-like panicle; bracts lanceolate and covered with erect hairs

Flowers: Ray-florets up to 1 mm wide, yellow, female, spread out and distinctly longer than the bracts; tubular disc-florets are hermaphrodite, with 5 stamens and an inferior ovary

Fruits: Achenes 1–2 mm long, hairy and with 8–10 grooves; pappus bristles 2–4 mm long; dispersal by wind

Habitat: In hill and mountain zones on marshes, wet meadows, wet woodland and moorland, on wet to intermittently moist clay or loam rich in bases and nutrients, neutral, and containing humus; colonises newly turned soil by means of runners

416 **Nodding bur-marigold – B. cérnua**
Plants 15–20 cm high, single-stemmed or branching annuals, with thin roots; flowering from July to Sept.

Bídens cérnua L.
Nodding bur-marigold

Asteráceae (Compósitae) – Sunflower family

Leaves: 5–15 cm long, lanceolate, with widely spaced teeth, pointed, sessile, opposite and slightly connate

Inflorescence: Composite flowers, drooping after flowering, 2–5 cm in diameter, usually solitary; outer bracts up to 3.5 cm long, lanceolate and with hairs on the edges; inner bracts

5–10 mm long and yellow

Flowers: Ray-florets oval, 10–15 mm long and up to 5 mm wide, sterile (no style) and yellow; disc-florets tubular, hermaphrodite and brownish-yellow; anthers rounded beneath; inferior ovary; fertilised by insects

Fruits: Flat achenes, with 4 angles and isolated hairs

Habitat: In hill (rarely mountain) zone on waste land, moors, ditches, and by ponds and riversides, especially in places with standing water in winter but not during the growing season; soil preferred is sand or clay rich in nutrients, nitrogen, and humus; accompanies cultivation; an early coloniser of muddy soil

417 **Giant golden rod – S. gigántea**
Plants 40–130 cm high, perennials with runners; flowering from August to October

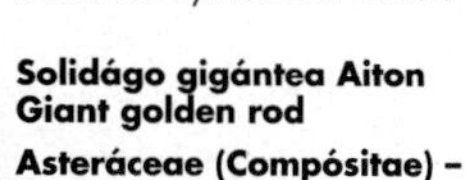

Solidágo gigántea Aiton
Giant golden rod

Asteráceae (Compósitae) – Sunflower family

Leaves: 8–15 cm long, lanceolate, coming to a long point at both ends, smooth or with short hairs along the vein beneath, entire or slightly toothed, and usually grey-green beneath

Inflorescence: Composite flowers 3–8 mm in diameter, on stems, in a branched panicle, with flowers on only one side of branches; receptacle lacks scales; bracts up to 4 mm long, smooth or with scattered hairs

Flowers: The 8–12 yellow ray-florets, which are female only, are little longer than the tubular florets (of the same yellow), which are hermaphrodite; inferior ovary

Fruits: Achenes 1–2 mm long, cylindrical, with a pappus of 1 or 2 rows of bristles

Habitat: In hill zone on waste land, in gravel pits, woodland, and thickets on the banks of streams, on wet loam or clay which is warm in summer, base-rich, rich in nutrients and with deep topsoil; an ornamental plant from the USA

418 **Alpine golden rod – S. virgaúrea ssp. minúta** Plants 10–40 cm high, perennials with pith inside the stems; flowering in July and August

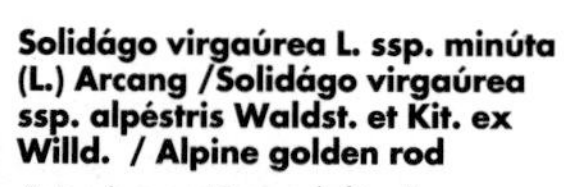

Solidágo virgaúrea L. ssp. minúta (L.) Arcang /Solidágo virgaúrea ssp. alpéstris Waldst. et Kit. ex Willd. / Alpine golden rod

Asteráceae (Compósitae) – Sunflower family

Leaves: Lanceolate, with stalks in lower and middle parts of plant, with only short stalks or sessile in upper parts, pointed at tips, usually entire and with scattered hairs or smooth at the edges

Inflorescence: Composite flowers laterally placed in leaf axils and at the end of the stem, forming a panicle which is not one-sided; composite flowers up to 20 mm long; bracts lanceolate, coming to gradual points and 6–10 mm long; receptacle without scales and smooth or with short hairs

Flowers: Ray-florets female and much longer than the hermaphrodite disc-florets, whose 5 petals are fused into a tube; 5 stamens, with anther halves rounded beneath; inferior ovary; fertilised by insects or self-fertile

Fruits: Achenes 4–6 mm long with pappus 5 mm long

Habitat: In subalpine and alpine zones in rough grassland, meadows, scrub and scrubby undergrowth, and on shingle by streams, on moderately moist to well-watered soils

419 **Fuchs's ragwort – S. fúchsii** Plants 50–130 cm high, with thin rhizomes; flowering from July to September

Senécio fúchsii Gmelin Fuchs's ragwort

Asteráceae (Compósitae) – Sunflower family

Leaves: Lanceolate, narrowing to a narrowly winged stalk, pointed, entire or finely toothed, smooth or somewhat hairy and often with a reddish main vein

Inflorescence: Numerous composite flowers forming an umbel-like panicle; outer 3–5 bracts narrow lanceolate; inner bracts (generally 8) often smooth, with dark brown points at the tips; receptacle slightly domed, without scales or hair

Flowers: Ray-florets female, up to 20 mm long, and yellow; disc-florets hermaphrodite, with 5 fused petals, 5 stamens and an inferior ovary; fertilised by insects

Fruits: Achenes up to 4 mm long, with 10–15 ribs and a pappus 8–12 mm long when ripe

Habitat: Usually in mountain zone, in sparse woodland, clearings, and thickets, in well-drained loam containing nutrients and humus and having a reasonable depth of topsoil; plants grow in light and half-shade; range extends northwards to Silesia

420 **Coltsfoot – T. fárfara** Plants 5–20 cm high, perennials with creeping rhizomes; flowering from March to May

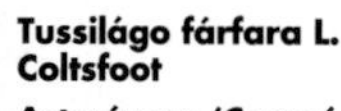

Tussilágo fárfara L. Coltsfoot

Asteráceae (Compósitae) – Sunflower family

Leaves: Basal leaves roundish to heart-shaped, usually with long reddish stalks flattened at the sides, deeply crenate at the edges, and also with fine teeth, with white or grey hairs beneath, and first appearing after flowering; stem leaves small, lanceolate, and red or brown

Inflorescence: Composite flower 2–3 cm across at the end of the stem; 1 row of bracts,

10–15 mm long, red or brown and surrounded by scale-like stem leaves; receptacle lacks hairs and scales

Flowers: Ray-florets female, numerous, up to 10 mm long; tubular disc-florets hermaphrodite, formed of 5 fused petals, 5 stamens and inferior ovary

Fruits: Achenes 3–5 mm long, cylindrical, with several rows of white pappus bristles; dispersal by wind

Habitat: From hill to subalpine zone on rock falls, by rubbish dumps, in gravel pits and excavations, on moist to well-watered, base-rich soil usually containing lime

421 **Alpine ragwort – S. alpínus** Plants 20–120 cm high, perennials with stems covered in scattered hairs; flowering in July and August

Senécio alpínus (L.) Scop. Alpine ragwort

Asteráceae (Compósitae) – Sunflower family

Leaves: Triangular or oval to cordate, on stalks, often heart-shaped at the base of the leaves proper, with irregular coarse teeth and with grey hairs beneath

Inflorescence: Long-stemmed composite flowers in an umbel-like panicle; outer bracts narrow lanceolate; inner bracts (about 20) up to 10 mm long with felted hairs at the base; receptacle rather domed and without scales or hair

Flowers: Ray-florets female, up to 15 mm long and yellow; disc-florets hermaphrodite, with 5 fused petals, 5 stamens and inferior ovary; fertilised by insects (flies and butterflies)

Fruits: Achenes 2–4 mm long, with 8 indistinct ridges, smooth, with a yellow pappus up to 5 mm long

Habitat: Generally in the subalpine zone, particularly by alpine huts, along streams, on alpine meadows, and in scrubby undergrowth, on moist to well-drained loam or clay which contains nutrients and generally lime and has deep topsoil; a good indicator of nitrogen

422 **Large-flowered leopard's bane – D. grandiflórum** Plants 20–50 cm high, perennials with hollow stems; flowering in July and August

Dorónicum grandiflórum Lam. Large-flowered leopard's bane

Asteráceae (Compósitae) – Sunflower family

Leaves: Basal leaves wide oval, on stalks, leaves proper flat or rather heart-shaped at the base, with round or blunt tips, generally crenate and rather hairy; lower stem leaves similar to basal, but semi-amplexicaul and with wider bases; top leaves sessile, with broad bases

Inflorescence: Usually 1 composite flower up to 4 cm across at the end of the stem; 20–30 bracts up to 20 mm long, lanceolate, green, and hairy

Flowers: Ray-florets female; disc-florets hermaphrodite, with 5 petals fused into a tube, 5 stamens, and inferior ovary; fertilised by insects

Fruits: Cylindrical achenes up to 4 mm long, with rough pappus hairs; dispersal by wind

Habitat: In alpine zone, widespread on loose scree on moist to well-watered, lime-rich, loose soils affected by seepage and with long periods of snow covering; plants send out creeping roots through scree; found in the Pyrenees and the Alps

423 **Tufted leopard's bane – D. clúsii** Plants 10–30 cm high, perennials with hollow, hairy stems; flowering in July and August

Dorónicum clúsii (All.) Tausch Tufted leopardsbane

Asteráceae (Compósitae) – Sunflower family

Leaves: Basal leaves narrow oval to oval, narrowing towards the stalks, crenate or entire and with a thick covering of hair, especially at the edges; stem leaves narrow oval to lanceolate and generally with narrower bases

Inflorescence: Usually 1 composite flower up to 6 cm across at the end of the stem; 20–30 bracts up to 15 mm long, lanceolate, green, and hairy

Flowers: Ray-florets female and up to 20 mm long; disc-florets hermaphrodite, with 5 petals fused into a tube, 5 stamens, and an inferior ovary; fertilised by insects

Fruits: Cylindrical achenes up to 4 mm long, with rough pappus hairs; dispersal by wind

Habitat: In alpine zone on moraines and scree on moist, lime-free, rock debris subject to long periods of snow cover; a central and southern European plant growing in the mountains of northern Spain, the Alps, and the Carpathians

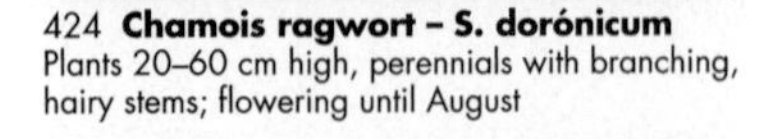

424 **Chamois ragwort – S. dorónicum**
Plants 20–60 cm high, perennials with branching, hairy stems; flowering until August

Senécio dorónicum L.
Chamois ragwort

Asteráceae (Compósitae) – Sunflower family

Leaves: Narrow oval to lanceolate, narrowing to a stalk, particularly in the lower stem, in the upper stem sessile and semi-amplexicaul, toothed, blunt, or pointed at the ends, often rather leathery and with felted grey hairs beneath

Inflorescence: Usually 1–5 composite flowers at the end of the stem; generally more than 10 outer bracts, small and narrow lanceolate; about 20 inner bracts, 10–15 mm long, with felted grey hairs and bearded at the tips; receptacle usually vaulted, and without scales or hairs

Flowers: Ray-florets female, 15–20 mm long, and yellow or orange-yellow; disc-florets hermaphrodite, with 5 petals fused into a tube, 5 stamens, and an inferior ovary; fertilised by insects

Fruits: Achenes 4–6 mm long, cylindrical, with 10–12 ridges, smooth, with white pappus 6–10 mm long in 1 row; dispersal by wind

Habitat: In subalpine and alpine zones in meadows, scrub, and rough pasture, on well-drained neutral soils containing lime

425 **Corn marigold – C. ségetum**
Annual plants with thin roots and simple or branching stems; flowering until August

Chrysánthemum ségetum L.
Corn marigold

Asteráceae (Compósitae) – Sunflower family

Leaves: Narrow oval to lanceolate, with irregular deep teeth on both sides or pinnately cut nearly to the middle, with stalks in lower part of plant, in upper part sessile or partly amplexicaul, smooth and blue-green

Inflorescence: Composite flowers 2–5 cm across at the end of the stem; bracts narrow oval, light green with paler margins but somewhat darker at the end and lying close to the flower head

Flowers: Ray-florets 10–20 mm long and golden-yellow; disc-florets with 5 petals fused into a tube, 5 stamens, and an inferior ovary; fertilised by insects (the plant attracts flies)

Fruits: Achenes 2–3 mm long, generally with 10 ridges and without toothed margins

Habitat: In hill zone in arable fields, waste places, rubbish dumps, and railway land, on well-drained, lime-poor, sandy or pure loam or clay rich in nutrients; a weed of cultivated land; also a garden plant; introduced from the Mediterranean area

426 **Marsh ragwort – S. aquáticus**
Plants 15–20 cm high, generally biennial, without runners; flowering from July to September

Senécio aquáticus Hill
Marsh ragwort

Asteráceae (Compósitae) – Sunflower family

Leaves: Basal leaves oval to lanceolate, notched or pinnately cut, on stalks, and yellow-green; stem leaves all pinnate, often sessile, with toothed lobes

Inflorescence: Composite flowers at the end of the stem forming an umbel-like panicle; side branches point obliquely outwards; outer bracts 2–4 mm long and narrow lanceolate; 12–15 inner bracts with 1–3 resin glands and scattered hairs on the backs when fruit is ripe; receptacle usually somewhat vaulted, without scales or hairs

Flowers: 12–15 ray-florets, female, up to 15 mm long; disc-florets hermaphrodite, with 5 petals fused into a tube, 5 stamens, and an inferior ovary

Fruits: Achenes 1–3 mm long, with up to 8 ridges and white pappus

Habitat: In hill zone, in marshes, wet meadows, moorland, ditches, and by springs, on wet or moist soils more or less rich in nutrients, usually lime-poor or leached of lime

427 **Ragwort – S. jacobaéa**
Plants 30–90 cm high, usually perennial, without runners; flowering from June to August

Senécio jacobaéa L.
Ragwort

Asteráceae (Compósitae) – Sunflower family

Leaves: Lower stem leaves lyre-shaped, middle ones pinnately cut almost to middle vein, smooth or with scattered hairs, with toothed lobes standing out at right angles; top stem leaves semi-amplexicaul

Inflorescence: Composite flowers at end of stem forming an umbel-like panicle; outer bracts 2–4 mm long and narrow lanceolate; 12–15 inner bracts with 1–3 resin glands and scattered hairs on the backs when fruit is ripe; receptacle usually somewhat vaulted, and without scales or hairs

Flowers: 12–15 ray-florets, female, up to 12 mm long; disc-florets hermaphrodite, with 5 petals fused into a tube, 5 stamens, and an inferior ovary

Fruits: Achenes 2–3 mm long, with a white pappus up to 5 mm long

Habitat: Found in hill and mountain zones as a weed of waste land, roadsides, thickets, and neglected or overgrazed pastures on all but the poorest soils

428 **Rock ragwort – S. rupéstris**
Plants 20–30 cm high, usually biennials with an unpleasaont smell; flowering from June to September

Senécio rupéstris W. et K.
Rock ragwort

Asteráceae (Compósitae) – Sunflower family

Leaves: Narrow oval to lanceolate, pinnately cut in the upper part of the leaf, with lobes pointing distinctly towards the tips, sessile, dark green above and grey-green beneath; lobes have teeth

Inflorescence: Composite flowers 2–3 cm across in irregular umbel-like panicles; outer bracts small, narrow lanceolate and ending in long black points; about 20 inner bracts 6–10 mm long, lanceolate, light green, and ending in black points

Flowers: 12–15 ray-florets, female, up to 15 mm long; disc-florets hermaphrodite, with 5 petals fused into a tube, 5 stamens, and an inferior ovary

Fruits: Achenes 2–3 mm long, with a pappus up to 6 mm long

Habitat: In mountain and subalpine zones in waste places, yards, on stony slopes, by the roadside and round cowsheds, on well-drained, base-rich, stony, and often lime-containing soils more or less rich in nutrients, which may or may not contain humus

429 **Yellow chamomile – A. tinctória**
Plants 20–50 cm high, annuals or perennials with rather hairy stems; blooming till August

Anthemis tinctória L.
Yellow chamomile

Asteráceae (Compósitae) – Sunflower family

Leaves: Oval, generally sessile, pinnate, covered with scattered hairs and grey-green beneath; lobes lanceolate with teeth pointing towards the tips

Inflorescence: Long-stemmed, solitary, composite flowers; receptacle hemispherical, with lanceolate scales; bracts covered in felted hairs

Flowers: 12–25 ray-florets up to 10 mm long and golden-yellow; disc-florets hermaphrodite, with 5 petals fused into a tube, 5 stamens, and an inferior ovary; fertilised by insects

Fruits: Achenes 2–3 mm long, smooth, with 5–7 longitudinal ridges on either side

Habitat: In hill and mountain zones in waste land, dry grassland, arable fields, by the roadside, ditches, and thickets, on dry, level, stony soils which are warm in summer and often poor in humus and topsoil; the plants prefer light; an early coloniser of newly turned soil; used in earlier times as a source of dye; a European and west Asian plant, with a range extending northwards to Scandinavia; naturalised in England and the USA

430 **Subalpine groundsel – S. abrotanifólius** Plants 10–40 cm high, perennials with rhizomes producing several shoots; flowering July to Sept.

Senécio abrotanifólius L.
Subalpine groundsel

Asteráceae (Compósitae) – Sunflower family

Leaves: Oval to narrow oval, doubly pinnately cut to the middle vein, with stalks in lower part of plant, in upper part sessile, smooth, or with short hairs and lanceolate lobes

Inflorescence: Solitary composite flowers arising from leaf axils, and 2–8 in an umbel-like panicle at the end of the stem; outer bracts lanceolate, 2–3 mm long and with scattered hairs; about 20 inner bracts 5–8 mm long, bearded at the base; receptacle rather vaulted, without scales or hairs

Flowers: Ray-florets female, 10–15 mm long, and yellowish or reddish-orange; disc-florets hermaphrodite, with 5 petals fused into a tube, 5 stamens, and inferior ovary; fertilised by insects

Fruits: Achenes 2–4 mm long, with 5–7 indistinct ridges, smooth and with a yellowish pappus 5–8 mm long; dispersal by wind

Habitat: In subalpine zone in meadows, mountain pine woods, and dry grassland on moderately damp, usually on soils poor in lime and containing some humus

431 **Alpine grey groundsel – S. incánus** Plants 5–15 cm high, perennials with stems covered in felted grey hair; flowering in July and August

Senécio incánus L.
Alpine grey groundsel

Asteráceae (Compósitae) – Sunflower family

Leaves: Oval to lanceolate, pinnately cut nearly to the middle vein, narrowing to wide stalks, and covered with grey or white felted hairs; lobes have blunt teeth or are deeply incised

Inflorescence: Up to 8 composite flowers at the end of the stem in an umbel-like raceme; outer bracts short and narrow lanceolate; 6–10 inner bracts lanceolate, up to 4 mm long, with reddish tips; receptacle rather vaulted, without scales or hairs

Flowers: 3–8 ray-florets female, 4–8 mm long, and dark yellow to orange; disc-florets hermaphrodite, with 5 petals fused into a tube, 5 stamens, and inferior ovary; fertilised by insects

Fruits: Achenes 2–4 mm long, with 9 ridges, often covered with sparse hairs and with a long yellowish pappus 3–5 mm long

Habitat: In alpine zone on scree and in stony grassland on well-drained to moderately dry loose soils poor in lime and containing humus

432 **Goat's beard – T. praténsis** Plants 30–60 cm high, biennials or perennials with remains of previous year's leaves at the base; flowering till July

Tragópogon praténsis L.
Goats's beard

Asteráceae (Compósitae) – Sunflower family

Leaves: Narrow triangular, going from a wide base to a long narrow point, sessile, entire, smooth, partially amplexicaul and with a reddish margin, especially in the lower part

Inflorescence: Composite flowers at end of stems; usually 8 bracts, lanceolate, fused at the base, smooth on the inner side, up to 45 mm long when fruit is ripe; receptacle smooth, without scales

Flowers: All hermaphrodite ray-florets, the 5 petals fused into a yellow ray; 5 stamens; inferior ovary of 2 cells

Fruits: Achenes with beaks up to 25 mm long; pappus of several rows and yellowish-white; pappus bristles have feathery hairs; dispersal by wind

Habitat: In hill and mountain zones in meadows, pastures, dunes, roadsides, and waste places, on damp to fairly dry loose loam or clay more or less rich in bases and nutrients, containing humus with a moderate depth of topsoil

433 **Shaggy hawkweed – H. villósum** Plants 10–30 cm high, perennials bearing 1–4 composite flowers, covered in white hairs; flowering July/ August

Hierácium villósum L.
Shaggy hawkweed

Asteráceae (Compósitae) – Sunflower family

Leaves: Basal leaves lanceolate to narrow oval, gradually narrowing towards the base, having no stalk or narrowing to a short one, entire or finely toothed and with long hairs; the 3–8 stem leaves are narrow oval, sessile or amplexicaul, pointed or blunt at the tip and thickly covered in hair

Inflorescence: Usually solitary flowers at the end of the stem; bracts lanceolate, pointed, up to 25 mm long, and with long hair, usually dark at the base

Flowers: All hermaphrodite ray-florets, the 5 petals fused into a yellow ray; 5 stamens; inferior ovary of 2 cells

Fruits: Achenes 30–45 mm long, dark brown with a yellowish-white pappus; dispersal by wind

Habitat: In subalpine and alpine zones in meadows and sunny scree, on well-drained, lime-rich, loose, stony soil with fairly deep topsoil; plants prefer light

434 **Mouse-ear hawkweed – H. pilosélla** Plants 5–30 cm high, perennials with thin runners; flowering from May to October

Hieráceum pilosélla L.
Mouse-ear hawkweed

Asteráceae (Compósitae) – Sunflower family

Leaves: Basal leaves lanceolate to narrow oval, gradually narrowing towards the base, blunt or pointed at the ends, usually entire, mid- to dark green above, with long hairs up to 7 mm long, and with grey or white felted hairs beneath; usually there are no stem leaves

Inflorescence: Solitary composite flowers at the ends of the stems; bracts narrow lanceolate, widest at the base and finely pointed at the end, and with stellate hairs, dark glandular hairs, and single hairs up to 7 mm long

Flowers: All hermaphrodite ray-florets, the 5 petals fused into a ray, usually pale yellow; 5 stamens; inferior ovary of 2 cells

Fruits: Achenes up to 2.5 mm long, black, with a yellowish-white pappus; dispersal by wind

Habitat: Usually in hill and mountain zones in grassy pastures and heaths, sparse pinewoods, banks, rocks, scree, and walls, on moderately dry, infertile, sandy, loose soils poor in lime

435 **Mouse-ear hawkweed – H. pilosélla** (blue-green variety) Plants 5–30 cm high, perennials with thin runners; flowering from May to October

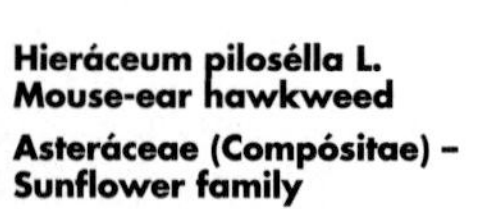

Hieráceum pilosélla L.
Mouse-ear hawkweed

Asteráceae (Compósitae) – Sunflower family

Leaves: Basal leaves lanceolate to narrow oval, gradually narrowing towards the base, blunt or pointed at the ends, usually entire, blue-green above (Fig. 435 shows a plant from a dry location), with long hairs up to 7 mm long, and with grey or white felted hairs beneath

Inflorescence: Solitary composite flowers at the ends of the stems; bracts narrow lanceolate, widest at the base and finely pointed at the ends, and with stellate hairs, dark glandular hairs, and single hairs up to 7 mm long

Flowers: All hermaphrodite ray-florets, the 5 petals fused into a ray, usually pale yellow; 5 stamens; inferior ovary of 2 cells

Fruits: Achenes up to 2.5 mm long, black, with a yellowish-white pappus; dispersal by wind

Habitat: Generally in hill and mountain zones in grassy pastures and heaths, sparse pinewoods, banks, rocks, scree, and walls, on moderately dry, infertile, sandy, loose soils poor in lime

436 **Wall hawkweed – H. murórum**
Plants 20–35 cm high, perennials with hairy stems; flowering from May to September

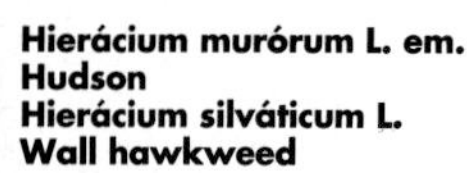

Hierácium murórum L. em. Hudson
Hierácium silváticum L.
Wall hawkweed

Asteráceae (Compósitae) – Sunflower family

Leaves: Basal leaves oval or narrow oval, on long stalks, bottom of leaves proper heart-shaped, notched in the lower part, blunt or pointed at the end, hairy, dark green above, sometimes with dark spots, and grey-green (rarely rather reddish) beneath; generally only one stem leaf

Inflorescence: Composite flowers in a panicle of up to 15 flowers; bracts lanceolate, pointed, 6–12 mm long, with numerous darker glandular hairs and distinctly longer simple hairs

Flowers: All hermaphrodite ray-florets, the 5 petals fused into a ray, generally pale yellow; 5 stamens; inferior ovary of 2 cells

Fruits: Achenes 3–4 mm long, black, with yellowish-white pappus

Habitat: From hill to subalpine zones in sparse, grassy, deciduous or coniferous forests, clearings, on shady rocks, walls, rock debris, and by the roadside, on well-drained, stony loam or clay rich in bases and nutrients

437 **Mountain hawk's beard – C. pontána**
Plants 25–50 cm high, perennials with long woody tap-roots; flowering in July and August

Crépis pontána (L.) D.T.
Crépis montána Tausch
Mountain hawkbeard

Asteráceae (Compósitae) – Sunflower family

Leaves: Broad lanceolate to narrow oval, narrowing to short stalks in the lower part of the plant and semi-amplexicaul in the upper part, with wide-spaced pointed teeth, pointed at the ends and with hairs, especially along the veins beneath

Inflorescence: Solitary; bracts lanceolate, 10–20 mm long, with numerous frizzy white hairs; receptacle covered with isolated hairs

Flowers: All hermaphrodite ray-florets, the 5 petals fused into a mid-yellow ray; 5 stamens; inferior ovary of 2 cells

Fruits: Achenes up to 12 mm long, with 17 unequal ridges, narrowing towards the end and with a yellowish-white pappus up to 9 mm long

Habitat: In subalpine zone in sunny grasslands on well-drained, base-rich, stony or pure loam or clay usually containing lime and humus

438 **Giant cat's ear – H. uniflóra**
Plants 15–50 cm high, perennials with woody rhizomes; flowering in July and August

Hypochoéris uniflóra VILL. s.l.
Giant cat's ear

Asteráceae (Compósitae) – Sunflower family

Leaves: Lanceolate to oval, sessile, blunt or round at the ends, with wide-spaced pointed teeth, without spots and with stiff, rough hairs on either side

Inflorescence: A solitary composite flower standing at the end of a distinctly thickened stem; outer bracts fringed; inner bracts up to 25 mm long, with black and white hairs and lanceolate; receptacle without hairs; scales with short hairs

Flowers: All hermaphrodite ray-florets, the 5 petals being into a pale to mid-yellow ray; 5 stamens; inferior ovary of 2 cells

Fruits: Achenes with beaks up to 20 mm long, brown, with indistinct ridges and a yellowish-white pappus; pappus bristles covered with feathery hairs

Habitat: In subalpine and alpine zones in sunny siliceous grassland and meadows on well-drained, base-rich, lime-free, acid loam or clay which contains humus and has a reasonable topsoil; an indicator of acidity and infertility

439 **Rough hawkbit – L. híspidus**
Plants 10–50 cm high, perennials with knotted rhizomes; flowering from June to August

Leoóntodon híspidus L. s.l.
Rough hawkbit

Asteráceae (Compósitae) – Sunflower family

Leaves: Basal leaves lanceolate to narrow oval, narrowing to the stalks, blunt or round at the end, entire or notched and either upright or recumbent; stem leaves scale-like in form

Inflorescence: Usually solitary; bracts 10–20 mm long, with short close hairs and frizzy hairs inside the points

Flowers: All hermaphrodite ray-florets, the 5 petals fused into a pale yellow ray; 5 stamens; inferior ovary of 2 cells

Fruits: Achenes up to 8 mm long, brownish, with 16–18 indistinct ridges, narrowing towards the end and with a yellowish-white pappus up to 9 mm long; the pappus bristles are covered in feathery hairs

Habitat: Found in all zones on meadows, pastures, grassy slopes, sparse woodland, moorland, water meadows, and on scree, on well-drained loamy or stony soil which is more or less base-rich and contains nutrients and humus

440 **Hawkweed ox-tongue – P. hieracioídes**
Plants 20–80 cm high, biennials or perennials with rhizomes producing several shoots; flowers July/Oct.

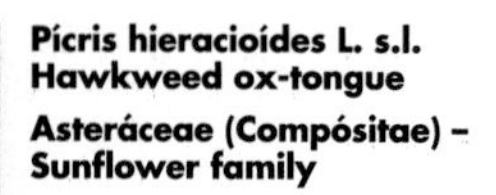

Pícris hieracioídes L. s.l.
Hawkweed ox-tongue

Asteráceae (Compósitae) – Sunflower family

Leaves: Lanceolate to narrow oval, in the lower part of the plant narrowing to winged stalks, in the upper part sessile and partially amplexicaul, pointed at the ends, entire or notched and covered in stiff hairs

Inflorescence: Composite flowers in umbel-like panicles; outer bracts numerous and arranged like tiles, erect and hairy; inner hairs up to 15 mm long, whitish-green and hairy

Flowers: All hermaphrodite ray-florets, the 5 petals fused into a mid-yellow ray often rather reddish in colour outside; 5 stamens; inferior ovary of 2 cells; fertilised by insects

Fruits: Achenes with beaks up to 5 mm long, rather bent into a sickle shape, brown or black and with longitudinal ridges

Habitat: In hill and mountain zones by ditches, on rocky slopes, in meadows and semi-dry grassland, and in gravel pits, on moderately well-watered base-rich soils containing nutrients and lime

441 **Hairy hawkweed – H. pilíferum**
Plants 5–20 cm high, perennials with distinctly hairy stems; flowering in July and August

Hiercium piliíferum HOPPE s.l.
Hierácium glandulíferum HOPPE s.l.
Hairy hawkweed

Asteráceae (Compósitae) – Sunflower family

Leaves: Basal leaves lanceolate to spathulate, gradually narrowing towards the base, entire, wavy-edged or with isolated teeth, distinctly long-haired, and blunt or rounded at the end; with 3–8 sessile stem leaves

Inflorescence: Composite flowers, generally solitary; bracts lanceolate, 10–15 mm long with single hairs up to 6 mm long and dark at the base

Flowers: All hermaphrodite ray-florets, the 5 petals fused into a pale to mid-yellow ray; 5 stamens; inferior ovary of 2 cells

Fruits: Achenes 2–3 mm long, dark brown with yellowish-white pappus; dispersal by wind

Habitat: In alpine (rarely subalpine) zone on moraines, rock debris, and siliceous grassland, on well-drained, base-rich, lime-poor, acid, stony loam or clay containing some humus; fairly common in the central and southern Alps

442 **Dwarf hawkweed – H. húmile**
Plants 10–30 cm high, perennials forking less than halfway up; flowering till August

Hierácium húmile Jacq.
Dwarf hawkweed

Asteráceae (Compósitae) – Sunflower family

Leaves: Basal leaves oval to lanceolate, narrowing abruptly to the stalks, usually pointed at tips and with irregular deep teeth; stem leaves narrow oval to lanceolate, narrowing towards bases and sessile

Inflorescence: Composite flowers, in groups of 2–8; bracts lanceolate, 10–15 mm long, with isolated glandular hairs and simple hairs up to 2 mm long; no stellate hairs

Flowers: All hermaphrodite ray-florets, the 5 petals fused into a pale to mid-yellow ray; 5 stamens; inferior ovary of 2 cells

Fruits: Achenes 3–4 mm long, dark brown or black, with a yellowish-white pappus; dispersal by wind

Habitat: In mountain and subalpine zones on stony slopes, rock fissures and rock debris, on well-drained to fairly dry, base-rich, stony soil usually containing lime; a typical plant rooting in rock fissures; it likes light and a little warmth

443 **Stem-clasping hawkweed – H. amplexicaúle**
Plants 10–40 cm high, perennials forking less than halfway up; flowering from June to August

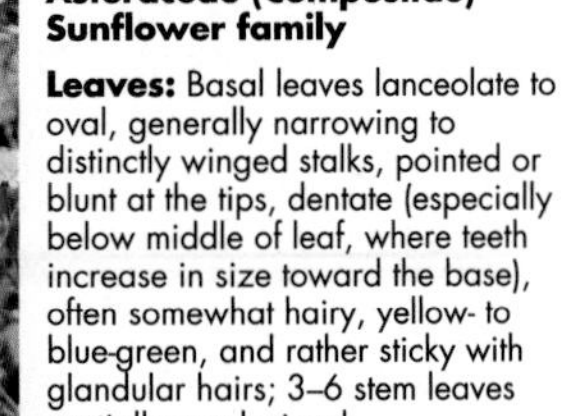

Hierácium amplexicaúle L.
Stem-clasping hawkweed

Asteráceae (Compósitae) – Sunflower family

Leaves: Basal leaves lanceolate to oval, generally narrowing to distinctly winged stalks, pointed or blunt at the tips, dentate (especially below middle of leaf, where teeth increase in size toward the base), often somewhat hairy, yellow- to blue-green, and rather sticky with glandular hairs; 3–6 stem leaves partially amplexicaul

Inflorescence: Composite flowers in groups of up to 12; bracts lanceolate, 10–15 mm long, with glandular hairs, bearded at the ends

Flowers: All hermaphrodite ray-florets, the 5 petals fused into a mid-yellow ray; 5 stamens; inferior ovary of 2 cells

Fruits: Achenes 3–4 mm long, black, with a yellowish-white pappus; dispersal by wind

Habitat: In mountain and subalpine (rarely hill) zones in crevices in rocks and walls, in scree and on rocky slopes, on moderately dry, lime-poor or lime-rich stony soils

444 **Pyrenean hawk's beard – C. pyrenáica**
Plants 20–50 cm high, perennials, **this specimen with many leaves all the way up plant;** flowers till Aug.

Crépis pyrenáica (L.) Greuter
Crépis blattarioides Vill.
Pyrenean hawk's beard

Asteráceae (Compósitae) – Sunflower family

Leaves: Basal leaves wither when flowers bloom; stem leaves narrow oval to lanceolate, sessile, amplexicaul with 2 pointed lobes, pointed at the tips and irregularly notched

Inflorescence: Usually with leaves almost as far as the flower heads; generally 2–8 composite flowers; outer bracts almost as long as inner, lanceolate and covered in hair; inner bracts lanceolate, 10–15 mm long and covered in yellowish to black hairs

Flowers: All hermaphrodite ray-florets; the 5 petals fused into a ray 25 mm long on the outer side of the flower head; corolla tube smooth; 5 stamens; inferior ovary

Fruits: Achenes 5–8 mm long, with about 20 ridges, narrowing towards the ends and with a white pappus

Habitat: In subalpine zone in scrubby undergrowth, rough grassland, and hayfields, on moist to well-drained soils usually lime-rich, base-rich, and containing nutrients and humus

445 **Pyrenean hawk's beard – C. pyrenáica**
Plants 20–50 cm high, perennials, **this specimen with leaves only part way up plant;** flowering till Aug.

Crépis pyrenáica (L.) Greuter
Crépis blattarioides Vill.
Pyrenean hawk's beard

Asteráceae (Compósitae) – Sunflower family

Leaves: Basal leaves wither when flowers bloom; stem leaves narrow oval to lanceolate, sessile, amplexicaul with 2 pointed lobes, pointed at the tips and irregularly notched

Inflorescence: Usually with leaves almost as far as the flower heads; generally 2–8 composite flowers; outer bracts almost as long as inner, lanceolate and covered in hair; inner bracts lanceolate, 10–15 mm long and covered in yellowish to black hairs

Flowers: All hermaphrodite ray-florets; the 5 petals fused into a ray 25 mm long on the outer side of the flower head; corolla tube smooth; 5 stamens; inferior ovary

Fruits: Achenes 5–8 mm long, with about 20 ridges, narrowing towards the ends and with a white pappus

Habitat: In subalpine zone in scrubby undergrowth, rough grassland, and hayfields, on moist to well-drained soils which are generally lime-rich, base-rich, and contain nutrients and humus

446 **Cat's ear – H. radicáta**
Plants 20–50 cm high, perennials with fleshy lateral roots and short rhizomes; flowering until October

Hypochoéris radicáta L.
Cat's ear

Asteráceae (Compósitae) – Sunflower family

Leaves: Basal leaves lanceolate to narrow oval, deeply dentate or pinnately cut, without stalks, narrowing towards the base, generally blunt at the tips, smooth or hairy, without spots and usually recumbent

Inflorescence: Solitary composite flowers on single stem or several flowers on forking stems; bracts lanceolate, 5–20 mm long, smooth, or rather hairy, and often rather darker at the tips; outer bracts much smaller than inner ones

Flowers: All hermaphrodite ray-florets; the 5 petals fused into a yellow ray; outer side of ray- floret often reddish or greenish; 5 stamens; inferior ovary of 2 cells

Fruits: Achenes with beaks up to 15 mm long, dark brown, with up to 15 distinct longitudinal ridges and with yellowish-white pappus

Habitat: In hill and mountain zones in rough grassland, meadows, parkland, heaths, and sparse woodland, on well-drained to moderately dry soils which are lime-poor and moderately base-rich and rich in nutrients

447 **Chicory hawkweed – H. intybáceum**
Plants 5–30 cm high, perennials with single or branching stems; flowering in July and August

Hierácium intybáceum All.
Chicory hawkweed

Asteráceae (Compósitae) – Sunflower family

Leaves: Stem leaves only, narrow lanceolate to lanceolate, narrowing towards the base, often amplexicaul, towards the ground often forming a quasi-rosette, pointed at the tips, with irregular teeth, generally light green, and sticky with numerous glandular hairs

Inflorescence: Up to 6 composite flowers; bracts narrow lanceolate, pointed and with numerous glandular hairs; hairy pits on the bottom of the receptacle

Flowers: All hermaphrodite ray-florets, the 5 petals fused into a yellowish-white ray; 5 stamens; inferior ovary of 2 cells; fertilised by insects

Fruits: Achenes 3–4 mm long, dark brown, generally with a yellowish-white pappus; dispersal by wind

Habitat: In subalpine and alpine zones on sunny rock faces, in rock debris, grassland, and scrub, on well-drained to fairly dry, loose, stony soils which are lime-poor but rich in silicates; found particularly in central and southern Alps

448 **Mountain hawkbit – L. montánus**
Plants 3–10 cm high, perennials with upright or curved stems; flowering till August

Leóntodon montánus Lam.
Mountain hawkbit

Asteráceae (Compósitae) – Sunflower family

Leaves: Basal leaves lanceolate, narrowing towards the stalks (often reddish) generally rounded at the tips, dentate or pinnately cut almost to the middle vein and smooth or rather hairy beneath; stem leaves, when present, only in the form of scales

Inflorescence: Usually solitary composite flower on stem, which thickens beneath it; bracts 10–15 mm long, with thick, black, erect hairs; receptacle without scales and usually without hairs

Flowers: All hermaphrodite ray-florets up to 20 mm long, the 5 petals fused into a yellow ray; 5 stamens; outer side of rays often with a reddish overlay; inferior ovary of 2 cells; fertilised by insects

Fruits: Achenes 4–8 mm long, with up to 20 ridges, brown, with a yellowish-white pappus

Habitat: Scattered in the alpine zone in scree on moist to well-drained, loose, often mobile, rough limestone or rock debris which is base-rich and contains lime

449 **Beaked hawk's beard – C. taraxacifólia**
Plants 20–70 cm high, annuals or perennials smelling of bitter almonds; flowering in May/June

Crépis taraxacifólia Thuill.
Beaked hawk's beard

Asteráceae (Compósitae) – Sunflower family

Leaves: Lanceolate to narrow oval, in lower part of plant narrowing to winged stalks, in upper part sessile or amplexicaul with small basal points, notched or pinnately cut and smooth or with sparse yellowish hairs

Inflorescence: Corymbs of composite flowers at ends of branches of forked stem; outer bracts lanceolate, pointed, with whitish margins; inner bracts with close hairs on the inside

Flowers: All hermaphrodite ray-florets, the 5 petals fused into a yellow ray; corolla tube hairy outside; 5 stamens; inferior ovary of 2 cells

Fruits: Achenes 5–9 mm long with beak, with 10 ridges and short hairs, and with a pappus 4–6 mm long

Habitat: In hill and mountain zones by roadsides, on walls and railway embankments, in waste places and thickets, on well-drained to moderately dry soils rich in bases and nutrients and usually containing lime and humus

450 **Dandelion – T. officinális**
Plants 5–30 cm high, perennials with leaves of many different shapes; flowering April/May to Oct.

Taráxacum officinále Weber s.l.
Dandelion

Asteráceae (Compósitae) – Sunflower family

Leaves: Lanceolate, usually pinnately cut to the middle vein, with triangular, often dentate lateral lobes and a large terminal lobe, on stalks and smooth; white hairs at the base of the leaf rosette

Inflorescence: Solitary flowers at the end of hollow, thin-walled stems; outer bracts narrow lanceolate, erect or bent back during flowering and without a pale margin; inner bracts lanceolate and longer than the outer ones

Flowers: All hermaphrodite ray-florets, the 5 petals fused into a yellow ray; corolla tube smooth; 5 stamens; inferior ovary of 2 cells

Fruits: Cylindrical brownish achenes with a whitish pappus; dispersal by wind

Habitat: From hill to subalpine zones in pastures, meadows, roadsides, and waste places on damp to moderately well-watered neutral soils rich in nutrients

451 **Smooth hawk's beard – C. capilláris**
Plants 15–80 cm high, annuals or biennials with thin tap-roots; flowering from June to September

Crépis capilláris (L.) Wallr.
Smooth hawk's beard

Asteráceae (Compósitae) – Sunflower family

Leaves: Narrow oval to lanceolate, in middle and lower parts of plant often with parallel edges, in upper part of plant sessile or amplexicaul, and in lower plant narrowing to winged stalks; leaves very variable in form

Inflorescence: Panicles of composite flowers; outer bracts narrow lanceolate, 2–4 mm long and covered in hair; inner bracts lanceolate, pointed, 4–8 mm long, with white and occasionally dark hairs and a few glandular ones

Flowers: All hermaphrodite ray-florets, the 5 petals fused into a yellow ray; corolla tube smooth; 5 stamens; inferior ovary of 2 cells; fertilised by insects

Fruits: Achenes 1–3 mm long, with 10 ridges, narrowing in the upper part; dispersal by wind

Habitat: In hill and mountain zones on grassland, arable fields, heaths, walls, waste places, and in ditches, on well-drained to fairly dry loamy soils which are lime-poor and fairly base-rich and contain nutrients and humus

452 **Smooth hawk's beard – C. capilláris**
Plants 15–80 cm high, annuals or biennials with thin tap-roots; flowering from June to September

Crépis capilláris (L.) Wallr.
Smooth hawk's beard

Asteráceae (Compósitae) – Sunflower family

Leaves: Narrow oval to lanceolate, **broadest in the upper half,** in upper part of plant sessile or amplexicaul, and in lower part narrowing to winged stalks; leaves very variable in form

Inflorescence: Panicles of composite flowers; outer bracts narrow lanceolate, 2–4 mm long and covered in hair; inner bracts lanceolate, pointed, 4–8 mm long, with white and occasionally dark hairs and a few glandular ones

Flowers: All hermaphrodite ray-florets, the 5 petals fused into a yellow ray; corolla tube smooth; 5 stamens; inferior ovary of 2 cells; fertilised by insects

Fruits: Achenes 1–3 mm long, with 10 ridges, narrowing in the upper part; dispersal by wind

Habitat: In hill and mountain zones on grassland, arable fields, heaths, walls, waste places, and in ditches, on well-drained to fairly dry loamy soils which are lime-poor and fairly base-rich and contain nutrients and humus

453 **Stinking aposeris – A. foétida**
Plants 5–25 cm high, perennials with mealy coating on the upper parts; flowering in June and July

Apóseris foétida (L.) Lessing
Stinking aposeris

Asteráceae (Compósitae) – Sunflower family

Leaves: Lanceolate, pinnate, or pinnately cut almost to the middle vein, with a large, dentate terminal lobe, and with lateral lobes progressively smaller towards the leaf bases, usually smooth and in a rosette

Inflorescence: Solitary composite flowers at the ends of the stems; outer bracts lanceolate, 1–2 mm long and somewhat hairy at the base; inner bracts lanceolate, 5–10 mm long, with some frizzy hairs at the base and looking as though dusted with meal

Flowers: All hermaphrodite ray-florets, the 5 petals fused into a yellow ray 15–20 mm long; 5 stamens; inferior ovary of 2 cells; fertilised by insects

Fruits: Egg-shaped, rather flat achenes 3–5 mm long, with short thick hairs, without beaks or pappus hairs; dispersal by ants

Habitat: In mountain and subalpine zones in shady meadows and sparse mountain forests on moderately moist to fairly well-drained loose soils which are base-rich and neutral and contain lime and humus

454 **Mountain hawkbit – L. montánus**
Plants 5–10 cm high, perennials with thick, knotted, short rhizomes; flowering until August

Leóntodon montánus Lam.
Mountain hawkbit

Asteráceae (Compósitae) – Sunflower family

Leaves: Basal leaves lanceolate, pinnately cut almost to the middle vein or only with widely spaced, fine teeth, and smooth on both sides or rather hairy beneath; lobes usually broadly triangular; stem leaves, when present, only in the form of scales

Inflorescence: Solitary composite flower on end of stem (larger than those of Fig. 448); bracts up to 15 mm long, with thick, dark, erect hairs

Flowers: All hermaphrodite yellow ray-florets up to 20 mm long, with 5 petals fused into a ray and 5 teeth at the end; 5 stamens; inferior ovary of 2 cells

Fruits: Achenes with white pappus

Habitat: In alpine zone on scree, on damp to well-drained loose, often mobile, rough limestone and rock debris which is base-rich and contains lime but no humus, and is covered by snow for long periods; a widespread and fairly common alpine plant; range extends eastwards to lower Austria

455 **Rough hawk's beard – C. biénnis**
Plants 30–100 cm high, biennials with woody tap-roots; flowering from May to July

Crépis biénnis L.
Rough hawksbeard

Asteráceae (Compósitae) – Sunflower family

Leaves: Narrow oval to lanceolate, irregularly notched or pinnately cut, with long, often rather yellowish hairs on either side of the main vein, narrowing to winged stalks at the base; upper leaves have narrower bases and are sessile or amplexicaul

Inflorescence: Terminal corymbs of numerous long-stemmed composite flowers; bracts in 2 rows of different lengths, up to 14 mm long, with short, white, close hairs; receptacle has scattered hairs

Flowers: Golden-yellow, lanceolate ray-florets formed of 5 fused petals; 5 stamens; inferior ovary; fertilised by insects or self-fertile

Fruits: Achenes 4–7 mm long, with up to 18 ridges, often with short hairs and a white pappus

Habitat: In hill and mountain zones in meadows and on roadsides, on damp or well-drained soils with medium or deep topsoil and containing nutrients and humus

456 **Pygmy hawk's beard – C. pygmaéa**
Plants 5–15 cm high, perennials with short rhizomes, usually with curved stems; flowering till Aug.

Crépis pygmaéa L.
Pygmy hawk's beard

Asteráceae (Compósitae) – Sunflower family

Leaves: 3–8 cm long, lanceolate to narrow oval, usually smooth, irregularly dentate or pinnately cut, with a large end lobe, often covered in felted white hairs, green-violet and with winged stalks

Inflorescence: Composite flowers with stems covered in felted white hairs; bracts in 2 rows of different lengths, up to 15 mm long, with white felted hairs, dark green or blackish with a whitish margin; receptacle has short hairs

Flowers: Yellow ray-florets of 5 fused petals, often with a reddish overlay on the outer side and up to 15 mm long; 5 stamens; inferior ovary of 2 cells

Fruits: Achenes up to 6 mm long, with up to 25 ridges, with a white pappus in 2–4 rows

Habitat: In alpine (rarely subalpine) zones on damp to well-drained, lime-rich, generally stable scree; a plant of the central and southern European mountains reaching to the Pyrenees

457 **Sow thistle – S. oleráceus**
Plants 20–100 cm high, annuals or perennials with thin tap-roots; flowering from June to October

Sónchus oleráceus L.
Sow thistle

Asteráceae (Compósitae) – Sunflower family

Leaves: Oval, rounded, or heart-shaped at the base or pinnately cut on both sides almost to middle vein; edges with sharp teeth, of unequal size; stem leaves sessile or amplexicaul with wide basal points

Inflorescence: Composite flowers on stems, forming a panicle; the topmost flower heads form an umbel; bracts of different lengths, 10–15 mm long, light green and with white down at the base

Flowers: Yellow ray-florets of 5 fused petals, often with a reddish overlay at the tip on the outer side and up to 2 cm long; 5 stamens; inferior ovary of 2 cells; fertilised by insects

Fruits: Achenes up to 3 mm long, brown, and the shape of an inverted egg

Habitat: In hill and mountain (rarely subalpine) zones in cultivated soil, waste land, and by walls, on well-drained to moderately dry soils containing nutrients and humus; roots up to 1 m deep; plant has accompanied cultivation from early times; formerly used as a vegetable

458 **Prickly lettuce – L. serríola**
Plants 30–140 cm high, annuals or biennials with tap-roots; flowering from July to September

Lactúca serríola L.
Prickly lettuce

Asteráceae (Compósitae) – Sunflower family

Leaves: Deeply dentate or pinnately cut, with bristly hairs on the middle vein beneath, blue-green and rigid; lobes irregular in shape, the base being arrow-shaped and amplexicaul

Inflorescence: Conical panicles with numerous composite flowers on stems; bracts lanceolate, of different lengths, light green, usually with white margins and darkish red at the tips

Flowers: Yellow ray-florets of 5 fused petals, 1–2 cm long and with 5 teeth on the flat tip; 5 stamens; inferior ovary of 2 cells; fertilised by insects or self-fertile

Fruits: Achenes up to 7 mm long, egg-shaped, with beaks up to 7 mm long, grey-brown and with 5–7 longitudinal ridges on each side

Habitat: Common in hill and mountain zones in sunny waste land, hedges, roadsides, railway embankments, etc., on moderately dry soil rich in bases and nutrients

459 **Fluellen – K. spúria**
Plants 10–30 cm high, annuals with many glandular hairs; flowering from July to October

Kíckxia spúria Dumortier
Linária spúria (L.) Miller
Fluellen

Scrophulariáceae – Figwort family

Leaves: Oval, rounded or heart-shaped at the base, with short stalk, alternate, entire, and blunt-ended

Inflorescence: Solitary flowers on long stems covered in shaggy hairs

Flowers: 5 stamens, oval, pointed, rather hairy and often reddish towards the tip; 5 petals, arranged into a top and bottom lip; upper lip violet inside; lower lip dark yellow; spur white and somewhat bent; 4 stamens enclosed in the corolla; superior ovary with 2 fused cells; fertisised by insects or self-fertile

Fruits: Capsules opening by pores with 2 deciduous lids

Habitat: In hill zone in cornfields and fallow land on fairly well-drained sandy soils which are base-rich, rich in nutriment, generally lime-poor, more or less neutral, and contain little humus

460 **Broom-leaved toadflax – L. genistifólia** Plants 30–80 cm high, annuals; flowering from June to September

Linária genistifólia (L.) Miller
Broom-leaved toadflax

Scrophulariáceae – Figwort family

Leaves: Alternate, 2–6 cm long, oval, pointed, without stalks, round at the base, without hairs, blue-green

Inflorescence: Plants branch plentifully in the upper part, with numerous terminal racemes bearing many flowers, opening from the bottom ones upwards

Flowers: 5 sepals, pointed and blue-green; 5 yellow petals divided into 2 lips, the upper with 2 lobes and the lower 3; spur very elongated; 4 stamens, enclosed in the corolla; superior ovary of 2 cells; fertilised by insects

Fruits: Dehiscent capsule with 2 flaps and many seeds

Habitat: Usually in hill zone, in dry grassland, docks, and waste land on dry, sandy, often gravelly soils; species shows great variability; introduced from south-east Europe; in the Alps the commonest subspecies are Linaria genistifólia ssp. dalmática (L.) Maire et Pet. and ssp. genistifólia (Mn.)

461 **Toadflax – L. vulgáris** Plants 20–60 cm high, perennials, erect but also branching from the base; flowering from June to September

Linária vulgáris Miller
Toadflax

Scrophulariáceae – Figwort family

Leaves: Linear lanceolate, narrowing in a wedge shape towards the base, 1.5–3 cm long, closely set, turned slightly downward at the edge, alternate and grey-green

Inflorescence: Stem ends in a dense raceme

Flowers: Stems of raceme and individual flowers generally bear glands; 5 sepals, with triangular tips and shorter than flower stem; petals 1–2 cm long, pale yellow with orange palate, and with two lips; spur somewhat longer than the corolla; 4 stamens, enclosed in the corolla; superior ovary of 2 cells; fertilised by insects or self-fertile

Fruits: Capsules opening by 4–10 large teeth; dispersal by ants and wind

Habitat: In hill and mountain zones in grassy and cultivated fields, roadsides, hedgerows, woodland clearings, quarries, and waste land, on fairly well-drained to dry soils rich in nutrients

462 **Wood cow-wheat – M. silváticum** Plants 5–25 cm high, annuals with erect rectangular stems; flowering from June to September

Melámpyrum silváticum L.
Wood cow-wheat

Scrophulariáceae – Figwort family

Leaves: Narrow linear, often with short stems, drawing to gradual points, entire, smooth or with scattered hairs and generally opposite; the leaves gradually become smaller going up to the inflorescence

Inflorescence: Solitary flowers in the axils of the top stem leaves; all flowers turned to the same side of the stem

Flowers: Calyx in 4 parts, 6–12 mm long, as long as or longer than the petals and pale green; 5 petals, arranged in 2 lips, yellow and with a bent corolla tube that has no ring of hairs; the lower lip has two humps pointing upwards at the base; 4 stamens, enclosed in the upper lip; superior ovary of 2 cells

Fruits: Capsules; dispersal by ants

Habitat: In mountain and subalpine zones in sparse woodland, thickets, meadows, spruce or mixed oak/spruce woods and scrub, on more or less well-drained, base-rich, lime-poor, neutral soil

463 **Radiate broom – G. radiáta** Shrub 30–70 cm high, much-branched, without thorns; flowering from May to July

Genísta radiáta (L.) Scop.
Cytísus radiátus (L.) M. et K.
Radiate broom

Fabáceae (Papilionáceae) – Pea family

Leaves: Opposite, on numerous cane-like twigs arranged on the stem; these display small, sheath-like fused stipules and 3 narrow lanceolate leaflets

Inflorescence: Short raceme

Flowers: 5 sepals, fused, thickly covered in hair, 3–6 mm long; yellow corolla 1–2 cm long, consisting of standard, keel, and wings, all of more or less equal length; standard rather hairy round middle vein; 10 stamens; all filaments fused together; superior ovary

Fruits: Pods 10–15 mm long, rather hairy at the base and containing numerous seeds

Habitat: In hill and mountain zones on dry warm hillsides, in sparse woodland, in meadows, and on rocks, on dry, lime-rich soils in warm places

464 **Hyssop-leaved iron woundwort – S. hyssopifólia** Plants 10–30 cm high, with underground woody stems; flowering from July to Sept.

Sideriítis hyssopifólia L.
Hyssop-leaved iron woundwort

Lamiáceae (Labiátae) – Mint family

Leaves: Oval to broad lanceolate, widest towards tip, entire or with isolated blunt teeth, often round at the ends, covered with scattered hairs, opposite, and having stalks only in bottom part of stem

Inflorescence: Whorls of up to 6 flowers set one above the other at end of stem to form an apparent spike

Flowers: 5 fused sepals, with teeth having awn-like spines, distinctly hairy, up to 10 mm long and with 5–10 veins; 5 fused petals, forming a top and bottom lip, pale yellow and 6–10 mm long; top part of upper lip slightly bent; lower lip clearly in 3 lobes; 4 stamens, not projecting beyond the corolla tube; anther halves spread out; superior ovary of 2 cells; pistil does not project from the corolla tube; usually self-fertile

Fruits: Egg-shaped nutlets, up to 2 mm long and glossy

Habitat: In subalpine zone on limestone rocks and debris, on dry soils in warm places, e.g. the Jura

465 **Yellow betony – B. alopecúros** Plants 20–50 cm high, perennials with knotty rhizomes; flowering from June to August

Betónica alopecúros L.
Stáchys alopecúros (L.) Bentham / Yellow betony

Lamiáceae (Labiátae) – Mint family

Leaves: Broad oval to cordate, on stalks, heart-shaped at the base, leaf proper 3–7 cm long, with coarse, blunt,or pointed teeth, and hairy

Inflorescence: Several whorls of up to 14 flowers set one above the other at end of stem to form an apparent spike

Flowers: 5 fused sepals 6–10 mm long, hairy and with teeth having awn-like bristles; 5 fused petals up to 15 mm long, pale yellow, with upper lip having 2 lobes and lower lip 3; 4 stamens; superior ovary of 2 cells; fertilised by insects

Fruits: Glossy nutlets up to 2.5 mm long

Habitat: In mountain and subalpine zones on waste land and rocky slopes, meadows and sparse woodland, on fairly moist to well-drained, loose clay or loam soils which are usually lime-rich and contain more or less humus; plants grow in light and half-shade; species shows great variety of form; found, for example, in the Bergamese Alps

466 **Small-flowered balsam – I. parviflóra**
Plants 15–80 cm high, smooth-leaved annuals with simple or branching stems; flowering June to Oct.

Impátiens parviflóra DC
Small-flowered balsam

Balsamináceae – Balsam family

Leaves: 4–12 cm long, oval to broad lanceolate, on stalks, pointed, alternate, with coarse teeth having distinct points

Inflorescence: Upright racemes

Flowers: Hermaphrodite, zygomorphic flowers, mid to pale yellow; 3 sepals, 2 atrophied, the third with a straight spur up to 1 cm long; 5 petals, the lateral ones fused in pairs, resulting in 3 apparent petals; largest up to 1 cm long; 5 stamens, alternating with the petals; anthers free-standing; superior ovary of 5 cells

Fruits: Club-shaped capsules, up to 2 cm long and springing open when touched

Habitat: In hill and mountain (rarely subalpine) zones in parkland, clearings in sparse woodland, gardens, hedgerows, roadsides, and on rubbish dumps, on well-drained to moist soils in humid locations, preferred soils being rich in nutrients, generally lime-poor, moderately acid, and containing humus

467 **Touch-me-not – I. nóli-tángere**
Plants 30–90 cm high, annuals, smooth, with explosive seed-pods; flowering from June to August

Impátiens nóli-tángere L.
Touch-me-not

Balsamináceae – Balsam family

Leaves: 3–12 cm long, narrow oval to oval, on stalks, with only short points at the tips, alternate, with very coarse blunt teeth often bearing fine points

Inflorescence: Solitary flowers or hanging racemes with few flowers

Flowers: Hermaphrodite, zygomorphic, and yellow; 3 sepals, 2 small, the third with a straight spur up to 3 cm long; 5 petals, the lateral ones fused in pairs, resulting in 3 apparent petals; largest petal 2–2.5 cm long, the same width, round, and with red dots at throat; 5 stamens, alternating with petals; anthers free-standing; superior ovary

Fruits: Spindle-shaped explosive capsules up to 3 cm long

Habitat: In hill and mountain zones in beech and spruce woods, woods in ravines and wet woodland, by springs and streams, and on woodland margins, on moist to wet loam or clay soils containing nutrients and humus; plants enjoy a humid atmosphere

468 **Monkey flower – M. guttátus**
Plants 25–60 cm high, perennials, often with isolated glandular hairs; flowering from July to Sept.

Mímulus guttátus DC.
Monkey flower

Scrophulariáceae – Figwort family

Leaves: Round or oval to broad lanceolate, with irregular teeth, on stalks in the lower part of the plant, in upper part without stalks and partially amplexicaul

Inflorescence: Solitary flowers on stalks arising from axils of upper stem leaves

Flowers: 5 sepals, fused into a tube, angled and 2-lipped; 5 fused petals, yellow and 2–4 cm long; upper lip 2-lobed, lower lip with 3 usually flat lobes and red-brown spots; 4 stamens (2 longer, 2 shorter); superior ovary of 2 cells; fertilised by bees

Fruit: Dehiscent capsules with 2 flaps; dispersal mostly by water

Habitat: In hill and mountain zones on gravel banks, banks of streams, in ditches and by springs, on wet, periodically flooded gravelly or sandy clay rich in nutrients and usually lime-poor; an early coloniser of newly turned soil; uncommon in wild, more common in gardens

469 **Perennial yellow woundwort – S. récta**
Plants 20–70 cm high, perennials with rhizomes or long tap-roots; flowering from June to October

Stáchys récta L.
Perennial yellow woundwort

Lamiáceae (Labiátae) – Mint family

Leaves: 2–5 cm long, lanceolate to oval, narrowing to short stalks or sessile, with teeth pointing towards the tips, and covered with scattered hairs or smooth

Inflorescence: Numerous whorls of 4–8 flowers placed one beneath the other

Flowers: More or less sessile; 5 fused sepals, 6–8 mm long, green, hairy and with spiny awn-like teeth; petals fused into 2 lips, whitish-yellow to yellow and 10–20 mm long; corolla tube slightly bent and hairy; upper lip distinctly or slightly domed, entire, hairy outside and often with dark red stripes; lower lip 5–8 mm long, with 3 lobes and brown marks; 4 stamens fused to the corolla tube; ovary superior, with 2 cells; fertilised by insects

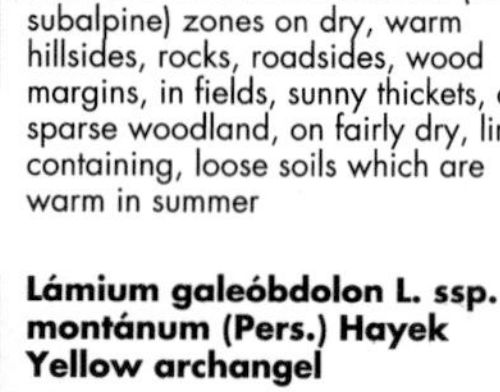

Fruits: 4 nutlets

Habitat: In hill and mountain (rarely subalpine) zones on dry, warm hillsides, rocks, roadsides, wood margins, in fields, sunny thickets, and sparse woodland, on fairly dry, lime-containing, loose soils which are warm in summer

470 **Yellow rattle – R. alectorólophus**
Plants 10–20 cm high, with branching 4-angled stems; flowering from May till August

Rhinánthus alectorólophus (Scop.) Pollich
Yellow rattle

Scrophulariáceae – Figwort family

Leaves: 2–6 cm long, narrow triangular to lanceolate, with irregular, blunt, or slightly pointed teeth pointing forwards, sessile, opposite, and hairy

Inflorescence: Solitary flowers in axils of top stem leaves (bracts)

Flowers: Sepals fused, flattened, and inflated, with 4 slightly pointed teeth, and covered with close hairs; petals 15–25 mm long, fused, with a long corolla tube and 2 lips; corolla tube slightly bent upwards; upper lip helmet-shaped, with a tooth up to 2.5 mm long and usually violet; lower lip shorter than top lip and with 3 lobes; 4 stamens; superior ovary

Fruits: Dehiscent capsules with 2 flaps

Habitat: From hill to subalpine zones on rather damp, neutral soils which are base-rich and contain lime, nutrients, and humus; an indicator of loamy soil; a great variety of forms

471 **Yellow archangel – L. galeóbdolon**
Plants 30–50 cm high, putting out runners at flowering or shortly after; flowering May to July

Lámium galeóbdolon L. ssp. montánum (Pers.) Hayek
Yellow archangel

Lamiáceae (Labiátae) – Mint family

Leaves: Topmost leaves narrow oval or lanceolate, opposite, on stalks, pointed at the ends and with sharply pointed teeth; those of the runners are wide lanceolate, usually with long points and white spots

Inflorescence: Numerous whorls of 4–8 flowers placed one beneath the other in the upper part of the stem

Flowers: Sessile; 5 fused sepals, 6–8 mm long, green, hairy, and with long teeth; petals fused into 2 lips and yellow; upper lip bent, entire, made of 2 petals and hairy outside; lower lip has 3 lobes, the middle one with brown spots and larger than the pointed lateral lobes; 4 stamens fused; ovary superior

Fruits: 4 smooth nutlets up to 3 mm long and smooth

Habitat: From hill to subalpine zones in damp, mixed or beech woods, and hedgerows on cool, moist, loose soils rich in nutrients and humus

472 **Dwarf eyebright – E. mínima**
Plants 2–10 cm high, annuals, rarely branching in lower parts; flowering from July to September

Euphrásia mínima Jacq. ex DC
Dwarf eyebright

Scrophulariáceae – Figwort family

Leaves: 1–2 cm long, roundish, hairy, with 2–4 teeth on either side, blunt or fine points but no spines, usually without stalks and opposite, particularly in the lower part of the plant

Inflorescence: Flowers springing from axils of top stem leaves

Flowers: No stems or only short ones; sepals 3–5 mm long, fused, with short glandular hairs and 4 pointed teeth; corolla 4–8 mm long with tube 2–4 mm long; upper lip yellowish, bluish, or reddish; lower lip yellowish, whitish, or lilac; 4 stamens; anthers pointed beneath; ovary superior; fertilised by insects or self-fertile

Fruits: Dehiscent capsules with 2 flaps

Habitat: In subalpine and alpine zones in rough pasture and alpine meadows, on well-drained, siliceous, acid, infertile loam with humus; this central and southern European mountain plant is variable in form

473 **Wolfsbane – A. vulpária**
Plants 30–80 cm high, perennials with turnip-like roots; flowering from June to August

Aconítum vulpária s.l.
Wolfsbane

Ranunculáceae – Buttercup family

Leaves: Basal leaves up to 15 cm in diameter, with 5–7 deeply cut lobes with scattered hairs on veins beneath; lobes wedge-shaped or rhomboid, usually with 3 secondary lobes, with pointed or blunt tips; the 1–3 stem leaves are similar to the basal leaves

Inflorescence: Usually branched, with curved upright loose racemes

Flowers: Zygomorphic, on stems, slightly hairy, with 5 sepals and yellow; top sepal shaped like a helmet constricted in the middle; other sepals round or oval; 2 nectaries; numerous stamens; 3–5 carpels; fertilised by bumble-bees

Fruits: Follicles containing many seeds

Habitat: In mountain and subalpine zones in mountain woodlands, scrubby undergrowth, wet woodlands, and damp deciduous woodlands, on cool, well-watered to wet loose loam or clay which is base-rich, rich in nutrients, moderately acid, and contains humus; plant grows in shade; also found in lower zones

474 **Kidney vetch – A. vulnerária**
Plants 10–60 cm high, usually with upright stems, often only biennial; flowering from May to August

Anthyllis vulnerária L.
Kidney vetch

Fabáceae (Papilionáceae) – Pea family

Leaves: Unequally pinnate or narrow oval, entire, lower leaves with stems; end leaflet much larger than the 2–6 lateral leaflets; these are oval to lanceolate and up to 6 cm long

Inflorescence: Flowers in thick cymes within fringed or lobed bracts

Flowers: Almost without stems; calyx often rather bulbous, with close or erect hairs and with 5 short pointed teeth; corolla 1–2 cm long, and golden-yellow; standard rather longer than wings and keel; 10 stamens; filaments usually fused together; superior ovary; fertilised by bumble-bees

Fruits: Usually single-seeded pods 4 – 6 mm long

Habitat: From hill to subalpine zones in meadows, sunny lime-containing rough pastures, quarries, sparse pine woods, and by the roadside, on moderately dry, base-rich, generally lime- and humus-containing loam or loess which is warm in summer

475 **Black medick – M. lupulína**
Plants 5–30 cm high, annuals or short-lived perennials; flowering from May to September

Medicágo lupulína L.
Black medick

Fabáceae (Papilionáceae) – Pea family

Leaves: The 3 leaflets are oval or rhomboid, finely toothed, especially in the upper part, with scattered close hairs and blunt-ended or rather flat; topmost leaflet has a stalk and is often rather larger than the lateral ones

Inflorescence: Spherical racemes on stems growing in axils of stem leaves

Flowers: On short stems; 5 sepals, green with narrow pointed tips; 5 petals forming standard, 2 wings and keel, yellow, up to 3.5 mm long and dropping off after flowering; 10 stamens, the top filament being free; superior ovary

Fruits: Kidney- or sickle-shaped pods

Habitat: In hill and mmountain zones by roadsides, in ditches, dry meadows, waste land, sunny thickets, and woodland margins, and rough pasture on limestone, on base-rich soils which contain lime and humus, have deep topsoil, and are warm in summer

476 **Brown clover – T. bádium**
Plants10–20 cm high, perennials without runners; flowering in July and August

Trifólium bádium Schreber
Brown clover

Fabáceae (Papilionáceae) – Pea family

Leaves: The 3 leaflets are oval or narrow oval, broadest in the middle, usually smooth, 1–2 cm long, finely toothed, and round-ended; stipules almost oval

Inflorescence: Spherical racemes containing 20–50 flowers at end of stem or growing in axils of top stem leaves

Flowers: On short stems; 5 fused sepals, with narrow triangular tips, smooth or with isolated hairs; petals golden-yellow, 5–9 mm long and turning dark brown after flowering; standard

2–3 times as long as the keel; 10 stamens, the top filament free; superior ovary; fertilised by insects

Fruits: Pods 2–4 mm long; dispersal by wind and animals

Habitat: In subalpine zone (and often washed down into the mountain zone) in meadows, snow valleys, and by camp sites, on well-drained to moist, base-rich, usually rather heavy loam or clay containing nutrients and humus; a good fodder plant

477 **Melilot – M. officinális** Plants 20–140 cm high, annuals or biennials distributed over almost all the world; flowering from June to October

Melilótus officinális (L.) Lam.
Common melilot

Fabáceae (Papilionáceae) – Pea family

Leaves: The 3 leaflets are oval (broadest towards tips) or lanceolate, generally entire in the lowest part, otherwise deeply toothed and blunt or rather flat at the tips; top leaflet has a longer stalk than the 2 lateral ones

Inflorescence: Dense racemes 4–10 cm long on stems bearing 30–70 flowers

Flowers: With thin stems; 5 sepals, green and with 5 narrow pointed tips; 5 petals, yellow and 5–7 mm long; standard and wings longer than the keel; 10 stamens; superior ovary

Fruits: Pods 3–4 mm long, smooth, containing 4–8 seeds and with longitudinal ridges

Habitat: In hill and mountain zones on crevices in the earth, by roadsides, banks of streams, waste places, ditches, and in quarries on moderately dry, neutral, and stony soils of all kinds which are warm in summer and base-rich and contain nutrients; plants attract bees; an early coloniser of newly turned soil; roots up to 90 cm deep

478 **Large yellow restharrow – O. nátrix**
Plants 2–40 cm high, perennials with woody, branching rhizomes; flowering in June and July

Onónis nátrix L.
Large yellow restharrow

Fabáceae (Papilionáceae) – Pea family

Leaves: On stalks, the 3 leaflets 1–3 cm long, narrow oval to oval, toothed, especially above the middle, and blunt or trimmed at the tips; the middle leaflet has a stem; bracts entire

Inflorescence: Flowers solitary or in groups of 2–3 in racemes on stems springing from the axils of the top stem leaves

Flowers: 5 fused sepals, 5–8 mm long, with glandular hairs and pointed tips; corolla 1.5–2.5 cm long and yellow; standard has distinct longitudinal stripes; keel distinctly bent upwards, narrowing into a beak; 10 stamens, all fused together; superior ovary; fertilised by insects

Fruits: Pods 1–2 cm long, containing many seeds and covered with glandular hairs

Habitat: In hill and mountain zones in warm places near dry grassland, stony hillsides, sunny roadsides, thickets and in limestone rough pasture, on warm, dry, base-rich clay containing humus

479 **Birdsfoot trefoil – L. corniculátus**
Plants 10–30 cm high, upright or recumbent, almost wholly smooth; flowering from May to July

Lótus corniculátus L.
Birdsfoot trefoil

Fabáceae (Papilionáceae) – Pea family

Leaves: Leaflets oval to narrow oval (broadest towards tips), 1–2 cm long, smooth or slightly hairy at the margins, and blunt or slightly pointed

Inflorescence: Umbels of 4–8 flowers

Flowers: 5 fused sepals, 5–7 mm long, generally smooth but often with hairy teeth, which meet before flowering; corolla is 10–15 mm long and yellow; keel usually pale yellow and bent upwareds; 10 stamens; upper filaments free-standing; superior ovary

Fruits: Pods containing several glossy seeds

Habitat: From hill to subalpine zones in pastures and grassy places, rough pasture on limestone, gravel pits, edges of thickets, on dry slopes and roadsides, on warm, moderately dry to well-drained, loose loam rich in bases and nutrients; seeds readily in fresh soil; fertiliser; attracts bees; a good fodder plant; roots up to 90 cm deep; this European plant is widespread and common

480 **Winged pea – T. marítimus**
Plants 5–30 cm high, perennials with scattered hairs and rhizomes; flowering from May to July

Tetragonólobus marítimus (L.) Scop.
Winged pea

Fabáceae (Papilionáceae) – Pea family

Leaves: On stalks; leaflets usually without stems, wedge-shaped or oval, broadest towards the tips, fleshy, slightly hairy, and blue-green; bracts rather smaller than leaves

Inflorescence: Solitary flowers on stems

Flowers: 5 fused sepals; calyx tube smooth, with hairy, pointed tips; corolla 2–3 cm long, pale yellow, often with red overlay in bloom; keel bent out and often dark; standard much larger than wings and keel; 10 stamens; topmost filaments free-standing; superior ovary

Fruits: Pods 4–5 cm long, with longitudinal wings up to 1 mm wide

Habitat: In hill and mountain zones in dry meadows, sparse pine woods, moorland, calcareous grassland and by springs on tuff, on intermittently flooded, base-rich, heavy clay, marl, and tuff which is warm and dry in summer and contains humus; an indicator of clay; attracts bumble-bees; a valuable fodder plant; stands high salt content

481 **Meadow vetchling – L. praténsis**
Plants 20–90 cm high, perennials with short runners; flowering in June and July

Láthyrus praténsis L.
Meadow vetchling

Fabáceae (Papilionáceae) – Pea family

Leaves: In pairs, with a branched or unbranched tendril, especially in upper part of plant; stalks without wings; both leaflets narrow oval to lanceolate, 1–4 cm long, with parallel veins, pointed and smooth or covered with short hairs

Inflorescence: Raceme with 3–12 flowers

Flowers: On stems; 5 fused sepals, 5–6 mm long, smooth or rather hairy and with 5 teeth of unequal length at tips; corolla 1–1.5 cm long and yellow; keel bent upwards; 10 stamens, all fused for much of their length; superior ovary; fertilised by insects

Fruits: Pods up to 4 cm long, with 5–12 seeds, which are glossy and up to 4 mm long

Habitat: In hill and mountain zones in pastures, moors and wet grassland, hedgerows, woodland margins and clearings, roadsides, and banks of streams, on moist or intermittently flooded, generally neutral loam or clay containing nutrients and humus; plant prefers high nitrogen content; deep roots

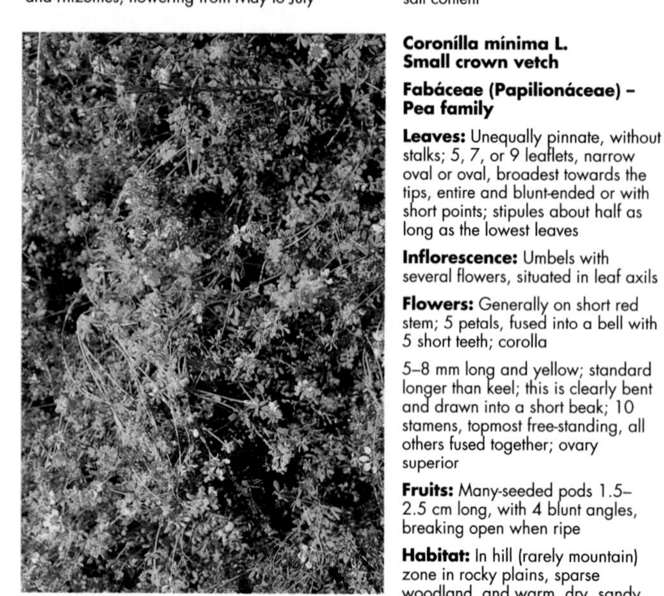

482 **Small crown vetch – C. mínima**
Plants 10–25 cm high, recumbent or climbing; flowering in June and July

Coronílla mínima L.
Small crown vetch

Fabáceae (Papilionáceae) – Pea family

Leaves: Unequally pinnate, without stalks; 5, 7, or 9 leaflets, narrow oval or oval, broadest towards the tips, entire and blunt-ended or with short points; stipules about half as long as the lowest leaves

Inflorescence: Umbels with several flowers, situated in leaf axils

Flowers: Generally on short red stem; 5 petals, fused into a bell with 5 short teeth; corolla

5–8 mm long and yellow; standard longer than keel; this is clearly bent and drawn into a short beak; 10 stamens, topmost free-standing, all others fused together; ovary superior

Fruits: Many-seeded pods 1.5–2.5 cm long, with 4 blunt angles, breaking open when ripe

Habitat: In hill (rarely mountain) zone in rocky plains, sparse woodland, and warm, dry, sandy places on dry, calcareous soils in warm places; uncommon, but found e.g. in the Valais

483 **Horseshoe vetch – H. comósa**
Plants 5–25 cm hight perennials, recumbent and spreading; flowering from May to July

Hippocrépis comósa L.
Horseshoe vetch

Fabáceae (Papilionáceae) – Pea family

Leaves: Unequally pinnate. on long stems, often covered with scattered hairs, and with 4–8 pairs of leaflets; these are short-stemmed or sessile, 5–15 mm long, oval or narrow oval, entire, rounded but often with sharp points

Inflorescence: Umbels on long stems, bearing 5–12 flowers

Flowers: Flower stem very short, green; 5 sepals, fused into a tube, with 5 narrow triangular points; corolla 8–12 mm long, and yellow; claws of petals 1–2 times as long as calyx; keel distinctly bent; 10 stamens, topmost free-standing; superior ovary

Fruits: Pods 1–3 cm long, with horeseshoe-shaped segments

Habitat: From hill to subalpine zones in warmer places in sparse woodlands, quarries, dry grassland, by the roadside, and in thickets, on fairly dry, base-rich, lime-free loam or loess which may or may not contain humus

484 **Small scorpion crown vetch – V. vaginális**
Plants 10–30 cm high, perennials, recumbent or climbing; flowering from May to July

Coronílla vaginális Lam.
Small scorpion crown vetch

Fabáceae (Papilionáceae) – Pea family

Leaves: Unequally pinnate, on short stalks, blue-green with 2–6 pairs of leaflets, which are sessile or on very short stems, 5–13 mm long, oval, rather fleshy, tough-rimmed and often mucronate

Inflorescence: Long-stemmed umbels bearing 3–10 flowers

Flowers: Short-stemmed; 5 fused sepals with triangular teeth; corolla 6–10 mm long and yellow; claws of petals longer than the calyx; keel points strongly upwards; 10 stamens, topmost free-standing; superior ovary

Fruits: 6-angled pods, 4 angles with wavy wings

Habitat: In mountain and subalpine (rarely hill) zones in warmer places in stony grassland, dry pinewoods, and on rocks, on generally dry, base-rich, stony or gravelly soils containing lime and humus

485 **Scorpion crown vetch – C. coronáta**
Plants 20–50 cm high, erect, smooth perennials; flowering in June and July

Coronílla coronáta L.
Scorpion crown vetch

Fabáceae (Papilionáceae) – Pea family

Leaves: Unequally pinnate, on very short stalks, with 3–6 pairs of leaflets on very short stems, 15–25 mm long, oval, rather fleshy, often mucronate and often blue-green beneath

Inflorescence: Long-stemmed umbels bearing 10–20 flowers

Flowers: Often with red stems; 5 sepals, with tapering triangular teeth at the tips; corolla

7–10 mm long and yellow; claws of petals as long as or longer than calyx; keel pointing strongly upwards; 10 stamens, the topmost free-standing; superior ovary

Fruits: Straight pods up to 3 cm long, clearly jointed; 4 blunt angles

Habitat: In hill and mountain zones in warm, half-shady locations on rocks, mountain woodland, woodland margins, in sparse oak and pine woods, on dry, base-rich loam or clay which is warm in summer; deep roots

486 **Mountain lentil – A. penduliflórus**
Plants 10–45 cm high, ascending or upright, branching; flowering in July and August

Astrágalus penduliflórus (Lam.)
Mountain lentil

Fabáceae (Papilionáceae) – Pea family

Leaves: Unequally pinnate, short-stalked, with 7–12 pairs of leaflets, short-stemmed, 5–20 cm long, blunt-ended or coming to short points and slightly hairy at the edges beneath; small stipules

Inflorescence: Racemes on stems, bearing 5–20 flowers

Flowers: Nodding and short-stemmed; stem often longer than calyx; 5 fused sepals, 5–7 mm long, with short, dark, close hairs; calyx teeth broad triangular; corolla 9–13 mm long and yellow; keel as long as wings; 10 stamens, topmost free-standing; superior ovary; fertilised by insects or self-fertile

Fruits: Nodding pods, swollen into a bladder shape, 2–3 cm long and thickly covered with dark hairs; dispersal by animals

Habitat: In subalpine and alpine zones on dry slopes, in larch woods, rocky alpine pastures, and sparse pine woods, on moderately dry to well-drained, base-rich, loose, neutral, stony loam or clay containing humus

487 **Yellow oxytropis – O. campéstris**
Plants 5–15 cm high, with basal leaf rosette, without glands; flowering from June to August

Oxytrópis campéstris L. (DC)
Yellow oxytropis

Fabáceae (Papilionáceae) – Pea family

Leaves: Unequally pinnate, on stalks and with 10–5 pairs of leaflets, 5–15 mm long, lanceolate, generally pointed, often with scattered hairs on both sides, and mainly (not always) opposite; stipules fused to the stems for almost half their length

Inflorescence: Upright, long-stemmed racemes with 8–18 flowers

Flowers: Pointing upwards on very short stems; 5 fused sepals with short black and white hairs; calyx teeth generally darker; corolla 15–20 mm long, yellowish-white, occasionally tinged with blue or purple; keel clearly mucronate at tip and with 2 deep purple spots on either side; 10 stamens, topmost free-standing; superior ovary

Fruits: Pod 1–2 cm long, swollen into a bladder shape, opening on top seam and covered with thick, partly dark, hairs

Habitat: Generally found in alpine zone on meadowland, on moderately dry, often calcareous soil

488 **Leafy lousewort – P. foliósa**
Plants 15–50 cm high, perennials, smooth or with scattered hairs; flowering in June and July

Pediculáris foliósa L.
Leafy lousewort

Scrophulariáceae – Figwort family

Leaves: Pinnately cut to middle vein, 10–25 cm long, up to 8 cm wide, smooth or with scattered hairs beneath, longer than the flowers, and with hairy stalks; leaflets also pinnate, with toothed points up to 2 cm long

Inflorescence: Dense racemes

Flowers: 5 fused sepals, 8–10 mm long, rounded at the base, hairy and with short, entire teeth; corolla 20–30 mm long and pale or sulphur yellow; lower lip as long as upper one, with 3 lobes and erect; upper lip slightly inflated and hairy; 4 stamens enclosed by upper lip; superior ovary; fertilised by bumble-bees

Fruits: Egg-shaped capsules, with crooked points and springing open on one side

Habitat: From mountain to alpine zones in scrubby undergrowth and grassy mountain slopes, on well-watered or intermittently well-watered, base-rich, usually calcareous loam or clay

489 **Tuberous lousewort – P. tuberósa**
Plants 10–20 cm high, with curved rising stems; flowering from June to August

Pediculáris tuberósa L.
Tuberous lousewort

Scrophulariáceae – Figwort family

Leaves: 3–15 cm long, narrow oval to lanceolate, and pinnately cut to the middle vein; leaflets also pinnate, with finely toothed points; leaves in the flower head are not longer than the flowers

Inflorescence: Short racemes, as wide as long when flowering starts

Flowers: At right angles to the stem; 5 fused sepals, 6–10 mm long, usually gradually narrowing towards the base, with narrow lanceolate, finely toothed points on the exterior, and margin covered with solitary hairs but smooth on the inside; corolla 15–20 mm long, pale yellow; upper lip without teeth and narrowing to a beak 3–4 mm long; lower lip as long as the upper one, smooth and only slightly erect; 4 stamens enclosed by the upper lip

Fruits: Capsules, longer than the calyx

Habitat: In subalpine and alpine zones in meadowland, on dry, lime-poor, acid loam containing humus

490 **Martagon lily – L. mártagon**
Plants 20–90 cm high, perennials with bulbs; flowering in June and July

Lílium mártagon L.
Martagon lily

Liliáceae – Lily family

Leaves: 4–15 cm long, up to 4 cm wide, narrow oval, widest in middle, pointed, sessile, in thick whorls in the middle part of the stem, alternate in top and bottom parts

Inflorescence: Raceme of 3–10 flowers

Flowers: Nodding, on long, oblique stems; 6 petals, narrow oval, up to 7 cm long, bent backwards from the middle, dull or pale purple with dark red spots; 6 stamens, shorter than the petals; superior ovary of 3 cells; fertilised by insects

Fruits: 3-celled capsules with many seeds; dispersal by wind

Habitat: From hill to subalpine zones in beech and mixed deciduous woodland and woods in ravines, in meadows and shrubby undergrowth, on water-bearing or wet loam or clay which is base-rich, contains humus and nutrients, and has more or less deep topsoil; grows in half-shade

491 **Snake's head fritillary – F. meleágris**
Plants 15–35 cm high, perennials with bulbs, not branching; flowering in April and May

Fritillária meleágris L.
Snake's head fritillary

Liliáceae – Lily family

Leaves: 4–6 on the stem, 5–15 cm long, up to 1 cm wide, linear, grass-like, with rather fleshy thickening, grooved, alternate, and grey-green

Inflorescence: Usually solitary terminal flowers, sometimes 2 or 3 in a raceme

Flowers: Up to 4 cm long and 3–4 cm across, nodding and bell-shaped; 6 petals, not fused, narrow oval, blunt-ended, purple-brown with chequered lighter markings; 6 stamens, shorter than the petals; superior ovary of 3 cells; fertilised by bees

Fruits: 3-celled capsules with many seeds

Habitat: In hill and mountain zones in wet meadows and along streams and rivers, on wet or intermittently flooded loam or clay close to ground-water level, neutral, and containing nutrients and humus; an indicator of wet ground; an ornamental garden plant; poisonous

492 **Spotted orchid – D. maculáta**
Plants 20–50 cm high, with palmately lobed root tubers; flowering in June and July

Dactylorhíza maculáta L.
Dactylórchis maculáta (L.) Soo
Spotted orchid

Orchidáceae – Orchid family

Leaves: Lower leaves 5–10 cm long, lanceolate, often pointed, not coming as high as the flower head and with dark brown or black markings above; top leaves small and narrow; topmost leaf well removed from flower head; 3–6 transitional bract-like leaves under the flower head

Inflorescence: Dense cylindrical spike 3–8 cm long

Flowers: 2 lateral petals, horizontal or drawn back; 3 petals bent together; lip 4–8 mm long; club-like or cylindrical spur, pointing straight downwards; inferior ovary of 3 cells

Fruits: Capsule opening by longitudinal slits to release many seeds

Habitat: From hill to subalpine zones in wet meadows, hedgerows, woodland and moorland, on moist, sometimes moist, or wet neutral to acid loam or clay containing humus

493 **Black vanilla orchid – N. nígra**
Plants 10–25 cm high, perennials with palmately lobed root tubers; flowering from July to August

Nigritélla nígra (L.) Rchb.
Black vanilla orchid

Orchidáceae – Orchid family

Leaves: 5–15 cm long, linear, with hollow grooves and pointing upwards

Inflorescence: Dense spike 1–2 cm long, club-shaped at the start of flowering, later long oval or spherical

Flowers: Usually very deep red, dark brown-red, or brilliant red, with a strong vanilla scent; petals 5–7 mm long, spread out in a star shape; inner petals only half as wide as outer; lip has rather hollow grooves at top of base; spur widens into a sack shape; inferior ovary of 3 cells; fertilised by insects, especially butterflies

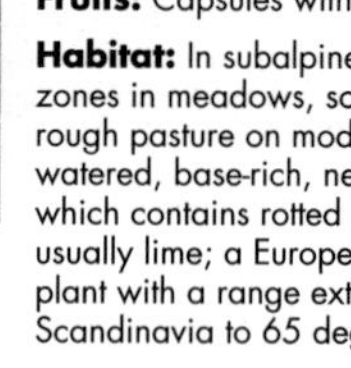

Fruits: Capsules with many seeds

Habitat: In subalpine and alpine zones in meadows, scrub, and rough pasture on moderately well-watered, base-rich, neutral loam which contains rotted humus and usually lime; a European mountain plant with a range extending in Scandinavia to 65 degrees

494 **Spring meadow saffron – B. vernum**
Plants 4–20 cm high, perennials with bulbs; flowering from February to May

Bulbocódium vernum L.
Cólchicum bulbocódium Ker-Gawl.
Spring meadow saffron

Liliáceae – Lily family

Leaves: 10 – 20 cm long, 1 to 1.5 cm wide, with folded grooves towards the top, pale or reddish brown, blunt-ended and appearing at the same time as the flowers in the spring

Inflorescence: Generally solitary flowers, rarely 2 or 3

Flowers: 6 petals; the outspread part towards the tip 2 – 5 cm long, up to 1 cm wide, with toothlike appendage at base,and generally pink; flower part consists of claws (not fused) in a tube, up to 10 cm long reaching deep below the soil; 6 stamens growing from base of petals; superior ovary with 3 styles fused almost as far as the stigmata

Fruits: 3-celled capsules with many seeds

Habitat: In subalpine (rarely hill or mountain) zone in meadows which are dry in summer but irrigated by meltwater in the spring, on dry sandy soils

495 **Autumn crocus – C. autumnále**
Plants 8–30 cm high, perennials with a visible tubers; flowering from August to October

Cólchicum autumnále L.
Autumn crocus

Liliáceae – Lily family

Leaves: 10–25 cm long, 2–5 cm wide, narrow oval or lanceolate, glossy, dark green, more or less flat and appearing together with the fruits in the spring

Inflorescence: Generally solitary flowers

Flowers: 6 petals 4–7 cm long. 1–1.5 cm wide, oval (widest towards tips) or longish, pink or lilac and without tooth-like appendage at base; lower part of petals is formed of a tube 8–25 cm long, reaching deep below the soil; 6 stamens, growing from the base of the petals; superior ovary of 3 cells; the 3 styles are not fused and are thickened at the tip; stigmata point downwards

Fruits: 3-celled capsules with many seeds

Habitat: From hill to subalpine zones in moist or wet meadows and woodlands on loam or clay which contains nutrients and humus and has a deep topsoil; poisonous; a weed of meadows and hayfields; commonly found; this central European plant's range extends northwards to Ireland and southwards to northern Spain

496 **Keeled garlic – A. pulchéllum**
Plants 25–60 cm high, with bulbs in tunics which disintegrate into fibres; flowering in July and August

Allium pulchéllum G.Don
Keeled garlic

Liliáceae – Lily family

Leaves: Narrow linear, smooth, 1–2 mm wide and very long

Inflorescence: Hemispherical or spherical umbels

Flowers: Flower stem 2–4 times as long as petals, and pale violet to pink; 6 petals, blunt-ended, 4–6 mm long and red to pinkish-violet; 6 stamens, clearly protruding beyond the petals; superior ovary of 3 cells; fertilised by insects

Fruits: Thin-skinned capsules with 3 compartments

Habitat: In hill zone on warm, dry, rough pasture, dry meadows, and in sparse thickets, on dry, lime-rich, stony or gravelly soil containing humus; the range of this plant, originating in the eastern Mediterranean region, extends northwards to the western and southern Alps, eastwards to the Caucasus; in central Europe it is found mainly in the Jura, in the southern Ticino, in the Veltlin and round Windsch

497 **Chives – A. schoenoprásum**
Plants 10–50 cm high, with long bulbs; flowering from May to August

Allium schoenoprásum L.
Chives

Liliáceae – Lily family

Leaves: All basal, standing 10–40 cm high, tubular, hollow, and smooth

Inflorescence: Many-flowered, dense spherical umbels; bracts shorter than the flower head

Flowers: Stems shorter than the petals; these are 7–15 mm long, elongated, coming to gradual points, pale to dark red with a darker central rib; 6 stamens, without lateral teeth at the tips and shorter than the petals; filaments widen slightly towards the base; superior ovary of 3 cells

Fruits: 3-celled capsules, usually with 2 seeds in each cell

Habitat: The wild varieties are found in the subalpine and alpine zones in moorland, wet meadows, and moist stony or rocky hillsides, on well-drained, compact, sandy or gravelly soils rich in nutrients; garden chives (illustrated here) prefer well-drained soil rich in nutrients and are often planted as an aromatic herb; it is also found wild on river banks

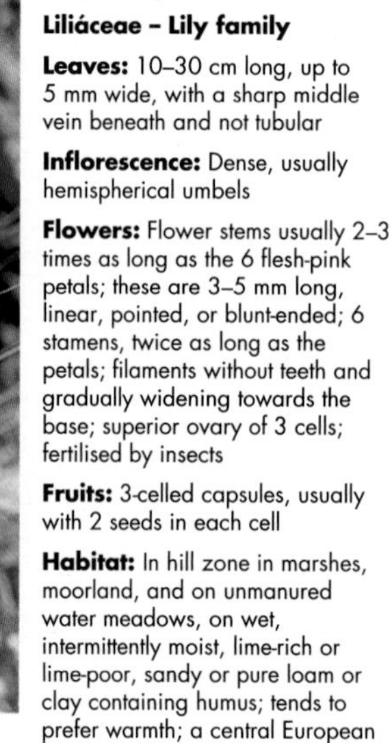

498 **Fragrant garlic – A. suaveólens**
Plants, 20–55 cm high, perennials with cylindrical bulbs; flowering from July to September

Allium suaveólens Jacq.
Fragrant garlic

Liliáceae – Lily family

Leaves: 10–30 cm long, up to 5 mm wide, with a sharp middle vein beneath and not tubular

Inflorescence: Dense, usually hemispherical umbels

Flowers: Flower stems usually 2–3 times as long as the 6 flesh-pink petals; these are 3–5 mm long, linear, pointed, or blunt-ended; 6 stamens, twice as long as the petals; filaments without teeth and gradually widening towards the base; superior ovary of 3 cells; fertilised by insects

Fruits: 3-celled capsules, usually with 2 seeds in each cell

Habitat: In hill zone in marshes, moorland, and on unmanured water meadows, on wet, intermittently moist, lime-rich or lime-poor, sandy or pure loam or clay containing humus; tends to prefer warmth; a central European plant; gregarious; found as far south as the Trieste region

499 **Round-headed leek – A. sphaerocéphalum**
Plants, 20–55 cm high, perennials with spherical bulbs; flowering in June and July

Allium sphaerocéphalum L.
Round-headed leek

Liliáceae – Lily family

Leaves: 30–80 cm long, semicircular in cross-section and with a deep groove on the upper side

Inflorescence: Spherical umbels with 2 small bracts

Flowers: Flower stem short or up to twice as long as petals; 6 petals, purple, 4–5 mm long, with central ridge and blunt-ended; 6 stamens, protruding beyond the petals when flower is open and with visible red anthers; superior ovary of 3 cells; fertilised by insects or self-fertile

Fruits: Thin-skinned capsules in 3 compartments

Habitat: In hill and mountain (rarely subalpine) zones, in scattered situations on rocky slopes in dry meadow land on warm, dry, base-rich, loose, lime-rich or lime-poor, sandy or loamy soils in hot places; this Mediterranean plant's range extends north to Belgium, east to Iran, and south to North Africa and Asia Minor; found in regions with little rainfall

500 **Asarabacca – A. europaéum**
Plants 5–10 cm high, with upward curving stems, with downy hairs; flowering in April and May

Asarum europaéum L.
Asarabacca

Aristolochiáceae – Birthwort family

Leaves: Round or kidney-shaped, heart-shaped at base, 2–8 cm in diameter, leathery, on long stalks and dark green; stalk has downy hair

Inflorescence: Solitary flowers, sessile or on short stems

Flowers: 3 petals, fused together, 1–1.5 cm long, bell-shaped, greenish outside and brown or dark purple inside, narrowing slightly below point where petals join, and hairy, especially outside; the 3 petals each have a point bent inwards; 12 stamens in 2 rows; inferior ovary with 6 cells; pollinated by small flies or beetles, but more often self-fertile

Fruits: Spherical capsules with 2–3 seeds in each compartment; dispersal by ants

Habitat: In hill and mountain zones in deciduous or coniferous woodland with good ground cover, on damp or moist neutral loam or clay containing lime, nutrients, and humus; an indicator of loam; plant good for binding soil together

501 **Flowering rush – B. umbellátus**
Plants 50–120 cm high, perennials with horizontal rhizomes; flowering in June and July

Butómus umbellátus L.
Flowering rush

Butomáceae – Flowering rush family

Leaves: Numerous, forming a basal rosette; 20–20 cm long, 6 –10 mm wide at base, linear, narrowing from a sheathed base, triangular in cross-section in lower part, flat in upper part, and dark or almost olive green

Inflorescence: Umbel of 20–50 flowers with stems of varying lengths; generally with a whorl of 3 narrow lanceolate bracts, 2–3 cm long

Flowers: 2–3 cm in diameter; 6 petals; inner petals oval, pink and white, with dark red veins; outer petals narrow triangular, dark red, with light margins and a little shorter than inner ones; 9 stamens in 3 rings; superior ovary

Fruits: 6 or 9 follicles up to 10 mm long, opening in the middle

Habitat: In hill zone in stagnant or slowly moving water with restricted movement, in ditches and on riverbanks; lately plant has begun to spread in reservoirs

502 **Field poppy – P. rhoéas**
Plants 20–70 cm high, erect or ascending, branching; flowering from May to September

Papáver rhoéas L.
Field poppy

Papaveráceae – Poppy family

Leaves: Doubly pinnate and on stalks in lower part of the plant, singly pinnate and sessile or only short-stalked in middle and upper parts; end lobes toothed; tips pointed

Inflorescence: Solitary flowers at end of stem, nodding before flowering starts

Flowers: 2 sepals, oval, blunt-ended, hairy, entire, enclosing the petals and dropping away on flowering; 4 petals, 2–4 cm long and red; many stamens; filaments dark red and thread-like; superior ovary with many ovules

Fruits: Smooth capsules, 10–20 mm long, with 8–18 stigmatic rays and usually as many indistinct longitudinal lines, opening by pore-like valves below disk

Habitat: In hill and mountain zones on waste places, in cornfields, by roadsides and railways, on well-drained to fairly dry neutral loam which is warm in summer, base-rich, and contains nutrients and humus

503 **Round-leaved penny cress – T. rotundifólium** Plants 5–12 cm high, perennials with long creeping stems; flowering in June and July

Thlaspi rotundifólium (L.) Gaudin ssp. rotundifolium
Round-leaved penny cress

Brassicáceae (Cruciferae) – Mustard family

Leaves: Opposite or alternate on sterile stems; on flower-bearing stems basal leaves form a rosette and leaves in upper part are generally alternate; leaves oval or round, with wide stalks on lower part of stem, in upper part sessile; no hairs, blue-green, and entire or slightly toothed

Inflorescence: Many-flowered umbel with greenish-red flower stems

Flowers: 4 sepals, 2–5 mm long, reddish-green, smooth, with white margins; 4 petals, up to 10 mm long, lilac, with rather dark veins and rounded or slightly incised at the end; 6 stamens; superior ovary; fertilised by insects or self-fertile

Fruits: Pods 3–8 mm long

Habitat: In subalpine and alpine zones in scree on damp, loose, mobile, generally coarse limestone debris which contains little humus or fine soil; found in the southern Carpathians, in the north and south Alps, and in the Bergamese Alps

504 **Perennial honesty – L. redivíva**
Plants 30–130 cm high, perennials with creeping root stocks; flowering in May and June

Lunária redivíva L.
Perennial honesty

Brassicáceae (Cruciferae) – Mustard family

Leaves: On stalks, up to 25 cm long, narrow heart-shaped, entire, toothed, and hairy; lower leaves almost opposite; upper leaves always alternate

Inflorescence: Many-flowered upright racemes

Flowers: Long stems with close hairs; 4 sepals with short hairs and 46 mm long; 4 petals, narrow oval, up to 20 mm long and white, lilac, or pale violet; 6 stamens, the 2 outer ones short and the 4 inner longer; ovary superior and 2-celled, the cells separated by a central wall; 2 stigmata; fertilised by moths and bees

Fruits: Shining white disc-like pods 3–9 cm long, 1 to 3 cm wide, narrow oval, on stems

Habitat: In mountain (rarely hill or subalpine) zone in humid conditions on steep wooded slopes and in shady woods in ravines, on damp or moist, base-rich, loose scree which contains nutrient, humus, and fine soil

505 **Honesty – L. ánnua**
Plants 20–90 cm high, with thick root stock, usually biennial; flowering in April and May

Lunária ánnua L.
Honesty

Brassicáceae (Cruciferae) – Mustard family

Leaves: Narrow oval, pointed, with rounded base and widely spaced fine teeth; lower leaves usually opposite, with stalks; upper leaves alternate and sessile

Inflorescence: Many-flowered upright racemes

Flowers: Long stems with short upright hairs; 4 sepals with short hairs, 7–12 mm long and with 2 appendages at the base on the inner side; 4 petals, 10–25 mm long and purple; 6 stamens, the 2 outer ones short and the 4 inner longer; ovary superior and 2-celled, the cells separated by a central wall; 2 stigmata; fertilised by moths and bees

Fruits: Oval or round translucent pods 4–4.5 cm long, with long stems; seeds 4–8 mm in diameter

Habitat: In hill zone in warm places on rubbish dumps, by the roadside, on bushy slopes, as an ornamental plant in country gardens, and wild in waste places; originating in south-eastern Europe and widely distributed as an escape from gardens

506 **Codlins and cream – E. hirsútum**
Plants 40–150 cm high, branching and covered with glandular hairs; flowering from June to Sept.

Epilóbium hirsútum L.
Codlins and cream

Onagráceae (Oenotheráceae) – Evening primrose family

Leaves: Narrow oval to lanceolate, 6–15 cm long, up to 3 cm wide, with many forward-pointing teeth 0.5–1 cm long, and usually semi-amplexicaul

Inflorescence: Corymbose racemes terminating main stem and branches

Flowers: On long stems; 4 sepals lying close to the petals, 8–10 mm long, gradually coming to points and with spines at the tips; 4 petals 10–18 mm long, entire, notched at the ends and purplish-rose; 8 stamens; inferior ovary; fertilised by insects

Fruits: Capsules splitting longitudinally; seeds have hairy plumes; on ripening fruits split from top to bottom and the 4 parts of the capsule wall are flung outwards; seeds dispersed by wind

Habitat: In hill and mountain zones in damp woodlands, ditches, weed beds, stream-banks, springs and scrub, on wet, base-rich clay containing nutrients and humus; plants grow in light and half-shade; a soil-binding plant; range extends north to Scotland and Scandinavia

507 **Broad-leaved willow-herb – E. montánum**
Plants 10–100 cm high, without runners and with few or no branches; flowering from June to August

Epilóbium montánum L.
Broad-leaved willow-herb

Onagráceae (Oenotheráceae) – Evening primrose family

Leaves: Narrow oval, generally rounded at base of leaf proper, pointed, with sharp or slightly rounded tips, 3–10 cm long, unequally toothed with teeth up to 1 mm long

Inflorescence: Terminal leafy racemes

Flowers: On long stems; 4 sepals, 3–5 mm long and pointed at the tips; 4 petals, 7–12 mm long, entire, distinctly notched at the tip, pale pink with deeper pink stripes on top; 8 stamens in 2 rings; inferior ovary; usually self-fertile

Fruits: Capsules splitting longitudinally, with close simple hairs and erect glandular hairs; seeds have hairy plumes; dispersal by wind

Habitat: From hill to subalpine zones in deciduous and coniferous mixed woodland with good ground cover, in hedgerows, gardens and parkland, by woodland paths and in clearings, on well-drained stony or loamy soils containing nutrient sand humus, in shady places; plant grows on rubbish dumps

508 **Subalpine willow-herb – E. alpéstre**
Plants 25–80 cm high, with short rhizomes, without branches or runners; flowering from June till August

Epilóbium alpéstre (Jacq.) Krocker
Subalpine willow-herb

Onagráceae (Oenotheráceae) – Evening primrose family

Leaves: In middle and upper parts of the plant in whorls of 3 (or less usually 4); in lower part usually in opposite pairs; leaves broad lanceolate to narrow oval, 4–10 cm long, widest towards base, rounded at base, sessile or on short stalks, and distinctly toothed

Inflorescence: Terminal leafy racemes

Flowers: 4 sepals, lying close to petals, 2–5 mm long, lanceolate, becoming narrower towards the tips, light green with tips often reddish, and distinctly covered with white hairs; 4 petals, up to 10 mm long, deeply notched, reddish, often with dark veins; 8 stamens with pale yellow anthers; inferior ovary; usually self-fertile

Fruits: Capsules, lightly covered with hair and splitting along their length

Habitat: In mountain and subalpine zones, in shrubby undergrowth, by camp-sites and in alder thickets, on wet to well-watered soils containing nutrients and generally lime

509 **Rose-bay willow-herb – E. angustifólium**
Plants 40–140 cm high, with long, horizontally spreading roots; flowering from June to August

Epilóbium angustifólium
Rose-bay willow-herb

Onagráceae (Oenotheráceae) – Evening primrose family

Leaves: Lanceolate, 5–15 cm long, up to 2 cm wide, narrowing gradually to rounded bases, pointed, with edges turned downwards, smooth, dark green above, blue-green beneath, with short stalks and prominent main and lateral veins

Inflorescence: Many-flowered upright terminal raceme; solitary flowers in leaf axils just below flower head

Flowers: 4 sepals, narrow lanceolate, often as long as petals, and usually dark red; 4 petals, 10–15 mm long, round or broad oval, rounded or slightly notched at the tips, spread out flat and purplish-red; 8 stamens; inferior ovary

Fruits: Capsules, lightly covered with hair and splitting along their length

Habitat: From hill to subalpine zones on waste ground, in scree, by the roadside, on river banks, in ground cover, on well-drained loam containing nutrients and humus; deep-rooting; range extends north to northern Norway and southern Finland

510 **Alpine willow-herb – E. fleischeri**
Plants 20–50 cm high, with many upright curving stems; flowering in July and August

Epilóbium fleischeri Hoechst
Alpine willow-herb

Onagráceae (Oenotheráceae) – Evening primrose family

Leaves: Linear to lanceolate, 3–5 cm long, up to 5 mm wide, smooth, with many tooth-like glands on margin, which is rarely rolled down; clearly visible middle vein but lateral veins hardly visible

Inflorescence: Usually solitary flowers springing from leaf axils

Flowers: On long stems; 4 sepals, narrow lanceolate, usually as long as petals, and brownish-red; 4 petals, up to 2cm long, oval, blunt-ended, and deep pink; 8 stamens; inferior ovary; pistil covered with hairs up to or above the middle; fertilised by insects

Fruits: Capsules, lightly covered with hair and splitting along their length; numerous seeds with hairy plumes; dispersal by wind

Habitat: In subalpine and alpine (rarely mountain) zones on moraines, scree, in gravel beds of mountain streams, on seasonally dry, base-rich, usually lime-poor, loose sand and gravel poor in fine soil; an early coloniser of alluvial sand; plant spreads by creeping roots and holds soil together; a fairly common alpine plant

511 **Lady's smock – C. praténsis**
Plants 15–60 cm high, branching freely; flowering from April to July

Cardámine praténsis L.
Lady's smock, cuckoo flower

Brassicáceae (Crucíferae) – Mustard family

Leaves: Basal leaves form a rosette and are unequally pinnate, with 2–7 pairs of leaflets and a larger end one; stem leaves have 7, 9, 11, or 13 narrow oval leaflets

Inflorescence: Terminal raceme with long-stemmed flowers

Flowers: 4 sepals, 2.5–4 mm long, and pale yellow or pale green; 4 petals, 7–14 mm long, narrow oval and violet, lilac, pink, or white; 2 outer, shorter stamens, 4 inner, longer ones; superior ovary with visible sheath; fertilised by insects

Fruits: Many-seeded, rod-shaped pods; ripe pods open suddenly by valves spiralling from base, propelling seeds some distance

Habitat: In hill and mountain zones in damp meadows and pastures, wet woodland, and by streams, on moist, slightly acid to neutral clay or loam containing nutrients and humus; also found on peat; an important indicator of fertile soil; range extends north to southern Sweden; also naturalised in USA

512 **Pyrenean rock beauty – P. pyrenáica**
Plants forming mats, perennials with thin branching rhizomes; flowering in June and July

Petrocállis pyrenáica (L.) R.Br.
Pyrenean rock beauty

Brassicáceae (Crucíferae) – Mustard family

Leaves: All in basal rosettes, wedge-shaped, with 3 (rarely 5) leaflets, 5–9 mm long, and hairy

Inflorescence: Solitary flowers or racemes with few flowers; stem and flower head practically identical

Flowers: 4 stamens, 2–3 mm long, narrow oval and reddish; 4 petals, 4–6 mm long, broad oval, rounded at the tips, pale lilac; 2 outer, shorter stamens, 4 inner, longer ones protruding clearly beyond the edge of the corolla; superior ovary, with sheath

Fruits: Pods 4–6 mm long, with networks of veins and short beaks; hairy stems

Habitat: In alpine (rarely subalpine) zone on scree, in rock crevices, stony slopes, on sunny limestone or dolomite poor in fine soil; found especially in the northern alpine ranges; not found in Switzerland on southern side of Alps; a central and southern European mountain plant found in the Pyrenees, Alps, and Carpathians

513 **Great burnet – S. officinális**
Plants 30–90 cm high, perennials, generally smooth; flowering from June to September

Sanguisórba officinális L.
Great burnet

Rosáceae – Rose family

Leaves: Pinnate; 7, 9, 11, 13, or 15 leaflets, narrow oval, rounded at tips or with fine points, on stalks, up to 5 cm long, with coarse pointed teeth, dark green above and blue-green beneath; basal leaves form a rosette; stem leaves have only a few leaflets

Inflorescence: Spherical or cylindrical dense raceme 1.5–5 cm long

Flowers: Each flower has 3 bracteoles; 4 petals, oval, up to 4 mm long and deep red; 4 stamens;, as long as the petals; single carpel with terminal style; fertilised by insects

Fruits: Single seed enclosed by hard calyx

Habitat: From hill to subalpine zones in damp grassland and by roadsides, on permanently or intermittently damp neutral or slightly acid loam or clay containing more or less nutrients and humus; deep-rooted; an indicator of moisture

514 **Hollow fumitory – C. cava**
Plants10–30 cm high, with spherical hollow tubers, flowering in March and April

Corydális cava (Mill.)
Hollow fumitory

Fumariáceae – Fumitory family

Leaves: On stalks, blue-green, ternate, with leaflets also on stalks; leaflets ternate with lobes divided and toothed

Inflorescence: Many-flowered racemes

Flowers: In axils of oval, entire bracts and 1.5–3 cm long; sepals very small; 4 petals, purple or rarely white; of 2 outer petals the upper one has a spur pointing backwards and curved down at tip, with the front widening and bending upwards, the lower petal widening at the front and pointing downwards; inner 2 petals are of similar form, joined at the front; 4 stamens; superior ovary of 2 cells; pollinated by bees

Fruits: Explosive capsules 2–2.5 cm long, holding many seeds

Habitat: In hill and mountain zones in parkland, beech and oak woods with good ground cover, wet woodland, orchards, and vineyards, on well-drained to moist base-rich loam containing nurients and humus and with deep topsoil; a good indicator of fertility and loam

515 **Shy stonecrop – S. anacámpseros**
Plants 10–25 cm high, perennials with upright stems; flowering from July to September

Sedum anacámpseros L.
Shy stonecrop

Crassuláceae – Stonecrop family

Leaves: 1–3.5 cm long, oval (often widest towards tip), flat, entire, blunt-ended or rounded, sessile, alternate, fleshy, with a spur-like stipule at the base; terminal leaves of sterile stems thickly bunched

Inflorescence: Many-flowered hemispherical cymes

Flowers: 5 sepals, 3–5 mm long, narrow oval, pointed, often with red spots on a pale green ground, and with tip bent inwards like a hook; 5 petals, up to 8 mm long, pointed, with greenish central ridge and otherwise dirty red or pink; red spots or longitudinal stripes on the upper side; 10 stamens in 2 rings; superior ovary

Fruits: Many-seeded follicles springing open along the inner side

Habitat: In subalpine and alpine zones on rocks, scree, and on rocky slopes; avoids lime; usually found on acid stony subsoil

516 **Soapwort – S. officinális**
Plants 30–70 cm high, perennials with creeping rhizomes; flowering from July to September

Saponária officinális L.
Soapwort

Caryophylláceae – Pink family

Leaves: 5–15 cm long, oval or lanceolate, pointed, entire, and alternate

Inflorescence: Compact terminal corymbs on main stem and branches

Flowers: Sepals 1.5–2.5 cm long, fused, smooth or with fine hairs and unequal teeth; 5 petals, pink, pale pink, or rarely white, narrow oval (broadest towards tips), rounded or slightly incised at tips and with 2 small blunt coronal scales at bases; 10 stamens; superior ovary

Fruits: Many-seeded capsules up to 20 mm long, opening by teeth

Habitat: In hill (rarely mountain or subalpine) zone by roadsides, ditches, gravel banks, waste ground, river banks, and in weed beds, on moderately dry to well-drained, moderately acid, stony, gravelly, or sandy soil containing nutrients and humus, in warmer locations; formerly used as a medicinal herb; root runners used in the manufacture of soap

517 **Cow-cockle – V. pyramidáta**
Plants 30–60 cm high, annuals with tap-roots; flowering in June and July

Vaccária pyramidáta Medicus
Vaccária hispánica (Mill.) Rauschert
Cow-cockle

Caryophylláceae – Pink family

Leaves: 5–10 cm long, cordate or lanceolate, sessile and fused at the base, entire, smooth, pointed, and blue-green

Inflorescence: Loose dichasia with long-stemmed flowers

Flowers: 5 fused sepals, 10–17 mm long, usually toothed with red margins, calyx tube with 5 sharp angles and inflated, particularly when ripe; 5 petals, pink, oval (broadest towards tips) clearly incised at tips, 15–20 mm long and with 2 transparent coronal scales on claws; 10 stamens; superior ovary; fertilised by butterflies or self-fertile

Fruits: Many-seeded capsule up to 1 cm long

Habitat: In hill and mountain zones on waste ground, in cornfields and arable land in warm places on dry, lime-rich, stony loam or clay containing more or less humus; roots up to 50 cm deep; not often found

518 **Houseleek – S. tectórum**
Plants 30–60 cm high, perennials with creeping rhizomes; flowering from July to September

Sempervívum tectórum L.
Houseleek

Crassuláceae – Stonecrop family

Leaves: Rosette leaves sharply pointed, greenish to pale red, usually smooth but distinctly hairy on the edges; stem leaves usually come to gradual points, green, often dark red at the tips beneath, and distinctly hairy at the edges

Inflorescence: Cymous panicle, generally with 30 or more flowers

Flowers: 12–16 sepals, usually fused at base; 12–16 petals, purplish-pink or dull red, 8–12 mm long, with distinct hairs and coming to fine points; twice as many stamens as petals, in 2 rings and with purplish-pink stamens; 12–16 free-standing carpels in a ring

Fruits: Many-seeded follicles

Habitat: From mountain to alpine zones in rocky areas and dry meadows in sunny places on lime-free substrates (also on dry silica rocks); a central and southern European mountain plant; has been planted for centuries on old walls, roofs, etc.

519 **Cobweb houseleek – S. arachnoídeum**
Plants 5–20 cm high, perennials with small rosettes; flowering from June to August

Sempervívum arachnoídeum L.
Cobweb houseleek

Crassuláceae – Stonecrop family

Leaves: Rosette leaves lanceolate, with reddish-brown tips and, in plants in sunny places, with a thick covering of hair; stem leaves narrow oval, coming to gradual points, rounded at bases, sessile, fleshy, covered with glands and yellowish or whitish

Inflorescence: Single umbels or cymous panicles

Flowers: 6–12 sepals, usually fused and half as long as petals, which are narrow oval, short, with sharp points, pale red to carmine, usually with deep red stripes and glandular hairs at the edges; twice as many stamens as flowers in 2 rings; free-standing carpels in a ring; fertilised by insects

Fruits: Many-seeded follicles

Habitat: In all zones on tops of boulders, in siliceous rough pasture, scree, and rocky fields, on dry, base-poor, lime-free soils which are warm in summer and contain few nutrients or fine soil

520 **Entire-leaved primrose – P. integrifólia** Plants 3–6 cm high, perennials with short vertical rhizomes; flowering in June and July

Prímula integrifólia L. em. Gaudin
Entire-leaved primrose

Primuláceae – Primrose family

Leaves: Elliptical, usually entire, up to 2.5 cm long and 1 cm wide, narrowing gradually towards the base, blunt-ended or round at the tip, with many short transparent glandular hairs on the edge and without a covering of mealy dust

Inflorescence: Erect umbels of 1–3 flowers

Flowers: 5 sepals, fused for much of length, up to 9 mm long and brownish-red; tips generally round; calyx and flower stem covered in colourless glandular hairs; 5 petals with fused claws, red-violet, with dark ring at the throat, with glandular hairs and distinctly incised; lobes 5–10 mm long and conical; 5 stamens; superior ovary

Fruits: Capsules 4–6 mm long

Habitat: Usually in alpine (rarely subalpine) zone on moorland, wet hollows, well-watered rough pasture, at the edge of snow valleys, on fairly moist acid ground rich in fine soil and covered by snow for long periods; found in Vorarlberg (Austria) and Switzerland

521 **Purple loosestrife – L. salicária** Plants 30–150 cm high, perennials, woody at the base; flowering in July and August

Lythrum salicária L.
Purple loosestrife

Lythráceae – Lossestrife family

Leaves: 5–15 cm long, lanceolate, entire, rounded or notched at the base, sessile and pointed at the tip; veins prominent; leaves in lower parts of plant opposite or in whorls of 3, in middle and upper parts opposite or alternate

Inflorescence: Spikes over 10 cm long, with flowers arranged in whorls in the axils of bracts

Flowers: Calyx tube 4–7 mm long, with 12 ridges and 6 teeth 2–3 mm long; 6 petals, deep purple (rarely white), lanceolate and 6–12 mm long; 12 stamens; superior ovary; fertilised by insects

Fruits: Many-seeded capsules, 3–6 mm long, enclosed in the calyx

Habitat: In hill (rarely mountain) zone in wet meadows rich in vegetation, ditches, moorland, reed-swamps at margins of lakes and slow-moving rivers and fens and marshes, on moist to wet, base-rich, moderately acid loam or clay soil containing nutrients and humus; deep roots

522 **Red campion – S. dioíca** Plants 20–90 cm high, with thick hair covering, one or more flowers in each head; flowering April to September

Siléne dioíca (L.) Clairv.
Red campion

Caryophylláceae – Pink family

Leaves: Basal leaves oval to broad lanceolate; stem leaves oval, entire, pointed, 5–10 cm long, hairy, and sessile

Inflorescence: Many-flowered loose terminal dichasia

Flowers: Scentless; 5 fused sepals, 10–15 mm long, hairy and brownish-red; teeth of calyx tube come to sharp points; 5 petals, incised from the tip almost to the middle, with rounded segments, pale purple and 10–25 mm long; 2 narrow scales up to 2 mm long at throat; flowers of one sex only, plant being dioecious; male flowers only have 10 stamens; female flowers have superior ovaries with 5 styles; plant fertilised by bumble-bees

Fruits: Capsules 8–14 mm long, opening by 10 teeth

Habitat: From hill to subalpine zones in mixed deciduous woodland, damp meadows, scrubby undergrowth, and by camp sites, on well-drained to moist, base-rich, loose clay or loam containing nutrients and humus; deep roots

523 **Flower of Jove – L. flos-jóvis** Plants 20–60 cm high, perennials, thickly covered with white hairs; flowering in June or July

Lychnis flos-jóvis (L.) Desr.
Siléne flos-jóvis Clairv.
Flower of Jove

Caryophylláceae – Pink family

Leaves: Narrow oval to elongated, 5–10 cm long, pointed, entire, covered in white hairs, sessile, and opposite

Inflorescence: Many-flowered terminal dichasia

Flowers: Calyx up to 15 mm long, covered with white woolly hairs, usually with 10 ridges darker in colour, and fine pointed teeth; 5 petals, 15–30 mm long, purple above, notched, with rounded segments, and scales up to 3 mm long, often toothed at throat; 10 stamens; superior ovary with 5 styles

Fruits: Capsules 10–15 mm long, on short stems and opening by 5 teeth; seeds up to 1 mm in diameter, flattened, dark-coloured, and with a granular surface

Habitat: In mountain and subalpine zones in sparse woodland and thickets, on loose, lime-poor, usually rather dry soils in warmer places

524 **Rock-rose – Heliánthemum hybrid** Plants 10–30 cm high, with glandular hairs, forming smallish mats; flowering from June to Aug.

Heliánthemum hybrid
Rock-rose (garden variety)

Cistáceae – Rock-rose family

Leaves: Narrow oval, on short stems, generally blunt-ended and opposite; stipules lanceolate and pointed

Inflorescence: Raceme-like cymes, frequently unilateral

Flowers: 5 sepals, of unequal size, the 3 inner being larger than the 2 others, narrow oval and generally mid-green; 5 (or 6) petals, wide oval, irregularly wavy at the edge, pale or dark yellow towards tip and deep yellow or reddish brown at the base; numerous stamens; 3 carpels, fused into a superior ovary; fertilisation by insects

Fruits: Many-seeded compartmented capsules; seeds mostly egg-shaped with indistinct ridges

Habitat: In hill and mountain zones, planted in gardens and parks; the genus includes over 80 species, mainly from the Mediterranean region, but with some from North and Central Europe. The garden rock-roses (illustrated here), with flowers of various colours, are hybrids derived from H. chamaecístus, H. appenínum, and allied species

525 **Glacier pink – D. glaciális** Plants 2–6 cm high, perennials with rhizomes giving rise to several stems, with a sterile leaf rosette; flowering till August

Diánthus glaciális Haenke
Glacier pink

Caryophyllaceae – Pink family

Leaves: Linear-lanceolate, 2–5 cm long, coming to a short point or blunt-ended, and sessile; leaf sheaths up to twice as long as the width of the leaf

Inflorescence: Solitary terminal flowers

Flowers: 5 sepals, fused into a calyx tube, up to 15 mm long and smooth; base of calyx enclosed by an epicalyx of 1–3 pairs of scales (bracteoles) at least as long as the corolla tube and gradually narrowing to points; 5 petals, broad oval, up to 10 mm long, pale violet to purple above, with dark red streaks and spots towards the throat, and distinctly toothed at the tip; 10 stamens; superior ovary with 2 styles; fertilised by insects

Fruits: Capsules opening by teeth

Habitat: In alpine zone in grassland and on ridges exposed to wind, on stony calcareous soils; the range of this plant, originating in the eastern Alps and Carpathians, extends west to the Tirol, Graubuenden, and into the Veltlin; rather a rare species

526 **Saxifrage tunic-flower – P. saxifrága**
Plants 10–25 cm high, perennials with short root stocks, usually smooth; flowering from June to Sept.

Petrorhágia saxifrága (L.) Link
Túnica saxifrága Scop.
Saxifrage tunic-flower

Caryophylláceae – Pink family

Leaves: 5–10 mm long, narrow lanceolate, entire, pointed, and opposite

Inflorescence: Loose panicles

Flowers: 5 sepals fused into bell-shaped calyx tube, greenish, 4–6 mm long, smooth, surrounded by 4 membranous bracts; calyx teeth about half length of tube, pointed and often reddish at tips; 5 petals, pale lilac or pink, oval (broadest towards tips), notched at tips, 6–10 mm long and with violet stripes towards base; no coronal scales; 10 stamens; superior ovary; fertilised by insects

Fruits: Many-seeded capsules up to 6 mm long

Habitat: In hill and mountain (rarely subalpine) zones on dry stony grassland, on warm, dry, lime-rich, neutral stony, sandy, or gravelly soil containing humus but no fine soil; a south European plant, with range extending north to central France and the French Jura

527 **Carthusian pink – D. carthusianórum**
Plants 10–40 cm high, perennials with branching rhizomes and sterile leaf rosettes; flowering until Oct.

Diánthus carthusianórum L.
Carthusian pink

Caryophylláceae – Pink family

Leaves: Narrow linear, at most 2 mm wide, coming to fine points, slightly drooping, opposite, and blue-green

Inflorescence: Flowers solitary or in small groups at end of stems; bracts lanceolate, awned and brown or yellowish-brown

Flowers: Generally 5 sepals, up to 18 mm long and smooth; bracts of epicalyx ending in awns; 5 petals, lobes oval (broadest towards tips), 5–15 mm long, dark purple, entire at sides but irregularly toothed at tip and with isolated hairs; 10 stamens, with dark violet anthers; superior ovary; fertilised by day-flying moths

Fruits: Capsules opening by 4 teeth; seeds up to 2.5 mm long

Habitat: From hill to subalpine zones in sparse woodland, dry meadows, calcareous rough pasture, on rocks, at the edges of woods and thickets, on fairly dry, warm, base- and lime-rich, preferably stony, loose loam or loess containing humus

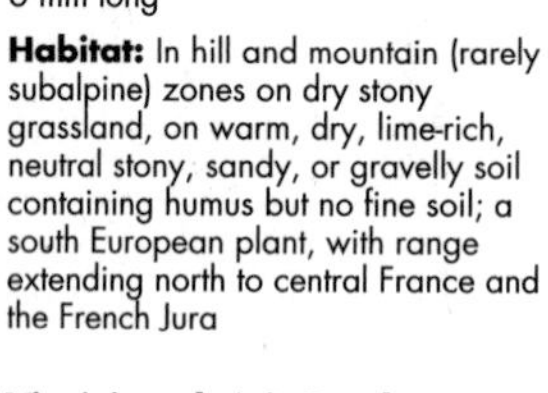

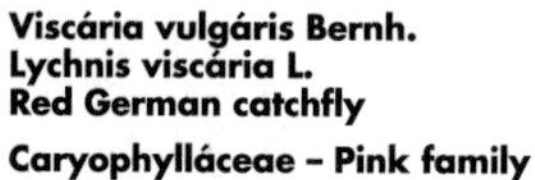

528 **Red German catchfly – V. vulgáris**
Plants 20–60 cm high, perennials with basal leaf rosettes; flowering from May till July

Viscária vulgáris Bernh.
Lychnis viscária L.
Red German catchfly

Caryophylláceae – Pink family

Leaves: 1–4 cm long, narrow lanceolate, entire, pointed and opposite

Inflorescence: Spike-like panicle of axillary cymes in whorls

Flowers: Short-stemmed; 5 sepals fused into calyx tube, reddish, 10–15 mm long, smooth, with 10 ridges; calyx teeth very short and pointed; 5 petals, spathulate, rounded (more rarely notched) at tips, and 15–20 mm long; coronal scales up to 3 mm long at the throat; 10 stamens; superior ovary; 5 styles; fertilised by day-flying moths

Fruits: Capsules up to 7 mm long, on stems, opening by 5 teeth; seeds up to 0.5 mm in diameter and dark-coloured

Habitat: In hill (rarely mountain) zone in sparse woodland, rough grassland, and heathland, on dry, lime-poor, sandy loam moderately base-rich and rich in nutrients, neutral to fairly acid and containing humus

529 **Corn cockle – A. githágo**
Plants 20–80 cm high, annuals with branching stems; flowering from June to August

Agrostémma githágo L.
Corn cockle

Caryophylláceae – Pink family

Leaves: 2–5 cm long, very narrow lanceolate, entire, pointed, and opposite; leaves lack bracts

Inflorescence: Long-stemmed solitary flowers

Flowers: 5 sepals, fused into a calyx tube 3–5 cm (rarely as much as 10 cm) long, hairy, with calyx teeth longer than the tube; 5 petals, oval, 2–4 cm long, dull purple, whitish towards the base, with dark red stripes, slightly notched at the ends, without coronal scales and shorter than the calyx teeth; 10 stamens; superior ovary with 5 styles and hairy stigmata; fertilised by butterflies and bees

Fruits: Many-seeded capsules; seeds up to 3.5 mm in diameter and poisonous

Habitat: In hill and mountain zones in waste places, cornfields and warmer places on well-drained to dry, rather acid, sandy or pure loam with humus; lately decreasing with use of selective weed-killers

530 **Deptford pink – D. arméria**
Plants 25–45 cm high, upper parts covered in rough hair; without sterile shoots; flowering June till Aug.

Diánthus arméria L.
Deptford pink

Caryophylláceae – Pink family

Leaves: Narrow lanceolate, pointed, rough-haired, opposite, and green or blue-green

Inflorescence: Short-stalked cymose clusters with 2 – 10 flowers; surrounded by narrow lanceolate bracts

Flowers: 5 sepals, up to 20 mm long, thickly set with short hairs and pointed; 5 petals; each lobe 3–6 mm long, narrow oval (broadest towards tip), irregularly toothed towards the tip, purple above, with numerous white dots and spots, darker towards the throat, with isolated pale hairs; 10 stamens with dark purple anthers; superior ovary; fertilised by day-flying moths

Fruits: Capsules opening by 4 teeth; seeds 1–1.5 mm long

Habitat: In hill (rarely mountain) zone in dry meadows, thickets, on roadsides, at margins of woods, and on waste ground, on moderately well-drained to fairly dry pure or rather sandy loam moderately base-rich and rich in nutrients and containing humus; plant likes warmth

531 **Mountain thrift – A. alpína**
Plants 10–20 cm high, perennials, with tap-roots and leaf rosettes; flowering in July and August

Arméria alpína Willd.
Státice montána Miller
Mountain thrift

Plumbagináceae – Thrift family

Leaves: Narrow linear, with 1–3 veins and parallel edges, pointed, 1–3 mm wide, narrowing a little towards the base, and smooth

Inflorescence: Flower head 2–3 cm in diameter, with many flowers, with blunt or pointed outer bracts and blunt inner bracts; flower stems 2–3 mm long, and white or green

Flowers: 5 sepals fused into a calyx tube, ribbed, with thick hairs on the 10 veins, whitish and translucent; 5 petals, fused only at base; lobe narrow oval (broadest towards tip), pink, often with dark red stripes; 5 stamens, protruding beyond petals; superior ovary of 5 cells

Fruits: Dry achenes opening in a ring at the base

Habitat: In alpine (rarely subalpine) zone, scattered in scree and rough pasture, on stony, lime-poor soils in warmer places; a mid and south European mountain plant found in the Alps and Pyrenees

532 **Ragged Robin – L. flos-cúculi**
Plants 20–80 cm high, perennials without branching stems; flowering from May to August

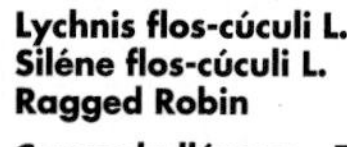

Lychnis flos-cúculi L.
Siléne flos-cúculi L.
Ragged Robin

Caryophylláceae – Pink family

Leaves: Narrow linear, 4–8 cm long, entire, generally rounded at the tips, opposite; there are both stem leaves and basal rosettes

Inflorescence: Flowers on stems in terminal and lateral dichasia

Flowers: 5 sepals fused into a calyx tube 5–9 mm long, with pointed triangular teeth and 10 dark red ridges; 5 petals, pink (rarely entirely white), 10–25 mm long and deeply incised into 4 segments, each narrow linear, pointed or blunt-ended; coronal scales up to 3 mm long at the throat, often with teeth; 10 stamens; superior ovary with 5 styles

Fruits: Capsules up to 8 mm long, opening by 5 teeth; seeds up to 1 mm in diameter, dark and with granular surfaces

Habitat: In hill and mountain zones in damp meadows, marshes, fens, moorland, and wet woods, on seasonally moist to wet, moderately acid loam or clay containing nutrients and humus

533 **Superb pink – D. supérbus**
Plants 20–60 cm high, perennials with thin rhizomes; flowering from June to September

Diánthus supérbus L.
Superb pink

Caryophylláceae – Pink family

Leaves: Narrow lanceolate, 3–5 cm long (rarely up to 15 cm), entire, coming to fine points, and opposite; leaf sheaths 1–2 times as long as width of leaf

Inflorescence: Flowers on stems, either solitary or in sparse racemes

Flowers: 5 sepals, fused into a calyx tube 20–30 mm long; bracts of epicalyx a third or a quarter length of calyx tube, with short awns; 5 petals, deeply incised more than halfway,

10–35 mm long, becoming greenish towards the throat, with dark hairs, and lilac or pale purple above; 10 stamens; superior ovary with 2 styles

Fruits: Capsules opening by 4 teeth; seeds 2–3 mm long, with thicker edges

Habitat: In subalpine (rarely mountain or alpine) zones on meadow land, moors, in sparse copses and by ditches, on seasonally wet or moist, often stony, neutral clay or peat containing nutrients and lime but poor in nitrogen

534 **Shaggy primrose – P. hirsúta**
Perennial plants with short, more or less vertical rhizomes; flowering in June and July

Prímula hirsúta All.
Prímula viscósa Vill.
Shaggy primrose

Primuláceae – Primrose family

Leaves: Oval to round, narrowing abruptly to short winged stalks, coarsely toothed towards the tips, sticky on both sides and whitish to mid-green

Inflorescence: Many-flowered umbels with flower stems 3–10 mm long

Flowers: 5 sepals, pale green and fused into a calyx tube up to 2.5 mm long; 5 petals, fused together and violet-pink, the outspread lobe oval (broadest towards tip), narrowing in a wedge shape, and with the tip distinctly incised; 5 stamens; superior ovary

Fruits: Capsules

Habitat: In subalpine and alpine zones on rocks, grassland, and stable scree, on stony, lime-poor, weakly acidic or basic soils; an Alpine and Pyrenean plant, found especially in central and north part of southern Alps; found in Pyrenees and east to Tauern Alps

535 **Bird's-eye primrose – P. farinósa**
Plants 5–20 cm high, perennials with short rhizomes; flowering from May to July

Prímula farinósa L.
Bird's-eye primrose

Primuláceae – Primrose family

Leaves: 2–8 cm long, up to 2 cm wide, long oval (widest towards tip), gradually narrowing to winged stalks, blunt-ended or rounded at the tips, dark green above, with white or sulphur-coloured meal beneath, smooth and entire or with irregular fine teeth

Inflorescence: Umbels with few or many flowers

Flowers: Bracts, flower stems, and calyx mealy; 5 sepals, fused into calyx tube 3–6 mm long, with teeth up to 2 mm long; 5 petals, fused in the lower region, with lobe clearly notched at the tip, lilac, rose, or pink, with a yellow spot with white edge at the throat

Fruits: Capsules 5–10 mm long

Habitat: In mountain, subalpine, and alpine zones in damp grassy and peaty places on moist to wet, infertile, neutral soil with lime

536 **Moss campion – S. acaúlis**
Plants 1–3 cm high, perennials forming thick, flat mats; flowering from June to August

Siléne acaúlis (L.) Jacq.
Moss campion

Caryophylláceae – Pink family

Leaves: Narrow lanceolate, 4–12 mm long, blunt-ended or pointed and with hairy edges

Inflorescence: Solitary flowers at ends of stems 1–3 cm long, without wings

Flowers: Hermaphrodite, male, and female flowers present; 5 sepals, fused for much of their length, 3–8 mm long, smooth, often reddish and with 10 veins and rounded teeth; 5 fused petals, corolla tube up to 12 mm long and usually white; outspread lobes narrow oval (broadest towards tips), purplish-red and distinctly notched at the tips; 10 stamens, with whitish or reddish filaments and pale yellow anthers; superior ovary; plant fertilised by moths

Fruits: Capsules 6–12 mm long, opening by 6 teeth; seeds up to 1 mm in diameter

Habitat: In alpine (rarely subalpine) zone on scree, ridges, rock debris and cliffs on moderately well-drained, stony, loose soils containing lime

537 **Cyclamen – C. purpuráscens**
Plants 5–15 cm high, with flat or spherical corms; flowering from June to October

Cyclamen purpuráscens Mill.
Cyclamen europaéum L.
Sowbread

Primuláceae – Primrose family

Leaves: Round, kidney-shaped or heart-shaped, 5–15 cm long, pointed or blunt at the tips, smooth, with blunt teeth or notched, dark green above with pale markings, and reddish beneath

Inflorescence: Solitary flowers

Flowers: On long stems and nodding; 5 sepals, fused to over half their length and with 5 triangular points; 5 petals, fused in bottom part into a short tube, white, pink, or carmine and with lobes 1.5–2.5 cm long, pointing backwards, and entire; 5 stamens; superior ovary; fertilised by insects

Fruits: Spherical capsules 0.5–1 cm in diameter; dispersal by ants

Habitat: In hill zone in deciduous woodlands, fairly well-drained, base-rich, neutral, loose and generally somewhat stony clay or loam containing nutrients, lime, and humus; prefers warmer places; plant grows in half or full shade; a mountain plant of central and southern Europe

538 **Marsh mallow – A. officinális**
Plants 50–120 cm high, perennials thickly covered with soft hairs; flowering in July and August

Althaéa officinalis L.
Marsh mallow

Malváceae – Mallow family

Leaves: Oval, rhomboidal, or triangular, pointed, with irregular coarse teeth and often slightly lobed, usually folded between the veins, and grey-green

Inflorescence: Flowers on short stems in upper leaf axils, forming an irregular raceme

Flowers: More than 5 epicalyx segments, 5–10 mm long, fused at the base and narrowly triangular; sepals 8–10 mm long; 5 petals, oval (broadest towards tips), narrowing in a wedge shape towards the base, up to 2.5 cm long, entire, flat, or irregularly notched at the tips and white or pink; numerous stamens, fused to each other in their lower parts and also to the petals; superior ovary

Fruits: Radially arranged nutlets

Habitat: In hill zone by ditches, salt and brackish marshes, and banks near the sea, in vineyards, fields, and waste land, on dry to moist, base-rich, thick, sandy ,or pure clay containing nutrients; cultivated from early times as a medicinal plant

539 **Alpine balsam – E. alpínus**
Plants 5–20 cm high, perennials forming loose mats; flowering in June and July

Erínus alpínus L.
Alpine balsam

Scrophulariáceae – Figwort family

Leaves: In basal rosette and alternate on stem; narrow oval to spathulate, 1–2 cm long, gradually narrowing to a short stalk, edge having short teeth, slightly notched at tip, with scattered hairs

Inflorescence: Single flowers in axils of top stem leaves, resembling a raceme

Flowers: 5 fused sepals, with 5 lobes incised almost to base, with glandular hairs and greenish; 5 petals, fused in lower part, oval (widest toward tip), 4–6 mm long, with corolla tube 5 mm long, lobes generally spread out flat, notched at tip, purple, with yellow spot at throat; 4 stamens, enclosed in corolla tube; superior ovary

Fruits: Egg-shaped capsules 3–4 mm long, splitting down middle into 2 flaps; many seeds

Habitat: In subalpine zone (also washed down to mountain zone) on alluvial soil and scree and in rock and wall crevices, on lime-rich stone; a plant of central and southern European mountains

540 **Purple saxifrage – S. oppositifólia**
Plants 1–6 cm high, perennials with upright flowering stems; flowering from May to July

Saxifrága oppositifólia L.
Purple saxifrage

Saxifragáceae – Saxifrage family

Leaves: Lanceolate to narrow oval, 2–4 mm long, with one lime-secreting cavity, hairy edges and usually straight points

Inflorescence: Solitary flowers

Flowers: 5 sepals, narrow oval, whitish to pale green, and usually with glandless hairs; 5 petals, oval, pointed, violet or lilac, each with 5 veins; 10 stamens in 2 rings, with grey-violet anthers and dark yellow filaments; intermediate ovary of 2 cells; fertilised by insects or self-fertile

Fruits: Capsules with many seeds

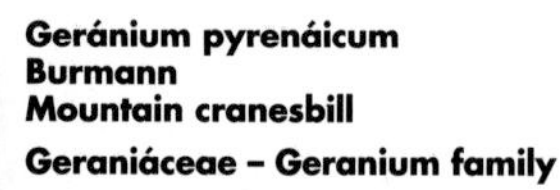

Habitat: In alpine zone (e.g. in the Valais up to 3800 metres above sea level), on rock slopes, debris, scree, rocks and ridges on moist to well-drained, base-rich, generally calcareous stony soil irrigated by meltwater; a polymorphous species with regard to leaves, habits, size, hairiness, and colour of petals; range extends to Spitzbergen

541 **Common mallow – M. sylvéstris**
Plants 20–100 cm high, biennials or perennials; flowering from June to September

Malva sylvéstris L.
Common mallow

Malváceae – Mallow family

Leaves: On long stalks, round, lobed, heart-shaped at base of leaf proper, with blunt teeth and rounded lobes

Inflorescence: Flowers in irregular clusters in top half of leaf stem

Flowers: 2–3 free-standing epicalyx segments; 5 sepals, fused as far as the middle and green; 5 petals, 2–3 cm long, oval (broadest at tips), fused at the base, with deep, broad notches at the tips, rose-purple to whitish with darker stripes; many stamens, with filaments fused into a tube; numerous superior carpels, each with seed cell; fertilised by insects

Fruits: 1-seeded kidney-shaped nutlets, disintegrating when ripe

Habitat: In hill (rarely mountain) zone in warm places in vineyards, fields, ditches, waste ground, walls, and rubbish dumps, and by the roadside, on clay, loam, or sand which is dry in summer and contains nutrients and humus; plant likes light and accompanies cultivation; used as a medicinal plant since the Stone Age, it has spread to both hemispheres and is also cultivated

542 **Mountain cranesbill – G. pyrenáicum**
Plants 20–60 cm high, perennials with thick tap-roots; flowering from May to August

Geránium pyrenáicum Burmann
Mountain cranesbill

Geraniáceae – Geranium family

Leaves: Roundish or kidney-shaped, 3–8 cm wide, soft to the touch, opposite and deeply incised, with 5, 7, or 9 lobes, incised to a lesser extent or toothed

Inflorescence: Long-stemmed flowers in pairs; inflorescence covered in glandular hairs

Flowers: 5 sepals, clearly covered in glandular hairs, narrow oval, not fused, blue-green, with short spiny points; 5 petals, violet, with deep, broad notches, 5–10 mm long and with hairs on the edges towards the base; 10 stamens, usually with violet anthers; superior ovary; usually fertilised by insects

Fruits: Capsules 1–2.5 cm long, splitting into 5 1-seeded sections; dispersal by explosive release and passing through animals

Habitat: In hill and mountain zones in meadows, waste places, by the roadside, and in thickets, on well-drained to moderately dry, humus-containing or fresh loam rich in nutrients

543 **Wood cranesbill – G. sylváticum**
Plants 20–60 cm high, perennials with thick rhizomes; flowering in June and July

Geránium sylváticum L.
Wood cranesbill

Geraniáceae – Geranium family

Leaves: Kidney-shaped to rectangular, up to 16 cm wide, with 5 or 7 lobes, with short scattered hairs on both sides; lobes with irregular secondary lobes and toothed; bracts pointed

Inflorescence: Flowers on stems in bunches arising from top leaf axils

Flowers: On stems; 5 sepals, narrow oval, greenish with a white border and spines 2–4 mm long; petals 10–18 mm long, broad oval, round at the tip, hairy towards the base, and reddish-violet; 10 stamens, with filaments widening to 1 mm at base and clearly hairy at the edges, and red-violet anthers; superior ovary

Fruits: Capsules 1–2.5 cm long, splitting into 5 1-seeded sections; flower stem upright when fruits ripe

Habitat: Usually in mountain and subalpine zones, in sparse copses, woodland, mountain meadows and undergrowth, on well-drained to moist, base-rich loam or clay containing humus and nutrients and with deep topsoil; roots up to 50 cm deep; range extends to Iceland

544 **Broad-leaved cranesbill – G. nodósum**
Plants15–40 cm high, perennials with thin rhizomes; flowering from May to August

Gerānium nodósum L.
Broad-leaved cranesbill

Geraniáceae – Geranium family

Leaves: Opposite, with short, close hairs on both sides; lower leaves pentagonal in outline and 4–12 cm wide, with irregularly toothed lobes; middle and upper leaves have 3 lobes, also with toothed lobes; bracts long and pointed

Inflorescence: Usually 2 flowers on each stem

Flowers: On stems; 5 sepals, narrow oval, 5–10 mm long, with spines 2–4 mm long at tips; flower stem and calyx covered in short hairs; 5 petals, oval (broadest towards tips), 10–18 mm long, with hairy edges towards base, notched at tips, pale to mid-violet with deep red stripes, often whitish towards base; 10 stamens; superior ovary

Fruits: Capsules 2–3.5 cm long, splitting into five 1-seeded sections; seeds with net-like pattern on surface

Habitat: In hill (rarely mountain) zone in shady places in mixed deciduous woodland, woodland margins, and hegerows, on well-drained to moist, loamy, base-rich soils with humus and nutrients; found in the southern Ticino, but native to southern Italy, Sicily, Serbia, and the Caucasus

545 **Aconite-leaved cranesbill – G. rivuláre**
Plants 15–50 cm high, perennials with forked stems; flowering in July and August

Gerānium rivuláre Vill.
G. aconitifólium
Aconite-leaved cranesbill

Geraniáceae – Geranium family

Leaves: 5–10 cm wide, with 5 or 7 lobes, with short stems in the upper part, irregular deep teeth on the lobes, and blue-green

Inflorescence: 2 or more flowers on stems arising from top leaf axils; stems remain upright when fruit is ripe

Flowers: 5 sepals, narrow oval, with both short and long hairs, 6–9 mm long, generally pale green, with points 1–2 mm long; 5 petals, oval (broadest towards tips), narrowing in a wedge shape towards base, 10–15 mm long, white or pale pink, with red veins; 10 stamens, with filaments somewhat wider towards the base and hairy edges; superior ovary of 5 carpels

Fruits: Capsules, splitting into five 1-seeded sections 2–4 cm long

Habitat: In subalpine zone on stream banks, in undergrowth, scrub, thickets and sparse coniferous woodland, on well-drained, lime-poor soils; range extends east to the soutern Tirol.

546 **Bloody cranesbill – G. sanguíneum**
Plants20–50 cm high, perennials with thick rhizomes; flowering from May to July

Gerānium sanguíneum L.
Bloody cranesbill

Geraniáceae – Geranium family

Leaves: Palmate, with 7 lobes, kidney-shaped in outline, generally with scattered hairs on both sides, and opposite; lobes have 2–4 entire points; bracts blunt-ended or pointed

Inflorescence: Usually solitary flowers with long stems, standing above neighbouring leaves

Flowers: Flower stem covered in short hairs; 5 sepals, hairy, 7–12 mm long and awned; 5 petals, 10–20 mm long, oval (broadest towards tips), covered in hairs on edges at base, irregularly notched at tips and purplish-crimson; 10 stamens, superior ovary

Fruits: Divided capsules 3–4 mm long, covered in glandular hairs in the lower part; seeds with net-like pattern on surface

Habitat: In hill and mountain zones in warm places in oak and pine woods, hedges, and on south-facing slopes, on dry, lime- or base-rich, loose, stony or deep loam or limy sand containing humus

547 **Herb Robert – G. robertiánum**
Plants 15–50 cm in height, annuals or biennials with tap-roots; flowering from May to October

Gerānium robertiánum L.
Herb Robert

Geraniáceae – Geranium family

Leaves: Triangular or pentagonal in outline, 3–8 cm wide, divided as far as the main vein into 3 or 5 leaflets, covered in sparse hairs on both sides, and opposite; segments have their own stalks and are also pinnate; secondary leaflets toothed

Inflorescence: Usually 2 flowers on each stem, arising from leaf axils, and not reaching above neighbouring leaves

Flowers: 5 sepals, 5–8 mm long, narrow oval, awned and covered in hairs; 5 petals, oval, entire, rounded (rarely notched) at the tips, on long claws, pink to deep pink and smooth-edged at the base; 10 stamens with orange anthers; superior ovary; fertilised by insects (especially moths) or self-fertile

Fruits: Capsules 1–2.5 cm long, in compartments, with 1 seed in each; seeds lack net-like pattern on surface

Habitat: In hill and mountain (rarely subalpine) zones in deciduous forests and hedges, on walls, waste ground, in scree and on rocks, on well-drained to moist, loose, loamy soils containing humus and nutrients in shady, humid places; an indicator of fertility

548 **Wormwood storksbill – E. absinthoídes**
Plants 20–50 cm high, recumbent or upright; flowering from May to August

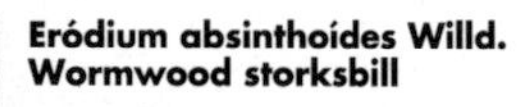

Eródium absinthoídes Willd.
Wormwood storksbill

Geraniáceae – Geranium family

Leaves: Narrow oval, doubly or trebly pinnate, on long stalks and opposite; in upper part of stem also short-stalked or sessile; primary leaflets alternate or opposite; last leaflets lanceolate, often somewhat incised and pointed

Inflorescence: Flowers in umbels on light or dark brown stems arising from top leaf axils; very short pale reddish-brown bracts at base of umbel

Flowers: 5 sepals, narrrow oval to oval, pale green to dull green with dark green veins, each with a long point; 5 petals, oval to round, pale to mid-pink or violet, with darker veins and not fused together; 5 stamens with deep yellow anthers and whitish or pink filaments; superior ovary; fertilised by insects

Fruits: Capsules with compartments holding single seeds

Habitat: Originating in Macedonia; occasionally found in gardens or parks

549 **Rose mallow – M. álcea**
Plants 30–110 cm high, perennials with downy stellate hairs; flowering from July to September

Malva álcea L.
Rose mallow

Malváceae – Mallow family

Leaves: Stem leaves deeply palmate, with 5–7 lobes (like a maple leaf), with roundish leaves proper on stalks; lobes in the upper part often incised and generally rounded; leaves take many shapes

Inflorescence: Single flowers on stalks in axils of leaves; 3–4 flowers in upper part may be close together

Flowers: 3 segments of epicalyx long oval or oval; 5 sepals, oval, pale green, with short points; 5 petals, not fused, oval, 20–35 mm long, deeply and irregularly incised at the tip, pale pink or white, with dark veins; numerous stamens fused into a tube; superior ovary; fertilised by insects

Fruits: Nutlets

Habitat: In hill and mountain zones by the roadside, in weed beds and thickets and by ditches, on well-drained, often sandy loam containing nutrients, lime, and humus; plants like light; used in earlier times as an ornamental and medicinal plant

Ranúnculus glaciális L.
Oxygráphis vulgáris Freyn
Glacier crowfoot

Ranunculáceae – Buttercup family

Leaves: Multiply ternate almost as far as the base, fleshy and dark green; lobes narrow oval, pointed or blunt-ended and often themselves with 2–5 lobes

Inflorescence: Long-stemmed flowers, singly or several to a stem

Flowers: White on first flowering, later tinged pink or brownish-red and 1–3 cm in diameter; usually 5 sepals, covered in reddish-brown hairs; 5 petals; many stamens; many superior carpels; receptacle without hairs

Fruits: Smooth achenes with straight beaks

Habitat: In alpine zone up to 4200 metres on moraines, in rock debris, rock crevice, and boulders, on damp, loose, more or less rough siliceous soils poor in nutrients and lime; an early coloniser of scree, spreading over it; an arctic/alpine plant, with a range extending to Iceland, east Greenland,and Spitzbergen; in the Alps this species is mainly found in the central ranges

550 **Glacier crowfoot – R. glaciális**
Plants 5–20 cm high, perennials with short rhizomes; flowering in July and August

551 **Water avens – G. rivále**
Plants 20–50 cm high, perennials with thick rhizomes; flowering from April to July

Geum rivále L.
Water avens

Rosáceae – Rose family

Leaves: Pinnate; end leaflet semicircular, 3–8 cm in diameter, often ternate towards the base, with toothed lobes; other segments smaller and also toothed

Inflorescence: 2–6 flowers on each stem in a narrow cyme

Flowers: Sepals narrow lanceolate, hairy and dark reddish-brown; petals as long (1–1.5 cm) as the inner sepals, erect, heart-shaped, narrowing towards bases, yellowish with reddish borders, usually arranged in a ring; many stamens and superior carpels

Fruits: Single-seeded hard achenes with straight beaks

Habitat: From hill to subalpine zones in riversides, marshes, wet moorlands, damp shrubby undergrowth, wet grassland, and wet mountain woodland, on moist or seasonally flooded, base-rich, calcareous loam or clay which may be weakly acidic; an indicator of fertility; in Mediterranean area grows only on mountains

Cómarum palústris L.
Poteniilla palústris Scop.
Marsh cinquefoil

Rosaceae – Rose family

Leaves: Pinnate, on long stems and with 5 or 7 leaflets, narrow oval, coarsely toothed, short-stalked, mid- to dark green above, smooth or with scattered hairs, and grey-green beneath with sparse, close hairs

Inflorescence: Loose terminal cyme with bracts at forks

Flowers: Parts in fives or rarely sevens, with double calyx; 5 outer sepals, narrow lanceolate, often somewhat fringed; 5 inner sepals, broad lanceolate with fine points; 5 petals, oval, 5–8 mm long, and dark red or dark purple, often lighter at tips; petals do not drop after flowering; many (about 20) stamens; 20–50 carpels on a high-vaulted, spongy, hairy receptacle; fertilised by insects

Fruits: Smooth, glossy, hard achenes with single seeds and beaks pointing sideways

Habitat: From hill to subalpine zones on fens, marshes, wet heaths, and moorland, on wet, seasonally flooded peat and moorland soils, moderately rich in nutrients and rather acidic; range extends north to Iceland

552 **Marsh cinquefoil – C. palústris**
Plants15–30 cm high, upright perennials; flowering from May to July

553 **Hellebore – Hellebórus hybrid**
Plants 15–35 cm high, evergreen perennials; flowering from December to April

Helleborus hybrid
Hellebore (garden variety)

Ranunculáceae – Buttercup family

Leaves: Basal leaves long-stemmed, kidney-shaped, or round, divided at base into 5, 7, or 9 lobes which are lanceolate to narrow oval, narrowing in a wedge towards the base, pointed, short-stemmed or sessile, entire and/or finely toothed; bracts have simpler structure

Inflorescence: Umbels with 1–2 flowers on each branch in axils of bracts

Flowers: 5 sepals, green or petaloid, wide oval or round, 1–3 cm long, wavy at edges, distinctly overlapping and reddish-green; numerous stamens; filaments generally pale yellow or white, anthers pale yellow; several superior carpels, fused at the base

Fruits: Many-seeded follicles opening along middle, with long beaks

Habitat: In the hill zone in parks and gardens. Forms of H. níger with spreading white or pinkish sepals bloom in winter and are grown as Christmas roses; later-flowering Lent roses are hybrids of other species, ranging in colour from white or cream to deep purple

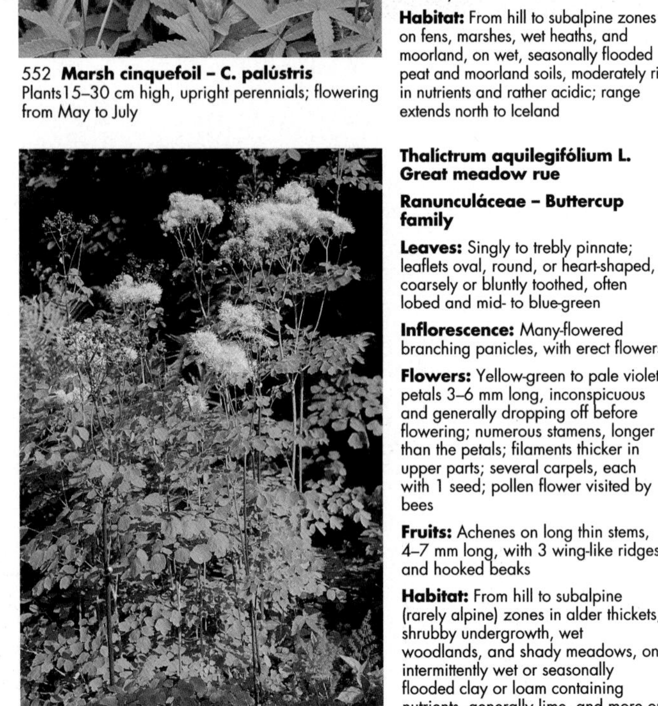

554 **Great meadow rue – T. aquilegifólium**
Plants 30–130 cm high, with leafy stems; flowering from May to July

Thalíctrum aquilegifólium L.
Great meadow rue

Ranunculáceae – Buttercup family

Leaves: Singly to trebly pinnate; leaflets oval, round, or heart-shaped, coarsely or bluntly toothed, often lobed and mid- to blue-green

Inflorescence: Many-flowered branching panicles, with erect flowers

Flowers: Yellow-green to pale violet petals 3–6 mm long, inconspicuous and generally dropping off before flowering; numerous stamens, longer than the petals; filaments thicker in upper parts; several carpels, each with 1 seed; pollen flower visited by bees

Fruits: Achenes on long thin stems, 4–7 mm long, with 3 wing-like ridges and hooked beaks

Habitat: From hill to subalpine (rarely alpine) zones in alder thickets, shrubby undergrowth, wet woodlands, and shady meadows, on intermittently wet or seasonally flooded clay or loam containing nutrients, generally lime, and more or less humus; an indicator of fertility and moisture; range extends north to northern Germany

555 **Burnet saxifrage – P. major**
Plants 30–100 cm high, perennials with ridged and forking stems; flowering from June to September

Pimpinélla major (L.) Hudson
Burnet saxifrage

Apiáceae (Umbellíferae) – Carrot family

Leaves: Unequally pinnate; leaflets narrow oval, pointed, without stems, irregularly toothed and heart-shaped or rounded at the base; end leaflet 3-lobed

Inflorescence: Terminal umbels with 20–30 flowers; no bracts

Flowers: 5 small, inconspicuous sepals; 5 petals, not fused, 1–1.5 mm long, oval, and white or often pink; 5 stamens between the petals; inferior ovary of 2 cells; fertilised by insects

Fruits: Dry schizocarps, splitting into 2 one-seeded carpels 2–3 mm long

Habitat: In mountain and subalpine zones on meadows, in shrubby undergrowth, on margins of woods, and in hedgerows, on well-drained base-rich loam containing nutrients and humus and with deep topsoil, in humid places; an indicator of fertility; only average value as fodder

556 **Wild angelica – A. sylvéstris**
Plants up to 2 m high, usually biennial, with hairs only near umbels; flowering from July to September

Angelica sylvéstris L.
Wild angelica

Apiáceae (Umbelliferae) – Carrot family

Leaves: Triangular in outline, leaves proper up to 45 cm long, doubly (rarely trebly) pinnate; last leaflets narrow oval to oval, with short points, up to 15 cm long, short-stalked or sessile, entire or in 2–3 lobes, often asymmetrical at base and singly or doubly toothed

Inflorescence: Primary umbels with 15–40 secondary umbels

Flowers: 5 petals, whitish or reddish; 5 stamens; inferior ovary of 2 cells; 2 styles; fertilised by insects

Fruits: Schizocarps 3–6 mm long, with ridges at edges, containing 2 single-seeded carpels

Habitat: From hill to subalpine zones in wet woodlands, shrubby undergrowth, water meadows, wet moorland, woodland clearings, and on river banks, on intermittently moist, loose, sandy or gravelly clay or loam containing nutrients and humus and with deep topsoil; used in earlier times as a medicinal plant; rooting up to 1 m deep; range extends north to Iceland and northern Finland

557 **Hairy chervil – C. hirsútum**
Plants 20–110 cm high, branching perennials, with thick stems; flowering from May to August

Chaeróphyllum hirsútum L.s.str.
Hairy chervil (white-flowered variety)

Apiáceae (Umbelliferae) – Carrot family

Leaves: Triangular in outline and doubly (rarely trebly) pinnate; last leaflets oval, pointed, without stalks, and toothed

Inflorescence: Primary umbels with 10–20 secondary umbels

Flowers: 5 petals, oval, entire, or incised or notched at tips, white or pink, and with distinct hairs; 5 stamens; inferior ovary of 2 cells; 2 styles with stigma often reddish; fertilised by insects

Fruits: Schizocarps up to 20 mm long

Habitat: In all zones in wet meadows, shrubby undergrowth, wet mountain woodlands, fir woods, by streams and springs, and in alder thickets on wet to moist, cool, clay with nutrients and humus in humid locations; a good indicator of nutrients; grows in light or half-shade; deep-rooted; a mountain plant of central and southern Europe (compare Fig. 558)

558 **Hairy chervil – C. hirsútum**
Plants 20–110 cm high, branching perennials, with thick stems; flowering from May to August

Chaeróphyllum hirsútum L.s.str.
Hairy chervil (pink-flowered variety)

Apiáceae (Umbelliferae) – Carrot family

Leaves: Triangular in outline and doubly (rarely trebly) pinnate; last leaflets oval, pointed, without stalks, and toothed

Inflorescence: Primary umbels with 10–20 secondary umbels

Flowers: 5 petals, oval, entire, or incised or notched at tips, white or pink, and with distinct hairs; 5 stamens; inferior ovary of 2 cells; 2 styles with stigma often reddish; fertilised by insects

Fruits: Schizocarps up to 20 mm long

Habitat: In all zones in wet meadows, shrubby undergrowth, wet mountain woodlands, fir woods, by streams and springs, and in alder thickets on wet to moist, cool, clay with nutrients and humus in humid locations; a good indicator of nutrients; grows in light or half -hade; deep-rooted; a mountain plant of central and southern Europe (compare Fig. 557)

559 **Alpine lovage – L. mutellína**
Plants 10–50 cm high, smooth perennials with rhizomes; flowering from June to August

Ligústicum mutellina Crantz
Meum mutellína Gaertner
Alpine lovage

Apiáceae (Umbelliferae) – Carrot family

Leaves: Basal leaves on long stalks: leaves proper triangular in outline, up to 10 cm long, multiply pinnate and blue-green; last leaflets pinnnately cut and pointed

Inflorescence: Stem bears 1–3 primary umbels and 1–2 stem leaves; primary umbels consist of 6–10 secondary umbels

Flowers: Hermaphrodite and male flowers; 5 petals, oval, usually notched or incised at tips and white or red; 5 stamens; where present, inferior ovary of 2 cells; fertilised by insects (usually flies)

Fruits: Schizocarps 4–6 mm long, up to 4 mm wide, with wing-shaped primary ridges

Habitat: In subalpine and alpine zones in wet meadows, shrubby undergrowth, and rough grassland, on wet, generally lime-poor, base-rich soils more or less rich in nutrients

560 **Bistort – P. bistórta**
Plants30–70 cm high, perennials with contorted rhizomes; flowering from May to July

Polygonum bistórta L.
Bistort, snake-root

Polygonáceae – Buckwheat family

Leaves: Oval, up to 20 cm long, widest at truncated or heart-shaped base of leaf proper, generally blunt-ended at tip, with winged stalk on lower leaves; upper stem leaves wide-spaced, sessile, and mostly pointed; sheathing stipules pointed and long

Inflorescence: Terminal cylindrical spike 3–8 cm long and up to 2 cm thick

Flowers: 5 petals, pink or reddish and 3–5 mm long; 4–8 stamens; superior ovary with pistil

Fruits: Triangular nutlets, 4–5 mm long, dark brown and glossy

Habitat: In mountain and subalpine zones in wet grassland, shrubby undergrowth, wet woodlands, alder thickets and river banks, on cool, wet or moist, seasonally flooded, lime-free loam or clay containing nutrients and humus; deep-rooted; attracts bees; an important fodder plant in wet meadows; often found in large patches, but never on dry soil

561 **Pale persicaria – P. lapathifólium**
Plants 20–80 cm high, annuals, recumbent or upright; flowering from July to October

Polygonum lapathifólium L.
Pale persicaria

Polygonáceae – Buckwheat family

Leaves: Lanceolate, 3–8 cm long, pointed, narrowing towards stem, with short stalks or sessile, green on both sides, generally with black spots, and smooth or slightly hairy beneath; sheathing stipules sessile and smooth or with short hairs on the edges; plant generally branches from ground level

Inflorescence: Terminal cylindrical spike up to 6 cm long and 1 cm thick

Flowers: Hermaphrodite; 4 or 5 petals, pink or pale pink and 2–3 mm long; superior ovary, with pistil

Fruits: Nutlets 2–3 mm long and flat or triangular; different shapes may be seen on one plant

Habitat: In hill and mountain zones in arable fields, waste places, mud patches, river banks, ditches and gardens on well-watered to moist, neutral or slightly acid sand, loam, or clay containing nutrients and humus; roots 30–40 cm deep; a weed of arable land; range extends worldwide except for arctic zones

562 **Willow grass – P. amphíbium**
Plants 20–30 cm long, semi-aquatic, with creeping stems; flowering from June to September

Polygonum amphíbium L.
Willow grass

Polygonáceae – Buckwheat family

Leaves: Up to 10 cm long, on stalks, narrow oval, pointed at tip, rounded or heart-shaped at base of leaf proper, and green on both sides; leaf stalks originate at or above middle of sheathing stipules

Inflorescence: Dense cylindrical terminal spikes

Flowers: 4 or 5 petals, pink or pale pink and 4 – 5 mm long; flowers are male, female or hermaphrodite; in hermaphrodite flowers there are 5 stamens and a superior ovary with 2 styles united ¼ the way up

Fruits: Lentil-shaped or triangular nutlets: fruit rarely formed

Habitat: In hill and mountain (rarely subalpine) zones in ponds and ditches; plant spreads vegetatively into large patches in wet meadow-land, wet waste ground, and on streambanks, on flooded or wet, often muddy clay and loam which is rich in nutriment, generally lime-free and neutral or moderately acid; plants grow in light; dispersion by water; aquatic forms have a smooth stem, terrestrial forms have bristly hairs; range extends over whole world except for tropics

563 **Milkwort – P. vulgáris**
Plants 10–30 cm high, perennials with branching rhizomes, thin tap-roots; flowering until July

Polygala vulgáris L.
Milkwort

Polygaláceae – Milkwort family

Leaves: Lower leaves alternate, without rosette; upper leaves 10 – 25 mm long, lanceolate, generally blunt-ended, sessile or with short stems, entire, generally widest at the middle and alternate

Inflorescence: Raceme of 5 – 30 red, blue or violet flowers on short stems; bracts 1 – 2.5 mm long

Flowers: Zygomorphic; 5 sepals, the 3 outer sepals being small, sepaloid, narrow triangular and brownish red; the 2 inner sepals form oval wings, 6 – 8 mm long and red, blue or violet, with much-branching lateral veins; corolla is grooved, with 2 outer petals united with lower (which is fringed) and joined to the staminal tube; 8 stamens; superior ovary, with simple style and stigma in 2 parts

Fruits: Capsule, compressed and generally heart-shaped

Habitat: In hill and mountain zones in rough grassland, heaths, roadsides and woodland margins, on well-drained to moderately dry loam which is poor in nutriment

564 **Tobacco flower – N. tabácum**
Plants 70–200 cm high, usually annuals with glandular hairs; flowering from July to October

Nicotiána tabácum L.
Tobacco flower

Solanáceae – Nightshade family

Leaves: Narrow oval to oval. entire, pointed, narrowing at base, may be more than 50 cm long and drooping, especially on lower stem

Inflorescence: Raceme- or panicle-like heads at ends of stem and branches

Flowers: On stems; 5 sepals fused into a tube, green, and with teeth narrow triangular, sharp pointed and unequally long; 5 fused petals, with lobes opening into a funnel, 4 – 5 cm long and rose pink; 5 stamens, as long as corolla tube; superior ovary in 2 cells

Fruits: Egg-shaped capsules springing open with 2 flaps; numerous very small seeds, with a warty surface

Habitat: In hill zone on waste places on previoiusly damp, loose soils in warmer locations; nowadays cultivated worldwide; different varieties are cultivated in Europe as a garden plant

565 **Purple gentian – G. purpúrea**
Plants 10–60 cm high, perennials with thick rhizomes; flowering in July and August

Gentiána purpúrea L.
Purple gentian

Gentianáceae – Gentian family

Leaves: Oval, entire, pointed, or blunt-ended, up to 25 cm long, 5–15 cm wide and with 5–7 veins; upper stem leaves narrow lanceolate

Inflorescence: In groups of 1–3 in axils of upper stem leaves and in a head at the end of the stem

Flowers: Calyx has 2 points and is incised down one side almost to the base; fused petals with 5–8 limbs extending well down towards the base, 2 –4 cm long, purplish-red, with distinct dark red dots and yellow inside; stamens fused to the corolla tube; superior ovary of 2 cells; fertilised by insects

Fruits: Capsules opening along the dividing wall of the ovary

Habitat: In subalpine and alpine zones on rough grassland, scrubby undergrowth, scrub, near alpine alders, on well-drained, base-rich, generally lime-poor clay soils which contain humus and are subject to lengthy periods of snow cover; plants grow in light or half-shade; a medicinal plant and herb; found in Scandinavia, the Alps, and the Apennines

566 **Comfrey – S. officinále**
Plants 20–120 cm high, thickly covered with erect hairs; flowering from May to August

Symphytum officinále L.
Comfrey

Boragináceae – Borage family

Leaves: Narrow oval (broadest towards base), gradually drawing to a point at tips, entire, gradually narrowing to stalk (which has wings extending down to the next leaf) and rough-haired

Inflorescence: Small, dense, terminal, forked cymes

Flowers: On short stems; 5 sepals, fused towards base, narrow triangular, hairy and often with darker points; 5 fused petals, yellowish, purple, or reddish-violet, 1–2 cm long, with short tips bent backwards, and often with darker stripes; corolla scales do not protrude from corolla; 5 stamens, with filaments fused and located roughly in middle of corolla tube; superior ovary of 2 cells; pistil protrudes beyond corolla tube

Fruits: 4 nutlets from each flower

Habitat: In hill and mountain zones in wet grassland, wet woodland, beside rivers and streams, and in ditches on moist or wet soil

567 **Tuberous comfrey – S. tuberósum**
Plants 10–40 cm high, perennials with stout, tuberous rhizomes; flowering from April to July

Symphytum tuberósum L.
Tuberous comfrey

Boragináceae – Borage family

Leaves: Narrow oval, 3–15 cm long, generally pointed and narrowing towards base; lower leaves long-stalked

Inflorescence: Small, dense, terminal, forked cymes

Flowers: On stems; 5 sepals, fused at base, narrow triangular, pointed and often darker towards base; 5 fused petals, yellowish-white or dull red, 1.5–2 cm long, with short tips bent backwards; corolla scales do not protrude from corolla; 5 stamens; superior ovary of 2 cells; fertilised by insects

Fruits: 4 nutlets, more or less warty with serrated edges round receptacles; dispersal by ants

Habitat: In hill and mountain zones in meadows, deciduous woodland, hedgerows, woodland margins, and shrubby undergrowth, on well-drained, base-rich, neutral loam or clay rich in nutrients and usually with deep topsoil; plant grows in shade or half-shade; a south European plant with a range extending to England

568 **Mountain lungwort – P. montána**
Plants 10–40 cm high, perennials with widely creeping rhizomes; flowering from March to May

Pulmonária montána Lejeune
Mountain lungwort

Boragináceae – Borage family

Leaves: Rosette leaves usually narrow oval to oval, pointed, 5–15 cm long, long-stalked, without spots, hairy on both sides and narrowing at the base of the leaves proper to stalks; stem leaves narrow oval, pointed, up to 10 cm long, hairy on both sides, often rounded at bases of leaves proper and partially amplexicaul

Inflorescence: Short-stemmed flowers in cymes

Flowers: 5 sepals fused into a bell, greenish or reddish-brown, with 5 pointed tips and numerous bristles; 5 fused petals, funnel-shaped in upper part, with 5 hairy tufts at entry to throat, and pale red at first, becoming violet; 5 stamens, not protruding from corolla; superior ovary of 2 cells

Fruits: Schizocarp with 4 nutlets; dispersal by ants

Habitat: In mountain and subalpine zones in sparse deciduous woodland, hedges, shrubby undergrowth, and at woodland margins, on moist or intermittently well-watered, base-rich, generally lime-poor, often sandy soils containing humus

569 **European scopolia – S. carniólica**
Plants 20–60 cm high, with branching stems and dense foliage; flowering in April and May

Scopólia carniólica Jacq.
European scopolia

Solanáceae – Nightshade family

Leaves: Oval, broadest towards base, 10–30 cm long, entire or slightly crenate, with prominent middle vein and narrowing towards stalks; stems covered in scale-like stipules at base

Inflorescence: Long-stemmed solitary flowers arising from leaf axils

Flowers: Nodding; 5 sepals fused into bell, green, and with triangular points; 5 petals fused into bell-shaped tube, 1.5 –2.5 cm long, glossy brown outside and matt olive-green inside; 5 petals, not protruding beyond the corolla; superior ovary of 2 cells

Fruits: Berries

Habitat: In hill (rarely mountain) zone on stony, bushy hills on previously dry, neutral, or weakly acid soils containing nutrients and humus, in warmer places; poisonous; occasionally escaped from gardens; range extends as far as Carinthia

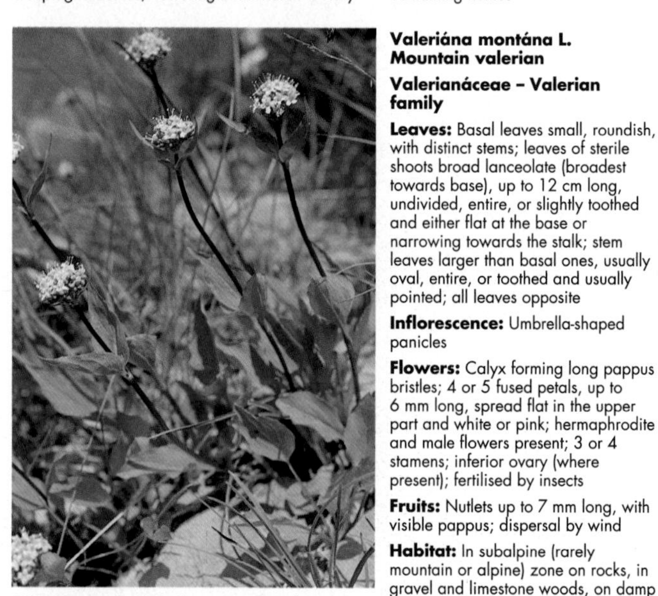

570 **Mountain valerian – V. montána**
Plants 10–50 cm high, perennials with several shoots; flowering from May to July

Valeriána montána L.
Mountain valerian

Valerianáceae – Valerian family

Leaves: Basal leaves small, roundish, with distinct stems; leaves of sterile shoots broad lanceolate (broadest towards base), up to 12 cm long, undivided, entire, or slightly toothed and either flat at the base or narrowing towards the stalk; stem leaves larger than basal ones, usually oval, entire, or toothed and usually pointed; all leaves opposite

Inflorescence: Umbrella-shaped panicles

Flowers: Calyx forming long pappus bristles; 4 or 5 fused petals, up to 6 mm long, spread flat in the upper part and white or pink; hermaphrodite and male flowers present; 3 or 4 stamens; inferior ovary (where present); fertilised by insects

Fruits: Nutlets up to 7 mm long, with visible pappus; dispersal by wind

Habitat: In subalpine (rarely mountain or alpine) zone on rocks, in gravel and limestone woods, on damp to well-drained, stony ground with fine soil, also on mobile limestone scree; plant grows in light

571 **Three-leaved valerian – V. trípteris**
Plants 10–50 cm high, perennials with woody stems at base; flowering from April to July

Valeriána trípteris L.
Three-leaved valerian

Valerianáceae – Valerian family

Leaves: On sterile shoots oval or round, with distinct, coarse, irregular teeth, blunt-ended or pointed, on long stalks and with bases heart-shaped; stem leaves generally ternate: side lobes toothed, lanceolate, and clearly smaller than the broad lanceolate middle lobe

Inflorescence: Umbrella-shaped panicles

Flowers: Calyx forming long pappus bristles; 4 or 5 fused petals, up to 6 mm long, spread flat in the upper part, white or pink and usually round at the end; 3 or 4 stamens; inferior ovary; fertilised by insects

Fruits: Nutlets up to 7 mm long, with visible pappus; dispersal by wind

Habitat: In mountain and subalpine zones on rock, in rock crevices and steep mountain forests, on moderately dry to well-drained, base-rich, calcareous and siliceous soils; plants grow in light or half-shade; a central and southern European plant

572 **Marsh valerian – V. dioíca**
Plants 10–20 cm high, perennials with rhizomes producing long runners; flowering from April to July

Valeriána dioíca L.
Marsh valerian

Valerianáceae – Valerian family

Leaves: Basal leaves oval or round, up to 25 mm long, entire or slightly incised or serrated, with stalks, and opposite; stem leaves, particularly in upper part, are pinnate, with 2–4 lateral leaflets on each side and a large, somewhat toothed terminal leaflet

Inflorescence: Umbrella-like terminal panicles with few flowers

Flowers: Calyx forming 10–15 long pappus bristles; 4 or 5 petals, those of male flowers 2–3 mm long, pink, with oval lobes spread out; those of female flowers 1–2 mm long and generally white; 3 or 4 stamens; inferior ovary; fertilised by insects

Fruits: Smooth nutlets 2–3.5 mm long; dispersal by wind

Habitat: In hill and mountain zones in ditches, wet moorland, streambanks, marshy meadows, fens and bogs, on wet or seasonally wet, moderately acid soils more or less rich in nutrients

573 **Valerian – V. officinális**
Plants 30–160 cm high, perennials with rhizomes but no runners; flowering from May to August

Valeriána officinális L.
Valerian

Valerianáceae – Valerian family

Leaves: Unequally pinnate; 5, 7, or 9 leaflets, lanceolate, 4–8 cm long, distinctly toothed, drooping at the leaf axil and covered with close hairs, particularly on the underside

Inflorescence: Umbrella-shaped panicles

Flowers: 5 petals, very small, present as a rolled-up point at blooming and so forming a thick ring; on fruiting these develop into about 10 feathery bristles 3–7 mm long; 5 fused petals,

3–6 mm long and pale pink; generally 3 stamens, rarely 4, fused at the base to the corolla tube; inferior ovary of 3 compartments, of which only 1 is fertile; fertilised by insects

Fruits: Nut-like fruits 2–4 mm long, usually smooth, rarely hairy

Habitat: From hill to subalpine zones in wet woodlands, hedges, ditches, wet meadows, wet moorland, and on stream and riverbanks, on intermittently moist to wet, more or less base-rich, neutral clay or loam containing humus and nutrients; also on limestone scree

574 **Alpine coltsfoot – H. alpína**
Plants 15–35 cm high, perennials with thin woolly rhizomes; flowering from May to July

Homógyne alpína (L.) Cass.
Alpine coltsfoot

Asteráceae (Compósitae) – Sunflower family

Leaves: Basal leaves round or kidney-shaped, long-stalked, leathery, smooth or slightly hairy, up to 3 cm in diameter, heart-shaped at base of leaf proper, with shallow teeth, and dark green above with clearly protruding veins; stem leaves scale-like

Inflorescence: Solitary composite flower on stem; flower head 1–3 cm long, with 10–20 greenish or reddish bracts with woolly bases

Flowers: Tubular disc-florets 6–10 mm long, reddish, with smooth corolla tube formed of 5 petals and purple tips; 5 stamens; inferior ovary of 2 cells; fertilised by insects or self-fertile

Fruits: Achenes 3–6 mm long, with up to 10 ridges and white pappus 6–10 mm long; dispersal by wind

Habitat: In subalpine and alpine (rarely hill) zones in fir woods, scrub, siliceous rough pasture and lowlands, on moist to well-drained, more or less base-poor soil containing rotted peat and humus

575 **Butterbur – P. hybridus**
Plants 10–50 cm high, perennials with thickened rhizomes, knobbly on top; flowering March/April

Petasítes hybridus (L.) G.M.Sch.
Butterbur

Asteráceae (Compósitae) – Sunflower family

Leaves: Basal leaves kidney-shaped, heart-shaped, or round, with shallow teeth, grooved red stalks, intitially downy but later becoming smooth, especially above, up to 60 cm in diameter and appearing at the end of flowering; stem has lanceolate, reddish, almost amplexicaul, scale-like leaves

Inflorescence: Numerous composite flowers, sweet-smelling, in short, thick racemes; bract shell of 2–3 rows of bracts surrounded by scales

Flowers: All tubular disc-florets; yellowish-white pappus; 5 petals fused into a tube, usually reddish; 5 stamens; inferior ovary of 2 cells

Fruits: Achenes 2–3 mm long with pappus 5–8 mm long

Habitat: In hill and mountain zones on stream and riverbanks, in alder thickets and wet meadows, on wet soils containing nutrients

576 **Grey wild lettuce – A. alliária**
Plants 50–150 cm high, perennials with grooved stems; flowering from June to September

Adenostyles alliária Kerner
Grey wild lettuce

Asteráceae (Compósitae) – Sunflower family

Leaves: Heart- or kidney-shaped, on long stalks, usually heart-shaped at base of leaf proper, irregularly toothed, downy beneath and usually with short points; stem leaves on short stems, sessile or amplexicaul with wide basal points

Inflorescence: Umbel-shaped panicles of many composite flowers; each head of 3–6 flowers with up to 6 lanceolate bracts 4–6 mm long, with hairy points; base of flower head without hairs

Flowers: Tubular florets all hermaphrodite, with pappus, 5 fused petals, pale pink; 5 stamens; inferior ovary; fertilised by insects

Fruits: Achenes 2–5 mm long, with pappus 4–8 mm long; dispersal by wind

Habitat: In mountain and subalpine (rarely alpine) zones in shrubby undergrowth, mixed mountain forests, woodland margins, and on stream banks, on wet, base-rich, neutral or weakly acidic, stony, and often lime-poor, loamy soil rich in nutrients; plants grow in light or shade

577 **Green wild lettuce – A. glábra**
Plants 30–90 cm high, perennials with knotted rhizomes; flowering from June till August

Adenostyles glábra (Miller) DC.
Green alpine lettuce

Asteráceae (Compósitae) – Sunflower family

Leaves: Heart- or kidney-shaped, on long stalks, irregularly heart-shaped at base of leaf proper, coarsely and usually regularly toothed, with coarse hair beneath and coming to very short points; stem leaves always distinctly stalked, without basal points

Inflorescence: Umbel-shaped panicles of many composite flowers; each head usually of 3 flowers, usually with 3 lanceolate or oval bracts, smooth, up to 6 mm long, reddish and without hairs at the tip

Flowers: Tubular disc-florets, all hermaphrodite, with pappus, 5 fused petals, pale pink or lilac; 5 stamens; inferior ovary; fertilised by insects

Fruits: Achenes 2–3 mm long, with pappus 3–5 mm long; dispersal by wind

Habitat: Usually in subalpine zone, in waste places, stony woodlands and meadows, on moist to well-drained, stony, loose, limestone scree containing more or less fine soil; plants grow in light or half-shade

578 **Cat's foot – A. dioica**
Plants 5–20 cm high, perennials with underground runners; flowering till July

Antennária dioica (L.) Gaertner
Cat's foot

Asteráceae (Compósitae) – Sunflower family

Leaves: In rosettes, narrow oval, narrowing in a wedge shape towards the base, initially downy on both sides and usually broadest towards the tips; stem leaves lanceolate and downy

Inflorescence: 3–12 short-stemmed composite flowers in terminal umbel; outer bracts very hairy and oval, inner ones mostly lanceolate; bracts on female flowers dark red, pink, or rarely white, on male flowers white, rarely pink or red; flower heads 4–8 mm long

Flowers: 5 petals, generally thread-like and whitish; 5 stamens (male flowers); inferior ovary (female flowers); fertilised by insects

Fruits: Achenes 1 mm long, with long white pappus 4–10 mm long; dispersal by wind

Habitat: In subalpine and alpine (rarely hill or mountain) zones on rough grassland, meadows, heaths and sparse woodland, on fairly well-drained to dry, usually lime-poor soils

579 **Hemp agrimony – E. cannabinum** Plants 70–150 cm high, perennials with knotty rhizomes, almost wholly covered in glands; flowering till Sept.

Eupatórium cannabinum L.
Hemp agrimony

Asteráceae (Compósitae) – Sunflower family

Leaves: Divided into 3 or 5 lobes, covered in scattered hairs, mostly short-stalked and opposite; lobes narrow oval to lanceolate, pointed and irregularly toothed with teeth pointing forwards

Inflorescence: Composite flowers in dense terminal corymbs, each head with 4–6 florets; bracts 4–6 mm long, blunt-ended, rather hairy and arranged like tiles

Flowers: All tubular florets; single-rowed pappus; 5 petals, fused into tube and pale red or pink; inferior ovary; fertilised by insects

Fruits: Achenes 2–3 mm long, with 5 ridges and a pappus 3–5 mm long; dispersed by wind

Habitat: In hill and mountain zones a common gregarious plant of marshes and fens, streambanks and moist woodlands, on wet to well-drained, base-rich, generally calcareous, loose loam or clay containing nutrients and humus; an indicator of moisture; a plant of Europe and Asia with a range extending west to Northern Ireland, north to southern Sweden, and south to Morocco

580 **Cotton burdock – A. tomentósum**
Plants 50–120 cm high, biennials with woody taproots; flowering in August and September

Arctium tomentósum Miller
Láppa tomentósa Lam.
Cotton burdock

Asteráceae (Compósitae) – Sunflower family

Leaves: Broad oval or heart-shaped, with blunt or pointed tip, usually rounded at base of leaf proper, entire or crenate, up to 45 cm long, covered in gossamer hairs beneath, and spirally arranged

Inflorescence: Composite flowers in upright or obliquely sloping panicles; bracts thickly covered in gossamer hairs; inner bracts have long red points

Flowers: All tubular florets, hermaphrodite and covered with numerous bristly scales at their base on the receptacle; whitish pappus; 5 petals fused into a long tube, in 5 parts towards the tip and generally purple-red; 5 stamens, with each half of anthers sharply pointed beneath; inferior ovary of 2 cells

Fruits: Achenes, narrow oval or oval, slightly flattened on the sides, smooth but with 3 ribs

Habitat: Generally in mountain zone in rich weed beds, thickets, waste places, stream banks and roadsides, on well-drained to moderately dry clay or loam soils

581 **Giant knapweed – R. scariósum**
Plants 30–140 cm high, perennials with stems up to 5 cm thick; flowering in June and July

Rhapónticum scariósum Lam.
Centaúrea rhapónticum L.
Giant knapweed

Asteráceae (Compósitae) – Sunflower family

Leaves: Narrow oval to lanceolate, widest towards base, coming to a gradual point, finely toothed or crenate, sessile on a narrow base, green above with isolated hairs, thickly covered with cottony hairs beneath; spirally arranged

Inflorescence: Very large solitary terminal composite flower; involucre up to 10 cm across, with numerous brown terminal appendages irregularly fringed at the margins

Flowers: Tubular florets of equal length, pink or purple and with a double pappus; 5 stamens; inferior ovary

Fruits: Brown cylindrical achenes up to 9 mm long with reddish pappus bristles

Habitat: Usually in subalpine zone, in moist meadows, shrubby undergrowth and thickets, on moist, base-rich, usually calcareous clay soil rich in nutrients; an alpine plant not often found

582 **Mountain thistle – C. personátus**
Plants 40–160 cm high, perennials with thick rhizomes and branching stems; flowering till August

Cárduus personátus (L.) Jacq.
Mountain thistle

Asteráceae (Compósitae) – Sunflower family

Leaves: Upper leaves narrow oval or oval, with long points, toothed, rather hairy or smooth above, with few hairs or thickly covered with grey cottony hairs and decurrent; lower leaves pinnately cut, coming to long points, with tips pointing forwards and slightly decurrent with narrow bases; leaves spirally arranged

Inflorescence: Composite flower heads solitary or in globular clusters at ends of main stem and branches; outer bracts, particularly upper ones, have non-prickly spines pointing upwards

Flowers: Rough pappus without feathery hairs; 5 petals, with petal tips as long as the corolla tube, and deep red; inferior ovary

Fruits: Achenes 2–4 mm long with pappus up to 15 mm long

Habitat: In mountain and subalpine zones on streambanks, thickets, shrubby undergrowth, alder thickets, damp waste land and wet meadows, on wet or moist, base-rich clay containing nutrients

583 **Creeping thistle – C. arvénse**
Plants 30–120 cm high, perennials with branching upper parts; flowering from July to September

Círsium arvénse (L.) Scop.
Creeping thistle

Asteráceae (Compósitae) – Sunflower family

Leaves: Basal leaves narrow oval to lanceolate, narrowed to a stalk-like base, rather rigid, smooth above, smooth or hairy beneath, more or less pinnately cut, with triangular teeth and lobes ending in strong spines; middle and upper leaves similar but sessile or semi-amplexicaul and more deeply pinnately cut; leaves spirally arranged

Inflorescence: Composite flower heads solitary or in clusters of 2–3 at ends of stem and branches; outer bracts have non-prickly spines

Flowers: All tubular florets; 5 fused petals, deeply divided and lilac; 5 stamens; inferior ovary

Fruits: Achenes 2–3 mm long, brownish and with pappus up to 3 mm long; dispersal by wind

Habitat: In hill and mountain zones in fields, roadsides, stream banks, forest clearings and waste land, on well-drained to fairly dry, lime-poor or lime-rich, stony, sandy, or loamy soil containing nutrients and humus and with deep topsoil; a very troublesome weed of cultivated land as it can regenerate from fragments of root; plant likes light; range extends north to 68 degrees 50 minutes; introduced to the USA

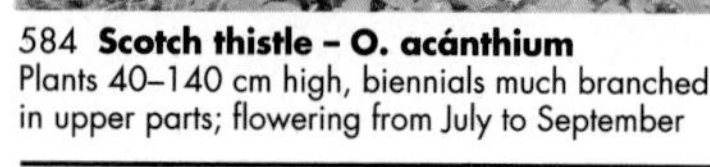

584 **Scotch thistle – O. acánthium**
Plants 40–140 cm high, biennials much branched in upper parts; flowering from July to September

Onopórdum acánthium L.
Scotch thistle

Asteráceae (Compósitae) – Sunflower family

Leaves: Narrow oval to oval, pinnately cut to the middle, dark green and hairy above, grey-green and with cottony hairs beneath, sessile and decurrent; leaflets triangular, ending in sharp spines

Inflorescence: Composite flower heads 2–4 cm in diameter, erect and solitary; bracts arranged into a sphere with a base 2–4 mm wide, dark-green below, ending in sharp spines

Flowers: Pappus has reddish bristles; petals fused into tubular florets, up to 2 cm long and purple; 5 stamens; inferior ovary

Fruits: Achenes 3–5 mm long, with longitudinal ridges and pappus 4–10 mm long, directly attached to the fruit

Habitat: In hill and mountain zones in fertile weed beds, rubbish dumps, waste land, ditches, camp sites, and roadsides, on dry, stony, sandy or pure clay or loam containing nutrients and humus

585 **Alpine thistle – C. deflorátus**
Plants 15–80 cm high, perennials with ascending or erect rhizomes; flowering till August

Cárduus deflorútus L.s.l.
Alpine thistle

Asteráceae (Compósitae) – Sunflower family

Leaves: Lanceolate, blunt-ended or pointed, generally smooth on both sides, simple with spiny teeth or pinnately cut with spines, sessile and slightly decurrent; spines not very prickly

Inflorescence: Solitary, spherical, composite flowers up to 3 cm across, usually nodding at flowering; bracts lanceolate, up to 10 mm long and narrowing to short spines

Flowers: Pappus rough, without feathery hairs; 5 fused petals to each tubular floret, up to 2 cm long and purple; inferior ovary

Fruits: Achenes 3–5 mm long and with pappus up to 15 mm long

Habitat: From mountain to alpine zones in scree, sunny rock slopes, dry woodlands and rough pasture, on well-drained to fairly dry, usually calcareous, loose, stony loam or clay containing nutrients and more or less humus; an early coloniser of fresh soil; binds soil together; a plant of central and southern European mountains; very variable in form

586 **Plume knapweed – C. nervósa**
Plants 10–35 cm high, perennials with thick grey hairs and blooming till August

Centaúrea nervósa Willd.
Centaúrea uniflóra ssp. nervósa Rouy
Plume knapweed

Asteráceae (Compósitae) – Sunflower family

Leaves: Wide lanceolate to narrow oval, entire or finely or coarsely toothed, with long points, sessile in upper part, in lower part narrowing to stalks or pinnately cut to middle; spirally arranged

Inflorescence: Solitary, terminal, composite flowers; bracts 1–3 mm wide, fringed and covered with brown appendages

Flowers: Tubular florets, outer longer than inner; 5 fused petals, generally dark purple; 5 stamens; inferior ovary

Fruits: Achenes 2–4 mm long, grey-brown, with pappus bristles up to 3 mm long

Habitat: In subalpine (rarely mountain or alpine) zone in moutain meadows, rough grassland and thickets, on damp, base-rich, usually lime-poor soil containing nutrients and humus; a plant of the central and southern European mountains

587 **Brown-rayed knapweed – C. jácea**
Plants 10–50 cm high, perennials with oblique stems; flowering from June to September

Centaúrea jácea L.s.l.
Brown-rayed knapweed

Asteráceae (Compósitae) – Sunflower family

Leaves: Lower leaves oval or lanceolate, blunt-ended, round or coming to short points, narrowing to short stalks,entire, finely toothed, or (especially towards base) irregularly lobed; upper leaves lanceolate, sessile and mostly pointed; all leaves spirally arranged

Inflorescence: Solitary, teminal, composite flowers; appendages to bracts up to 4 mm long, light or dark brown and largely covering green part of bracts

Flowers: Tubular florets, outer longer than inner; 5 fused petals, purple; 5 stamens; inferior ovary

Fruits: Achenes up to 3 mm long, without pappus

Habitat: From hill to subalpine zones in grassland, moors, waste land, and thickets, on damp to moderately dry, sometimes seasonally flooded, base-rich, calcareous loam which contains nutrients and humus and usually has deep topsoil

588 **Saw-wort – S. tinctória**
Plants 20–90 cm high, perennials branching in upper parts; flowers from July to September

Serrátula tinctória L.
Saw-wort

Asteráceae (Compósitae) – Sunflower family

Leaves: Lower leaves oval to lanceolate, whole, on stalks, sharply pointed and toothed; middle and upper leaves pinnately cut almost to the middle and sessile or with short stalks; segments sharply pointed; end leaflet larger than lateral ones

Inflorescence: Loose corymbs of composite flowers, usually with cylindrical involucre; male and female heads present, female larger than male; bracts greenish or brownish with violet tips

Flowers: 5 petals in tubular florets, in 5 parts towards tip and purple; female florets have corolla swollen in middle, white abortive anthers, and spreading style branches; male florets lack swollen corolla, have dark-blue anthers and abortive style branches; inferior ovary

Fruits: Greenish achenes 3–6 mm long, with pappus up to 8 mm long; dispersal by wind

Habitat: In hill and mountain zones in wood margins, clearings, and rides, grassland, moors, ditches and reed-beds, on moist or seasonally dry, base-rich, usually calcareous, often peaty loam or clay with nutrients and rotted humus; once used as a dye plant; plant grows in light or half-shade; range extends to southern Scandinavia

589 **Marsh thistle – C. palústre**
Plants 30–120 cm high, biennials with short rhizomes; flowering from June to October

Círsium palústre (L.) Scop.
Marsh thistle

Asteráceae (Compósitae) – Sunflower family

Leaves: Rigid, narrow oval to lanceolate, pinnately cut by wide bays, dark green above with few hairs, grey-green beneath and cottony (especially when young), sessile and amplexicaul with long lobes; leaflets end in short spines

Inflorescence: Crowded clusters of composite flowers; bracts narrow oval, pointed, dark green, often with white margins and dark points

Flowers: Tubular florets; 5 petals, divided into 5 parts towards tip and purple; 5 stamens; inferior ovary

Fruits: Yellowish-brown achenes 2–3 mm long, without spots and with pappus 5–10 mm long

Habitat: In hill and mountain zones in marshes, moors, wet grassland, hedgerows, and woodland on wet or seasonally moist, base-rich, neutral or rather acid, sandy or pure clay or loam with nutrients and humus; plant grows in light or half-shade; an indicator of moisture

Círsium acaúle Scop.
Stemless thistle

Asteráceae (Compósitae) – Sunflower family

Leaves: In a rosette, rigid, narrow oval, pinnately cut to near middle vein, rather hairy on veins; leaflets have spiny points

Inflorescence: Solitary, teminal, composite flower almost without stem; bracts lanceolate, green, nearly without spines but pointed

Flowers: Tubular florets; 5 fused petals up to 35 mm long, divided into 5 towards tip and purple; 5 stamens; inferior ovary; fertilised mainly by bumble-bees

Habitat: In mountain and subalpine zones in rough pasture and meadows, on seasonally moist or moderately dry, rather stony clay or loam containing lime, humus, and moderate amounts of nutrients; a weed in grazing land; an indicator of loam; a central European plant with a range extending to southern Sweden

590 **Stemless thistle – C. acaúle**
Plants 5–20 cm high, perennials with rhizomes; flowering from July to September

591 **Woolly thistle – C. erióphorum**
Plants 40–140 cm high, biennials with thick tap-roots; flowering from July to September

Círsium erióphorum (L.) SCOP.s.l.
Woolly thistle

Asteráceae (Compósitae) – Sunflower family

Leaves: Rigid, pinnately cut nearly to middle vein, dark green above and covered with rough hairs, thickly covered in white hairs beneath, with edges often rolled down to some extent; leaflets narrow lanceolate, ending in long, yellow spines; veins white or pale yellow; leaves arranged spirally

Inflorescence: Solitary, terminal, composite flowers 4–8 cm wide on stem and branches; bracts thickly covered in woolly hairs and with long spines

Flowers: Tubular florets; 5 petals, fused towards base, 2–4 cm long and purple or reddish- violet; 5 stamens; inferior ovary; fertilised by insects

Fruits: Achenes 4–6 mm long, yellow streaked with black, with pappus 3 cm long

Habitat: In mountain and subalpine zones in grassland, open scrub, roadsides, stream banks, and forest clearings, on well-drained or fairly dry, base-rich clay or loam containing nutrients, lime, and humus

592 **False saw-wort – C. vulgáris**
Plants 15–60 cm high, annuals with oblique branches and tap-roots; flowering from May to July

Crupína vulgáris Cass.
False saw-wort

Asteráceae (Compósitae) – Sunflower family

Leaves: Pinnately cut to the middle vein; leaflets narrow lanceolate, finely toothed, and covered with glands

Inflorescence: Composite flowers on stems in loose racemes or panicles; involucre up to 2 cm long, cylindrical, narrowing somewhat towards the top, and with few bracts, lanceolate, translucent white at the margins, arranged like tiles, and with globular glands; receptacle covered in scales

Flowers: Tubular florets, 3–8 on each flower head, neuter at edge and hermaphrodite inside; pappus of several rows of rough dark bristles; 5 fused petals, separating to form a 5-part cone at the tip; 5 stamens; inferior ovary; fertilised by insects

Fruits: Dark brown achenes 3–5 mm long with pappus of equal length

Habitat: In hill (rarely mountain) zone on dry, warm, stony hills and on rocks, on dry stony soil

593 **Mountain aster – A. alpínus**
Plants 5–20 cm high, perennials with thin rhizomes; flowering from June to August

Aster alpínus L.
Mountain aster

Asteráceae (Compósitae) – Sunflower family

Leaves: Narrow oval to lanceolate, broadest towards tips, entire, gradually narrowing to stalks, but in upper part of stem sessile and rounded or blunt-ended at tips

Inflorescence: Solitary, terminal, composite flowers 3–5 cm wide ; bracts lying close, hairy at margins, and 6–12 mm long

Flowers: Ray-florets lanceolate, spread out flat in a single row and usually bluish-violet; inner tubular florets yellow; 5 stamens; inferior ovary; fertilised by insects

Fruits: Achenes up to 3 mm long, rather hairy and with yellowish pappus up to 6 mm long; dispersal by wind

Habitat: In subalpine and alpine zones in sunny scree and meadows, on well-drained to dry , flat, stony clay or loam containing lime and humus; a common arctic/alpine plant

594 **Attic fleabane – E. átticus**
Plants 15–50 cm high, perennials with slightly oblique stems; flowering from July to September

Erígeron átticus Vill.
Attic fleabane

Asteráceae (Compósitae) – Sunflower family

Leaves: Narrow oval to lanceolate, entire, generally broadest towards tips, narrowing to winged stalks, and on upper stem generally sessile

Inflorescence: 3–10 composite flowers in a loose panicle; flower heads up to 35 mm in diameter, with involucre 5–10 mm long; bracts lanceolate, pointed, greenish, and thickly covered in glandular hairs

Flowers: Ray-florets clearly upright, 3–6 mm long and purplish-red; inner ray-florets atrophied into threads; disc-florets brownish at start of flowering, later yellow; 5 stamens; inferior ovary; fertilised by insects

Fruits: Achenes 2–3 mm long, covered in hairs, with pappus 4–6 mm long; dispersal by wind

Habitat: In subalpine (rarely alpine) zone on moraines, in stony grassland, and meadows, usually on dry, stony, lime-poor soil in warmer places

595 **Single-flowered fleabane – E. uniflórus**
Plants 3–10 cm high, perennials with stems thickly covered in red hairs; flowering till September

Erígeron uniflórus L.
Single-flowered fleabane

Asteráceae (Compósitae) – Sunflower family

Leaves: Basal leaves narrow oval, narrow oblong, or spathulate, with blunt or rounded tips, generally smooth above, hairy beneath, with long hairs on edges; stem leaves sessile and spirally arranged

Inflorescence: Solitary, terminal, composite flower; bracts lanceolate, brownish or reddish, pointed and covered in thick, white, woolly hair

Flowers: Ray-florets spread flat and white, lilac or pink, without thread-florets; tubular florets with 5 fused petals and yellowish; 5 stamens; inferior ovary; fertilised by insects

Fruits: Achenes 2–3 mm long, hairy, with pappus 3–5 mm long

Habitat: In alpine (rarely subalpine) zone on rock ledges and wind-blown ridges on well-drained or seasonally dry, base-rich, usually lime-poor, neutral or weakly acid, flat and stony loam or clay rich in humus

596 **Yarrow – A. millefólium**
Plants 15–40 cm high, with scattered hairs, branching in upper parts; flowering June to October

Achilléa millefólium L.
Yarrow, milfoil (pink-flowered variety)

Asteráceae (Compósitae) – Sunflower family

Leaves: Lanceolate to linear lanceolate and doubly or trebly pinnately cut as far as middle vein, with up to 50 leaflets on each side; last leaflets narrow lanceolate; there is a sterile basal rosette

Inflorescence: Composite flowers in terminal corymbs, with numerous brown, blunt bracts up to 6 mm long overlapping like tiles

Flowers: Ray-florets white or pink, female and only slightly drawn back after flowering; tubular florets whitish, hermaphrodite, and without spurs; fertilised by insects or self-fertile

Fruits: Achenes up to 2 mm long, flat oval, without toothed margins or pappus bristles; dispersal by wind and insects

Habitat: From hill to subalpine zones in meadows, arable fields, woodland, by the wayside and on exposed sand, on well-drained to moderately dry sandy, stony or loamy soil which is rich in nutriment

597 **Purple lettuce – P. purpúrea**
Plants 20–120 cm high, perennials with branching stems; flowering from July to Seprember

Prenánthes purpuúea L.
Purple lettuce

Asteráceae (Compósitae) – Sunflower family

Leaves: Narrow oval or lanceolate, notched or pinnately cut, pointed, with heart-shaped base, parially amplexicaul, smooth, mid-green above and grey-green beneath, and in lower part of plant narrowing to winged stems; upper leaves sessile

Inflorescence: Numerous panicles containing 2–5 composite flowers; outer bracts much smaller than inner; both small and green

Flowers: Only ray-florets present, each of 5 fused petals, with 5 teeth, and violet or purple; 5 stamens, reddish and white; inferior ovary

Fruits: Achenes 3–5 mm long, with 3–5 ridges and white pappus

Habitat: From hill to subalpine zones in woodlands with good ground cover, shrubby undergrowth, and along forest paths, on damp to well-drained, generally lime-poor, rather acid loam containing nutrients and humus, in humid conditions; plant grows in shade or half-shade

598 **Orange hawkweed – H. aurantiacum**
Plants 15–40 cm high, perennials with runners both above and below ground; flowering till August

Hierácium aurantiácum L.
Orange hawkweed

Asteráceae (Compósitae) – Sunflower family

Leaves: Narrow oval to lanceolate, narrowing towards base, entire or with isolated fine points, generally with short hairs on both sides and green

Inflorescence: Panicles, often umbrella-shaped, containing 2–12 composite flowers; bracts lanceolate with dark-coloured, stellate, glandular and simple hairs up to 3 mm long

Flowers: Only ray-florets present, yellow-orange to brownish-red; 5 stamens; inferior ovary

Fruits: Dark-coloured achenes up to 2.5 mm long, with yellowish-white pappus; dispersal by wind and vegetative reproduction by runners

Habitat: In mountain and subalpine zones in siliceous rough pasture and grassland on well-drained or seasonally well-watered, lime-poor, often rather acid clay or loam containing humus; also cultivated as a garden plant; plant grows in light; a European mountain plant, often found wild in rough grassland in parks and by roadsides

599 **Golden hawk's beard – C. aúrea**
Plants 5–25 cm high, perennials with rhizomes, upright; flowering from June to August

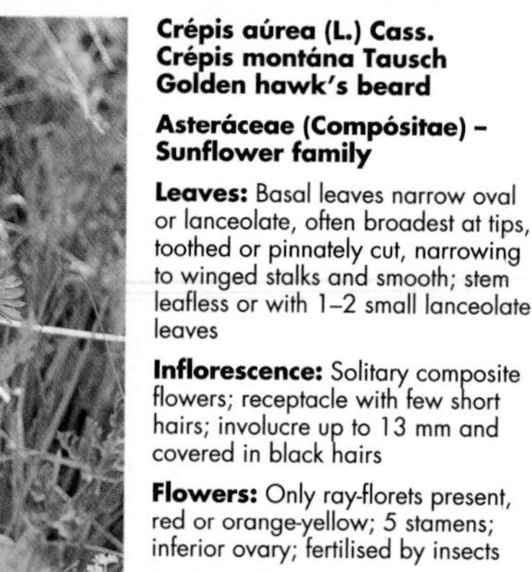

Crépis aúrea (L.) Cass.
Crépis montána Tausch
Golden hawk's beard

Asteráceae (Compósitae) – Sunflower family

Leaves: Basal leaves narrow oval or lanceolate, often broadest at tips, toothed or pinnately cut, narrowing to winged stalks and smooth; stem leafless or with 1–2 small lanceolate leaves

Inflorescence: Solitary composite flowers; receptacle with few short hairs; involucre up to 13 mm and covered in black hairs

Flowers: Only ray-florets present, red or orange-yellow; 5 stamens; inferior ovary; fertilised by insects

Fruits: Achenes 4–6 mm long, with up to 20 ridges, narrowing towards top and with pappus 5–7 mm long; dispersal by wind

Habitat: In subalpine and alpine (rarely mountain) zones round huts, by paths and camp sites and in fertile meadows, on well-drained, neutral or weakly acid loam rich in bases and nitrogen and containing nutrients and humus; a good indicator of nitrogen

600 **Red clover – T. praténse**
Plants 10–25 cm high, usually perennials with upright stems; flowering from May to October

Trifólium praténse L.
Red clover

Fabáceae (Papilionáceae) – Pea family

Leaves: Ternate, on stems; leaflets oval, blunt-ended, round, rather pointed or shallowly notched at tips, usually entire, often with lighter green, whitish, or reddish markings, somewhat hairy and up to 3 cm long

Inflorescence: Egg-shaped or spherical raceme, solitary or in small groups at ends of stem and branches

Flowers: 5 fused sepals, whitish-green and rather hairy, with thread-like greenish points; 5 red petals 5 times as long as the calyx tube (10–15 mm long); standard significantly longer than wings or keel; 10 stamens; superior ovary; fertilised by insects

Fruits: Pods

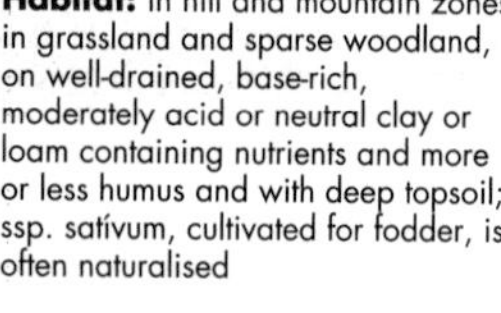

Habitat: In hill and mountain zones in grassland and sparse woodland, on well-drained, base-rich, moderately acid or neutral clay or loam containing nutrients and more or less humus and with deep topsoil; ssp. sativum, cultivated for fodder, is often naturalised

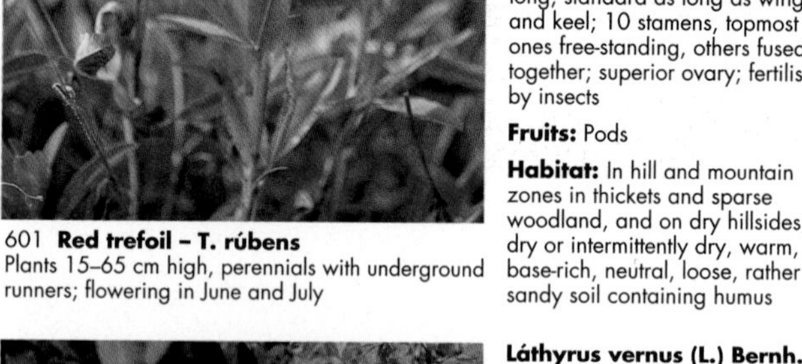

601 **Red trefoil – T. rúbens**
Plants 15–65 cm high, perennials with underground runners; flowering in June and July

Trifólium rúbens L.
Red trefoil

Fabáceae (Papilionáceae) – Pea family

Leaves: Ternate, on short stalks, and usually smooth; leaflets lanceolate, 3–6 cm long, blunt-ended, pointed or somewhat notched at tips, and finely toothed; stipules plentiful and very long

Inflorescence: Egg-shaped or cylindrical terminal flower heads on stems 2–6 cm long

Flowers: 5 fused sepals, greenish, clearly covered in hairs in upper part, and with a distinct ring of hairs on the inner rim of the calyx tube; 5 purple petals up to 15 mm long; standard as long as wings and keel; 10 stamens, topmost ones free-standing, others fused together; superior ovary; fertilised by insects

Fruits: Pods

Habitat: In hill and mountain zones in thickets and sparse woodland, and on dry hillsides, on dry or intermittently dry, warm, base-rich, neutral, loose, rather sandy soil containing humus

602 **Alpine clover – T. alpínum**
Plants 5–15 cm high, perennials with brown fibres on the rhizomes; flowering from June to August

Trifólium alpínum L.
Alpine clover

Fabáceae (Papilionáceae) – Pea family

Leaves: Ternate and on stalks; leaflets linear lanceolate to narrow oval, usually pointed, 2–7 cm long, mostly entire, smooth and broadest towards the base or in the middle

Inflorescence: Terminal racemose flower heads with up to 12 flowers

Flowers: On stems and scented; 5 fused sepals, whitish green, smooth and with narrow, triangular, smooth point; 5 petals, flesh-pink and up to 8 times as long as calyx tube; standard clearly longer than wings or keel; 10 stamens; topmost stamens free-standing, others fused together; superior ovary; fertilised by insects

Fruits: Pods

Habitat: In subalpine and alpine zones in scrub, meadows, and rough grassland, on generally dry, lime-poor soils in places warm in summer; roots more than 1 m deep; a mountain plant of central and southern Europe

603 **Spring vetchling – L. vernus**
Plants 15–30 cm high, perennials with short branching rhizomes; flowering in April and May

Láthyrus vernus (L.) Bernh.
Spring vetchling

Fabáceae (Papilionáceae) – Pea family

Leaves: On stalks, with 4–8 leaflets and often with an awned tip; leaflets linear lanceolate to oval, with long points, usually without stems, broadest towards the base, often covered in short hairs and up to 7 cm long

Inflorescence: Racemes on long stems, with 3–7 flowers

Flowers: 5 fused sepals, whitish green, often with a reddish overlay and with calyx teeth of unequal length; 5 petals, up to 2 cm long, with standard, 2 wings, and a keel, initially red, then blue and finally greenish-blue; 10 stamens, with all filaments fused together; superior ovary; fertilised by insects (especially bees)

Fruits: Smooth pods up to 6 cm long

Habitat: In hill and mountain zones in beech and pine woods with good ground cover, on moist to well-drained, base-rich, calcareous, loose, clay or loam of moderate depth containing humus and nutrients; plant grows in shade or half-shade

604 **Mountain milk-vetch – O. jacquínii**
Plants 5–20 cm high, perennials with branching rhizomes; flowering in July and August

Oxytrópis jacquínii Bunge
Oxytrópis montána DC.
Mountain milk-vetch

Fabáceae (Papilionáceae) – Pea family

Leaves: Narrow oval or lanceolate, unequally pinnate and with reddish-brown stalks; 13–20 pairs of leaflets, narrow oval, pointed, or blunt-ended, and covered in scattered hairs or smooth

Inflorescence: Long-stemmed spherical or egg-shaped racemes

Flowers: Stems lacking or very short; 5 fused sepals, reddish, with short black hairs and calyx points narrow triangular and shorter than corolla tube; 5 petals, generally purple-violet, up to 1.4 cm long; keel has a short narrow point; standard much larger than keel or wings; 10 stamens, 9 fused together; superior ovary

Fruits: Single-celled cylindrical pods 2–3 cm long, with dark hairs

Habitat: In alpine (rarely subalpine) zone on rubbish dumps, banks, scree, and cliffs, on moderately dry calcareous soils

605 **Crown vetch – C. vária**
Plants 40–80 cm high, perennials with recumbent or upright stems; flowering until August

Coronilla varia L.
Crown vetch

Fabáceae (Papilionáceae) – Pea family

Leaves: Narrow oval, unequally pinnate, and on very short stalks; 6–12 pairs of leaflets, short-stalked, narrow oval, entire, and with short narrow points

Inflorescence: Long-stemmed racemes, generally spherical and with up to 20 flowers

Flowers: Short-stemmed; 5 fused sepals, yellowish-green and with short calyx teeth drawing to points; corolla of lilac standard, 2 whitish or pink wings, and a white keel with a dark purple, bent point; 10 stamens, topmost one not fused with the rest; superior ovary

Fruits: Pods up to 8 cm long, incised and later breaking up into 1-seeded sections

Habitat: In hill and mountain zones in banks, thickets, quarries, rough pasture, dry meadows, by the roadside, and in ditches, on generally dry, calcareous, base-rich soil in warmer places; extends northwards to England

606 **Mountain kidney-vetch – A. montána**
Plants 5–20 cm high, perennials, recumbent or upright; flowering in June and July

Anthyllis montána L.
Mountain kidney-vetch

Fabáceae (Papilionáceae) – Pea family

Leaves: Narrow oval, unequally pinnate, and on short stalks; 6–13 pairs of leaflets, narrow oval, sessile, with thick silky hairs on both sides, and up to 1 cm long

Inflorescence: Dense, round racemes surrounded by radial bracts

Flowers: Short-stemmed; 5 fused sepals, whitish or pale green, with thick hairs and with calyx points narrow triangular, often reddish and 3–5 mm long; 5 petals, generally purple, divided into standard, 2 smaller wings, and a small keel; 10 stamens, filaments nearly all fused together; superior ovary; fertilised by insects

Fruits: Pods 3–6 mm long

Habitat: From hill to subalpine zones on scree, rocky slopes, and in dry grassland, on dry, stony, calcareous soil in warmer places; a plant of central and southern European mountains with a range extending to the north Spanish mountains and the Pyrenees; rare

607 **Mountain sainfoin – O. montána**
Plants 10–30 cm high, with recumbent or upright stems; flowering in July and August

Onóbrychis montána DC.
Mountain sainfoin

Fabáceae (Papilionáceae) – Pea family

Leaves: Narrow oval or oval, unequally pinnate and on stalks; pinnae lancolate or narrow oval, sessile or short-stalked, pointed, flat or rounded at tip, and often with a short narrow point

Inflorescence: Dense, round, long-stemmed racemes

Flowers: Generally without stems; 5 fused sepals, greenish or reddish, with calyx teeth up to twice as long as calyx tube; corolla 10 to 15 mm long and purplish red; keel generally longer than banner; wings about as long as calyx; superior ovary; fertilised by insects

Fruits: Pods 5 – 8 mm long, with teeth up to 1.5 mm long; dispersion by animals

Habitat: In subalpine (rarely mountain) zone in meadows, calcareous rough grassland and stony ground, on sunny, dry, base-rich, lime-rich soil containing humus; useful fodder plant in meadowland; deep-rooted; mountain plant of Central and Southern Europe

608 **Sainfoin – O. viciifólia**
Plants 25–30 cm high, perennials without runners, generally with upright stems; flowering till August

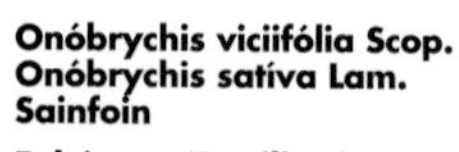

Onóbrychis viciifólia Scop.
Onóbrychis sativa Lam.
Sainfoin

Fabáceae (Papilionáceae) – Pea family

Leaves: Narrow oval, unequally pinnate and with stalks; 6–13 pairs of leaflets, lanceolate, short-stalked, and rounded, flat or provided with a narrow point at tip

Inflorescence: Narrow oval or triangular many-flowered raceme

Flowers: Generally without stems; 5 fused sepals, greenish, with reddish, hairy calyx rim; calyx teeth up to 3 times as long as calyx tube; corolla 10–15 mm long and generally pink, rarely whitish; keel often as long as standard; wings shorter than calyx; 10 stamens; superior ovary; fertilised by insects

Fruits: Pods 5–8 mm long, with teeth 0.5–1 mm long

Habitat: In hill and mountain (rarely subalpine) zones in ditches, thickets, dry meadows, and by the roadside, on warm, moderately dry or dry, infertile, base-rich, loose soil containing humus and with deep topsoil; a useful fodder plant on dry grassland and often cultivated

609 **Hungarian sainfoin – O. arenária**
Plants 20–50 cm high, perennials, usually with curved ascending stems; flowering in June and July

Onóbrychis arenária (Kit.)Ser.
Hungarian sainfoin

Fabáceae (Papilionáceae) – Pea family

Leaves: Narrow oval, unequally pinnate and stalked; 5–12 pairs of leaflets, lanceolate, often short-stalked, and mostly pointed

Inflorescence: Narrow oval, often elongated, many-flowered racemes

Flowers: With no or only short stems; 5 fused pale green sepals; calyx teeth long and narrow triangular, dark green, with whitish or pale green edges; corolla 8–12 mm long; standard and keel pale pink or white, generally with red veins and equal in length; wings much shorter than calyx; 10 stamens; topmost stamen free-standing; superior ovary; fertilised by insects

Fruits: Pods 3–6 mm long, with teeth up to 1.5 mm long; dispersal by being caught in animals' fur

Habitat: In hill and mountain zones on rocky ground, dry grassland, calcareous rough pasture, dry pinewoods, and dry, warm hills, on dry, base-rich, stony loam or calcareous sand which is warm in summer; roots more than 1 m deep; rare

610 **Alpine sainfoin – H. hedysaroídes**
Plants 15–25 cm high, perennials with underground runners; flowering in July and August

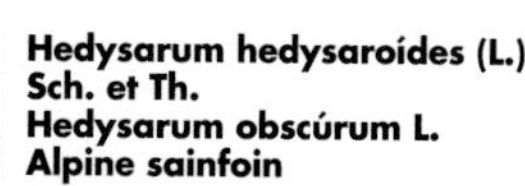

Hedysarum hedysaroídes (L.) Sch. et Th.
Hedysarum obscúrum L.
Alpine sainfoin

Fabáceae (Papilionáceae) – Pea family

Leaves: Oval to rectangular, unequally pinnate and on stalks; 5 – 9 pairs of leaflets, narrow oval, short-stalked, or sessile, 1–3 cm long, entire and with round or blunt tips

Inflorescence: Racemes with up to 35 flowers on dark red stems

Flowers: On short stems, bending down slightly; 5 fused sepals, oval, with scattered hairs and dark red; calyx teeth 1–3 mm long and reddish-black; corolla 1.5–2.5 cm long and purple; keel much longer than wings and standard; 10 stamens; superior ovary; fertilised by insects

Fruits: Pods 15–40 mm long, usually smooth, and with 3–6 ridges; dispersal usually by wind

Habitat: In subalpine and alpine zones in meadow land, rough grassland, scree, cliffs, and scrub, on well-drained soils which are more or less base-rich and rich in nutrients

611 **Common vetch – V. satíva**
Plants 15–80 cm high, more or less covered in hairs; flowering from May to October

Vícia satíva L. s.l.
Common vetch

Fabáceae (Papilionáceae) – Pea family

Leaves: Oval, pinnate, with terminal tendril(s) and usually on stalks; 4–7 pairs of leaflets, lanceolate, short-stalked or sessile, gradually narrowing towards base, round at tips and usually with short narrow points; stipules half-arrow-shaped and toothed

Inflorescence: Flowers solitary or in pairs or threes, usually sessile in leaf axils

Flowers: 5 sepals fused in reddish-green calyx tube; calyx teeth as long as tube and pointing straight forward; corolla up to 3 cm long and often multicoloured; standard pale pink to pale red-violet and pointing straight up; keel smaller and greenish-white to reddish; wings often rather darker than standard; 10 stamens; superior ovary;

Fruits: Pods 35–70 mm long

Habitat: In hill and mountain zone in cornfields, arable fields, by roadsides, and on waste ground, on well-drained calcareous loam containing nutrients; ssp. satíva is cultivated as a fodder plant

612 **Bush vetch – V. sépium**
Plants 20–50 cm high, perennials with upright or climbing stems; flowering till June

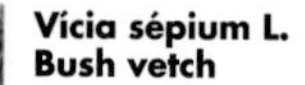

Vícia sépium L.
Bush vetch

Fabáceae (Papilionáceae) – Pea family

Leaves: Oval, pinnate, with branching terminal tendrils and short-stalked; 4–7 pairs of leaflets, long oval, short-stalked, up to 3 cm long and blunt-ended or pointed; stipules often have nectar glands on the outside, visited by bees

Inflorescence: Short racemes with 3–6 flowers

Flowers: 5 sepals fused into a dark red-brown calyx tube and hairy; calyx teeth 10–15 mm long and pink to brownish-violet; standard and wings usually much bigger than keel; 10 stamens; superior ovary; fertilised by insects

Fruits: Flat, smooth pods 20–35 mm long

Habitat: In hill and mountain zones in thickets, meadows, woodland, along paths and at wood margins, on moist to well-drained, base-rich, calcareous, loose loam or clay with nutrients and humus; a good indicator of fertility; plant grows in light or half-shade

613 **Round-leaved rest-harrow – O. rotundifólia**
Plants 10–35 cm high, perennials with woody rhizomes; flowering from May to July

Onónis rotundifólia L.
Round-leaved rest-harrow

Fabáceae (Papilionáceae) – Pea family

Leaves: Oval, unequally ternate, on long stalks , with finely toothed stipules opening flat; lateral lobes oval to round, toothed, 1–3 cm long and usually sessile; end lobe broad oval or round, notched and long-stalked

Inflorescence: Up to 3 flowers on long stems rising from leaf axils

Flowers: 5 fused sepals, greenish and rather hairy; calyx teeth narrow triangular and dark green; corolla 15–25 mm long and pink; standard much larger than keel and wings and often rather hairy; 10 stamens, all fused together; superior ovary

Fruits: Pods 20–35 mm long, usually flat and containing 9 seeds

Habitat: In hill and mountain (rarely subalpine) zones on cliffs, and in sparse pine woods, on dry, stony, calcareous soil in warmer places; a mountain plant of central and southern Europe

614 **Rest-harrow – O. spinósa** Plants 25–60 cm high, perennials with woody rhizomes, covered in glandular hairs; flowering till September

Onónis spinósa L.s.l.
Rest-harrow

Fabáceae (Papilionáceae) – Pea family

Leaves: Usually triangular, unequally ternate, and short-stalked; lobes narrow oval to oval, generally all stalked, toothed, and whitish-green; stipules toothed; thorns present in leaf axils on branches

Inflorescence: Solitary short-stemmed flowers in axils of upper stem leaves

Flowers: 5 fused sepals, green and with glandular hairs; calyx teeth longer than calyx tube; corolla 10–25 mm long, pink to violet-pink, often whitish; standard rather hairy outside and about as long as keel; 10 stamens; superior ovary; fertilised by insects

Fruits: Pods, inflated, covered in glandular hairs and up to 10 mm long

Habitat: In hill and mountain zones on dry meadows, roadsides, in thickets, and on moorland, on moderately dry, base-rich, usually calcareous, neutral loam or clay with humus; an indicator of infertility; a weed on pasture

615 **Whorled lousewort – P. verticilláta**
Plants 5–25 cm high, perennials with scattered hairs; flowering from June to August

Pediculáris verticilláta L.
Whorled lousewort

Scrophulariáceae – Figwort family

Leaves: Stem leaves in whorls of 3–4, pinnately cut to middle vein, smooth or slightly hairy and 3–6 cm long; leaflets irregularly toothed; leaf-bracts in flower spikes rather smaller and covered with scattered hairs

Inflorescence: Dense racemes

Flowers: 5 fused sepals 4–6 mm long, rounded at the base, divided only on the underside, with hairs on veins and with short teeth; corolla 10–18 mm long and purple; lower lip protruding and as long as upper lip; this is helmet-shaped, notched on the front edge and not toothed; 4 stamens enclosed by upper lip; the two longer filaments are hairy; superior ovary; fertilised by bumble-bees

Fruits: Egg-shaped capsules up to 12 mm long, with crooked points

Habitat: In subalpine and alpine zones on meadows, in calcareous rough grassland and wet moorland, on moist to well-drained, base-rich, generally calcareous, neutral, stony clay or loam containing humus; also found as a marsh plant

616 **Rhaetian lousewort – P. kérneri**
Plants 5–15 cm high, with recumbent or ascending stems; flowering in July and August

Pediculáris kérneri D.T.
Rhaetian lousewort

Scrophulariáceae – Figwort family

Leaves: Pinnately cut to middle vein, often rather hairy and dark green or reddish-brown; leaflets irregularly toothed

Inflorescence: 1–3 flowers on end of shoot

Flowers: 5 fused sepals, 5–9 mm long, narrowing towards base, generally reddish and often hairy; corolla 15–20 mm long and dark purple; upper lip bent forwards and elongated; lower lip protruding and smooth; 4 stamens enclosed by upper lip; superior ovary; fertilised by insects (bumble-bees)

Fruits: Egg-shaped capsules up to 15 mm long, with crooked points

Habitat: In alpine zone on rough grassland, meadows, and scree, on well-drained, stony, calcareous, rather acid soils containing humus; a plant of the Alps and Pyrenees, with a range extending east to Salzburg; fairly common in Switzerland in the central and southern Alps; less often found in northern Alps; a semi-parasitic plant

617 **Wild thyme – T. serpyllum**
Plants 5–25 cm high, perennials, woody in lowest parts; flowering from May to October

Thymus serpyllum L.s.l.
Wild thyme

Lamiáceae (Labiátae) – Mint family

Leaves: Oval or round, sessile or short-stalked, entire, blunt-ended or rounded at tips, and with edges flat or slightly rolled downwards

Inflorescence: Whorls forming cylindrical or spherical heads

Flowers: Short-stemmed; 5 sepals fused into 2-lipped calyx; corolla 2–6 mm long, pink or purple (rarely white); upper lip has 3 teeth and lower lip 2 elongated teeth; 4 stamens; superior ovary of 2 cells; fertilised by insects

Fruits: 4 nutlets

Habitat: From hill to subalpine zones on banks, cliffs, grassland, sparse pine woods, and rocks, on dry, base-rich, neutral, sandy ground lacking fine soil; the precise range of individual species in this genus awaits clarification

618 **Summer savory – S. horténsis**
Plants 10–25 cm high, annuals, not woody, with branching roots; flowering till September

Satureja hortensis L.
Summer savory

Lamiáceae (Labiátae) – Mint family

Leaves: Lanceolate or narrow lanceolate, 1–3 cm long, with short stalk, entire, with short hairs or smooth, and blunt or pointed at tip

Inflorescence: Solitary or in half-whorls of 2 or 3 flowers in axils of top stem leaves

Flowers: Short-stemmed; calyx in 5 parts, bell-shaped, 10-nerved, up to 6 mm long, often rather hairy and with 5 narrow, elongated teeth; calyx tube smooth inside; corolla up to 8 mm long and whitish or lilac; 4 stamens; superior ovary; style-branches of unequal length

Fruits: 4 nutlets

Habitat: In hill and mountain zones on waste ground, cliffs, and ditches, on moist to well-drained, loose soils with nutrients in warmer places; also forms weed beds; often planted in gardens as a pot-herb and medicinal plant; originally an eastern Mediterranean plant, its range has spread by cultivation; strongly scented, especially leaves

619 **Alpine calamint – A. alpínus**
Plants 10–25 cm high, perennials with woody rhizomes; flowering from July to September

Acinos alpínus (L.) Moench
Saturéja ácinos (L.) Scheele
Alpine calamint

Lamiáceae (Labiátae) – Mint family

Leaves: Elliptical or oval, short-stalked, often slightly toothed or entire, without rolled-down edges, 10–20 mm long, usually blunt-ended and opposite

Inflorescence: Usually 3 flowers in whorls in axils of topmost leaves of stem

Flowers: Short-stemmed; calyx in 5 segments, forming a bell, pouched in lower part, usually with 13 nerves, 4–7 mm long, usually pale green, hairy and with 5 long narrow teeth, those of lower lip being longer than those of the upper one; corolla 10–20 mm long and purple or violet; 4 stamens; superior ovary; fertilised by insects

Fruits: 4 nutlets

Habitat: Usually in subalpine zone, on rocks, dry grassland, stony ground, and sparse pine woods, on moist to dry, stony, generally calcareous, base-rich clay or loam containing humus; a mountain plant of central and southern Europe with a range extending to Sicily

620 **Marjoram – O. vulgáre**
Plants 20–50 cm high, perennial aromatics; flowering from July to September

Oríganum vulgáre L.
Marjoram

Lamiáceae (Labiátae) – Mint family

Leaves: Usually oval (broadest towards base), on stalks, 10–40 mm long, entire or obscurely toothed, dark green above, grey-green beneath, with blunt or rounded tips

Inflorescence: Panicles of dense cymes, with narrow oval bracts 3–6 mm long

Flowers: Short-stemmed; 5 fused sepals, pale or dark green, usually with red teeth; corolla pink or flesh-pink and 3–8 mm long; 4 stamens, protruding beyond corolla tube; 4 stamens with diverging halves; superior ovary; fertilised by insects; attracts bees

Fruits: 4 spherical or egg-shaped nutlets; dispersal by wind

Habitat: In hill and mountain (rarely subalpine) zones in dry pastures, hedgerows, wood margins, and sparse pine woods, thickets, and scrub, on fairly dry, base-rich, usually calcareous soil

621 **Wild basil – C. vulgáre**
Plants 15–60 cm high, perennials with underground runners; flowering from July to September

Clinopódium vulgáre L.
Saturéja vulgáris (L.) Fritsche
Wild basil

Lamiáceae (Labiátae) – Mint family

Leaves: Oval, 20–45 mm long, short-stalked, entire or slightly toothed, often covered thickly in hairs; rarely dotted with glands and with a blunt or rounded tips

Inflorescence: Calyx in 5 segments, forming a narrow calyx tube, 13-nerved, up to 10 mm long, 2-lipped and hairy inside; corolla carmine and 10–20 mm long; upper and lower lip with 3 and 2 teeth respectively; 4 stamens; superior ovary; fertilised by insects

Flowers: Calyx in 5 parts, narrowly tubular, with 13 veins, up to 10 mm long and divided into two lips; calyx tube hairy inside; corolla carmine abd 10–20 mm long; upper and lower lips with 3 and 2 teeth respectively; 4 stamens; superior ovary; fertilised by insects

Fruits: 4 nutlets; dispersal by wind and by being attached to animals' fur

Habitat: In hill and mountain zones in thickets, hedges, wood borders and scrub, by roadsides and in sparse oak and pine woods, on well-drained, base-rich, often calcareous, loose loam or clay soil containing nutrients and humus; plant grows in light or half-shade; range extends over all Europe except the Arctic

622 **Lesser calamint – C. nepéta**
Plants 20–60 cm high, perennials with erect or climbing stems; flowering till September

Calamíntha nepéta (L.) Scheele s.l.
Saturéja calamíntha Scheele
Lesser calamint

Lamiáceae (Labiátae) – Mint family

Leaves: Oval or round, on stalks, blunt or rounded at tips, hairy on both sides, 10–20 mm long and notched, toothed, or rarely entire

Inflorescence: Half-whorls of 3–5 flowers in leaf axils of upper stem leaves

Flowers: Short-stemmed; 5 fused sepals, with 13 nerves and pointed teeth; corolla 10–20 mm long and lilac or pale violet; upper lip with 3 teeth, lower lip with 2 teeth; 4 stamens; superior ovary; fertilised by insects

Fruits: 4 oval nutlets up to 1.5 mm long

Habitat: In hill (rarely mountain) zone by paths, in thickets, on rocks, walls, and scree, on moderately dry or dry, base-rich, calcareous soil containing humus but lacking fine topsoil in warmer places; a Mediterranean plant, with a range extending north to southern England and south to Algeria

623 **Bastard balm – M. melissóphyllum**
Plants 15–60 cm high, perennials with creeping rhizomes; flowering in May and June

Melíttis melissóphyllum L.
Bastard balm

Lamiáceae (Labiátae) – Mint family

Leaves: Oval or heart-shaped, 4–9 cm long, on stalks, generally round at base of leaves proper, with regular coarse teeth and covered in scattered hairs

Inflorescence: Solitary flowers or 2–3 flowers in whorls, on stems, in axils of upper stem leaves

Flowers: 5 sepals fused into a bell-shaped tube, with 10 nerves, pale to mid-green, often with glandular hairs on the nerves and 15–20 mm long; corolla 30–40 mm long, purple, pink, or white and with 2 lips; lower lip has red-violet spots and is in 3 segments, the middle one broad; upper lip entire and covered in fine glandular hairs; 4 stamens; superior ovary; fertilised by insects

Fruits: Egg-shaped nutlets, more or less triangular and up to 5 mm long

Habitat: In hill and mountain zones in sparse deciduous woodland and thickets, on moderately well-drained to fairly dry, base-rich, generally calcareous, loose, often stony soils containing humus

624 **Alpine bartsia – B. alpína**
Plants 10–25 cm high, perennials with underground creeping rhizomes; flowering till August

Bártsia alpína L.
Alpine bartsia

Scrophulariáceae – Figwort family

Leaves: Oval, pointed, sessile, with round or heart-shaped bases, regularly toothed and with dark violet overlays, particularly in the upper part of the plant

Inflorescence: Solitary flowers without stems in axils of top leaves

Flowers: 4 sepals fused into a bell, 5–8 mm long, with numerous glandular hairs, pale green and each with a violet tooth; 5 fused petals, covered in glandular hairs and 15–25 mm long; upper lip helmet-shaped, usually entire and longer than lower lip, which has 3 segments without fringed tips; 4 stamens included in corolla tube; anthers hairy in lower part; superior ovary; fertilised by insects (bumble-bees)

Fruits: Pointed exploding capsule with 2 flaps, containing many seeds

Habitat: In subalpine and alpine zones in meadows, wet moorland, rock ledges, and scrub, on wet or moist, base-rich soils fairly rich in nutrients

625 **Hedge woundwort – S. sylvática**
Plants 25–90 cm high, perennials with long underground runners; flowering till September

Stáchys sylvática L.
Hedge woundwort

Lamiáceae (Labiátae) – Mint family

Leaves: Oval to narrow heart-shaped, on long stalks, 4–10 cm long, pointed, coarsely toothed, hairy on both sides and with a heart-shaped base

Inflorescence: Whorls, generally with 6 flowers, forming a broken terminal spike

Flowers: 5 fused sepals, covered thickly in hairs, greenish-red and each with an awned point; 5 fused petals, 10–15 mm long, brownish red and 2-lipped; upper lip entire and often hairy; lower lip longer, with pale or dark marking and often 3-lobed; 4 stamens, protruding beyond the upper lip; superior ovary; fertilised by insects and self

Fruits: 4 nutlets 10–20 mm long

Habitat: In hill and mountain zones in wet woodlands, hedgerows, waste land, and meadows on damp to wet clay or loam, neutral and rich in nutrients and humus

626 **Water mint – M. aquática**
Plants 15–50 cm high, with shoots also bearing leaves in water; flowering from July to October

Mentha aquática L.
Water mint (dark green variety)

Lamiáceae (Labiátae) – Mint family

Leaves: Oval, on stalks, generally pointed, unequally toothed, 3–8 cm long, dark green above, whitish-green beneath, and hairy

Inflorescence: Terminal heads 2 cm across of 1–3 whorls, and usually 1–3 axillary whorls below

Flowers: On stems; 5 fused sepals, greenish or reddish, hairy, up to 5 mm long and each with a narrow triangular tooth; corolla 4–7 mm long, pink or lilac, with similarly shaped points and corolla tube smooth inside; 4 stamens, protruding beyond corolla tube; superior ovary; fertilised by insects

Fruits: 4 egg-shaped nutlets, up to 1 mm long and pitted

Habitat: In hill and mountain zones in swamps, marshes, fens, and wet woods and by rivers and ponds, on seasonally flooded or wet, base-rich, often peaty soil containing nutrients and rotted humus; plant grows in light or half-shade; a garden escape in the USA

627 **Water mint – M. aquática**
Plants 15–50 cm high, with shoots also bearing leaves in water; flowering from July to October

Mentha aquática L.
Water mint (pale green variety)

Lamiáceae (Labiátae) – Mint family

Leaves: Oval, on stalks, usually pointed, unequally toothed, 3–8 cm long, pale or mid-green, and hairy on both sides

Inflorescence: Terminal heads 2 cm across of 1–3 whorls, and usually 1–3 axillary whorls below

Flowers: On stems; 5 fused sepals, greenish or reddish, hairy, up to 5 mm long and each with a narrow triangular tooth; corolla 4 to 7 mm long, pink or lilac, with similarly shaped points and corolla tube smooth inside; 4 stamens, protruding beyond corolla tube; superior ovary; fertilised by insects

Fruits: 4 egg-shaped nutlets, up to 1 mm long and pitted

Habitat: In hill and mountain zones in swamps, marshes, fens, and wet woods and by rivers and ponds, on seasonally flooded or wet, base-rich, often peaty soil containing nutrients and rotted humus; plant grows in light or half-shade; a garden escape in the USA

628 **Spotted dead nettle – L. maculátum**
Plants 15–40 cm high, perennials with long runners; flowering from April to September

Lámium maculátum L.
Spotted dead nettle

Lamiáceae (Labiátae) – Mint family

Leaves: Broad heart-shaped or triangular, usually on red stalks, pointed and notched or toothed

Inflorescence: Dense axillary whorls among top pairs of leaves

Flowers: Sessile; 5 fused sepals, 7–12 mm long, greenish or violet, without spots, rather hairy and each with a narrow triangular tooth; corolla 20–30 mm long, generally purple and with a bent corolla tube; upper lip bent and with short hairs on the margin; lower lip often with a dark or whitish marking, and a lanceolate tooth for each lateral segment; 4 stamens, rising into the upper lip; anthers violet-brown, bearded, and with orange-yellow pollen

Fruits: 4 smooth nutlets up to 3 mm long; dispersal by ants

Habitat: From hill to subalpine zones in woodlands, hedges, thickets, by waysides, by fences and on stream banks, on damp to well-drained, generally calcareous, neutral soils containing nutrients

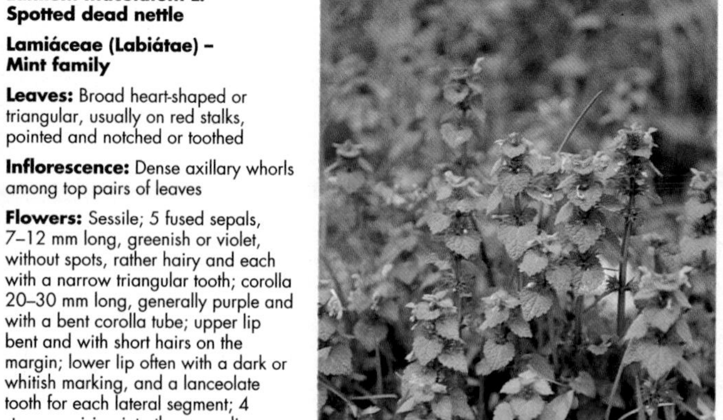

629 **Red dead nettle – L. purpúreum**
Plants 5–25 cm high, annuals or bienniaks, often branching at the base; flowering March to October

Lámium purpúreum L.
Red dead nettle

Lamiáceae (Labiátae) – Mint family

Leaves: Triangular or heart-shaped, generally on green stalks, rounded, blunt, or pointed at the tips, with blunt teeth, somewhat hairy and tinged with red on top; upper leaves sessile

Inflorescence: Dense axillary whorls among top pairs of leaves

Flowers: Sessile; 5 fused sepals, 5–8 mm long, whitish, pink, and greenish, with scattered hairs, each with a deep red, narrow, triangular tooth; corolla 8–12 mm long, rose or purple and with a straight corolla tube; upper lip bent and with short hairs on margin; lower lip with dark spots and a narrow lanceolate tooth on each lateral segment; 4 stamens; anthers violet-brown, bearded, with orange-yellow pollen

Fruits: 4 smooth nutlets 2–3 mm long; dispersal by ants

Habitat: From hill to subalpine zones in vineyards, arable fields, gardens, waste land, and on roadsides, on fresh, light, neutral soil rich in nutrients

630 **Betony – B. officinális**
Plants 20–60 cm high, perennials with knotty rhizomes; flowering from July to September

Betónica officinális L.
Stáchys officinális (L.) Trevisan
Betony

Lamiáceae (Labiátae) – Mint family

Leaves: Narrow oval, on stalks in middle and upper parts of plant, heart-shaped at bases of leaves proper, pointed, 4–12 cm long and covered in scattered hairs

Inflorescence: Compact heads with 8–14 flowers

Flowers: More or less sessile; 5 fused sepals, 5–8 mm long, dark red or greenish, hairy (especially in lower part) and each with an awned tooth up to 3 mm long; corolla 10–15 mm long, whitish to deep pink and with bent corolla tube; upper lip generally entire and bent forwards or backwards; lower lip in 3 segments, only slightly longer than upper one; 4 stamens; superior ovary; fertilised by insects or self-fertile

Fruits: 4 smooth nutlets up to 2.5 mm long

Habitat: In hill and mountain zones in open woodlands, hedgerows, wet grassland, heaths, and deciduous woodlands, on mostly lime-poor, neutral or moderately acid loamy or peaty soil which is seasonally wet but dry in summer; an indicator of infertility; formerly used as a medicinal plant

631 **Self-heal – P. vulgáris**
Plants 5–25 cm high, perennials with runners above ground; flowering from June to September

Prunélla vulgáris L.
Self-heal

Lamiáceae (Labiátae) – Mint family

Leaves: Basal leaves narrow oval, 10–35 cm long, entire or shallow-toothed, on stalks and generally blunt-ended; leaves on lower stem on stalks, but on upper stem short-stalked or sessile

Inflorescence: Spikes of numerous whorls, generally 6-flowered, set above each other

Flowers: 5 fused sepals, 5–8 mm long, with 10 irregular nerves and distinctly hairy; corolla 7–16 mm long, blue-violet to purple and with corolla bent like a knee; upper lip protruding, rounded in upper part, entire and with 2 slightly developed lateral teeth; middle segment of lower lip has toothed edge; 4 stamens, rising under the upper lip, with stamens elongated into teeth; superior ovary

Fruits: 4 nutlets, up to 2 mm long and often triangular

Habitat: From hill to subalpine zones in grassland, sparse woodland, on stream banks and along woodland paths, on well-drained neutral soils containing nutrients

632 **Narrow-leaved hemp nettle – G. angustifólia**
Plants 10–60 cm high, annuals or biennials without thickened stems; flowering from June to September

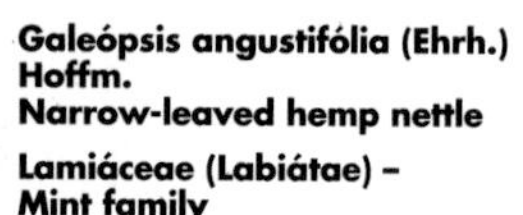

Galeópsis angustifólia (Ehrh.) Hoffm.
Narrow-leaved hemp nettle

Lamiáceae (Labiátae) – Mint family

Leaves: Narrow oval to lanceolate, entire or with at most 4 teeth, short-stalked or sessile and pointed at tips

Inflorescence: Spikes of numerous whorls of up to 12 flowers, set above each other

Flowers: More or less sessile; 5 fused sepals, 3–7 mm long, thickly covered in close hairs and each with a finely pointed tooth; corolla 10–25 mm long, pale or mid-carmine and with a straight hairy corolla tube; upper lip bent forward and covered in short hairs; lower lip usually broader and longer than upper lip; 4 stamens; superior ovary; fertilised by insects or self-fertile

Fruits: 4 nutlets up to 3 mm long

Habitat: From hill to subalpine zones in scree, rubbish dumps, arable fields, waste ground, ditches, railway embankments, and gravel pits, on dry, base-rich, loose, stony soil containing little humus or topsoil; roots up to 1 m deep; an early coloniser of newly turned soil

633 **Common hemp nettle – G. tétrahit**
Plants 20–80 cm high, perennials with thickened stems below leaf nodes; flowering from June to Oct.

Galeópsis tétrahit L.
Common hemp nettle

Lamiáceae (Labiátae) – Mint family

Leaves: Broad lanceolate, 4–12 cm long, rounded at bases of leaves proper or narrowing in a wedge shape to the stalks, pointed at tips, smooth or with scattered hairs and notched

Inflorescence: Spikes of numerous whorls of up to 16 flowers, set above each other; awned bracts

Flowers: 5 fused sepals, 8–15 mm long, whitish or pale green, each with a narrow triangular dark green tooth; corolla up to 20 mm long, pink or blue-violet, with a white corolla tube; upper lip helmet-shaped; lower lip in 3 lobes, with a violet mark and a yellow spot at base; 4 stamens; superior ovary; fertilised by insects or self-fertile

Fruits: 4 nutlets 2–3 mm long; dispersal by sticking to animals' fur

Habitat: From hill to subalpine zones in waste places, arable fields, grassland, forest clearings, by roadsides, fences, and in scree, on moist to fairly dry soils

634 **Alpine skullcap – S. alpína**
Plants 10–25 cm high, perennials with woody underground stems; flowering from June to August

Scutellária alpína L.
Alpine skullcap

Lamiáceae (Labiátae) – Mint family

Leaves: Oval, 2–4 cm long, on short stalks, flat or heart-shaped at base, blunt-ended, with scattered hairs and irregular blunt teeth

Inflorescence: Spike-like head of flowers in axils of greenish or violet bracts

Flowers: 5 sepals fused into a bell, in 2 lips, 2–3 mm long, hairy and with a round scale (scutellum) 3–5 mm long on the back of the upper lip; corolla blue-violet, 20–30 mm long and with a whitish or pale violet lower lip; upper lip has 3 lobes, the middle one shaped like a helmet; 4 stamens, reaching under the top lip; superior ovary

Fruits: 4 nutlets 1–2 mm long and covered with grey fibres

Habitat: Usually in subalpine zone, on rocky or stony slopes and in rough pasture on calcareous ground lacking topsoil but warm and dry in summer; a plant of southern and central European mountains, with a range extending south to the Spanish mountains and those of the Balkan Peninsula

635 **Wall Germander – T. chamaédrys**
Plants 10–30 cm high, perennials with lower stems woody; flowering till August

Teúcrium chamaédrys L.
Wall Germander

Lamiáceae (Labiátae) – Mint family

Leaves: Oval or narrow oval, often narrowing towards stalks, 10–25 mm long, with irregular teeth on both sides, round or blunt-ended at tips, usually hairy, dark green above and whitish-green beneath

Inflorescence: Solitary flowers or whorls of 2–6 in axils of top leaves

Flowers: On stems; 5 fused sepals, greenish or reddish, with 10 dark nerves, 5–8 mm long, rather hairy and each with a narrow triangular short-awned tooth; corolla 10–15 mm long and pink; no upper lip; lower lip in 5 lobes, with a large, entire, rather toothed middle lobe; 4 stamens, protruding from the corolla tube; superior ovary

Fruits: 4 egg-shaped or spherical nutlets 1–2.5 mm long, with veined surface

Habitat: From hill to subalpine zones in dry grassland, sparse oak and pine woods, on moderately dry or dry base-rich, loose, stony soil with humus; grown in gardens, and sometimes naturalised on old walls

636 **Motherwort – L. cardiáca**
Plants 30–120 cm high, generally with branching stem; flowering from June to September

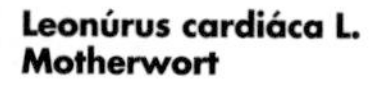

Leonúrus cardiáca L.
Motherwort

Lamiáceae (Labiátae) – Mint family

Leaves: Lower and middle stem leaves round, on red stalks, hairy, with 5–7 lobes like a maple leaf, flat or heart-shaped at base; lobes have sharp points or are themselves lobed; topmost stem leaves lanceolate, smaller, and with fewer teeth

Inflorescence: Terminal spike of dense-flowered axillary whorls

Flowers: 5 fused sepals, dark red or blackish, 4–8 mm long, and with awned teeth; corolla

7–12 mm long and pale purple to flesh-pink; upper lip swollen or helmet-shaped, entire and distinctly hairy outside; lower lip 3-lobed, with wide middle lobe; 4 stamens, reaching under top lip and usually clearly visible; superior ovary; fertilised by insects

Fruits: 4 hairy nutlets

Habitat: In hill and mountain zones in thickets, by the roadside, on waste ground, in gardens, on walls and by fences, on well-drained, neutral, loamy soil containing nutrients and humus

637 **Black columbine – A. atráta**
Plants 20–70 cm high, perennials, often with glandular hairs on stems; flowering from June to July

Aquilégia atráta Koch
Black columbine

Ranunculáceae – Buttercup family

Leaves: Basal and lower stem leaves doubly ternate, primary segments on stalks and secondary ones incised or 3-lobed; tips blunt-ended or rounded; top stem leaves entire, narrow oval, without stalks and only slightly toothed

Inflorescence: Solitary flowers or few-flowered racemes

Flowers: 5 sepals, deep violet, 15–25 mm long, pointed and often with white tips; 5 nectaries, formed like petals, deep violet, each with a long, hollow, backward-pointing spur; numerous stamens, the innermost sterile staminodes; generally 5 carpels

Fruits: Many-seeded follicles, 2–3 cm long, often with glandular hairs

Habitat: From hill to subalpine zones in sparse woodlands, thickets, fens, scrub, and woodland margins, on fairly dry calcareous soils with nutrients

638 **Policeman's helmet – I. glandulífera**
Plants 1–2 m high, smooth annuals originating in the Himalayas; flowering from July to September

Impátiens glandulífera Royle
Policeman's helmet

Balsamináceae – Balsam family

Leaves: Lanceolate, 10–25 cm long, on stalks, narrowing in a wedge shape towards stalks, pointed at tips, sharply and finely toothed, and opposite in lower and middle parts of stem; in top part leaves are whorled; stalked glands present on the stem

Inflorescence: Flowers 5–10 together in racemes on long stems in leaf axils of upper whorls

Flowers: Zygomorphic; sepals fused and reddish; of the 5 purplish-pink petals the lateral ones are fused in pairs; spur short and bent back; 5 stamens, with free-standing filaments; anthers fused and joined to the ovary; superior ovary of 5 cells; fertilised by insects

Fruits: Many-seeded club-shaped capsules 20–50 mm long; dispersal as capsules explode

Habitat: In hill zone on riverbanks, wet woodlands, and waste land, on moist, sandy, loamy or clay soil containing nutrients and humus; plant grows in shade or half-shade

639 **Balfour snapweed – I. balfoúrii** Plants 50–100 cm high, annuals, widespread as ornament but often escaping from gardens; flowering till Oct.

Impátiens balfoúrii Hooker
Balfour snapweed

Balsamináceae – Balsam family

Leaves: Wide lanceolate, 5–15 cm long, often narrowing towards stalks, coming to short points, toothed, alternate, grey-green beneath and without glands on stalks

Inflorescence: Racemes of at most 10 flowers

Flowers: Zygomorphic; sepals fused; of the 5 petals the lateral ones are fused in pairs, the top lobe red and the bottom one white; spur long and straight or slightly bent upwards; 5 stamens with free-standing filaments; anthers fused and joined to the ovary; superior ovary of 5 cells; fertilised by insects

Fruits: Many-seeded explosive capsules

Habitat: In hill zone along roadsides and on wood margins, in thickets, rubbish dumps, and gardens, on moist to well-drained, sandy, loamy, or clay soil containing nutrients and humus; plant originally a native of the Himalayas

640 **Alpine toadflax – L. alpína** Plants 5–15 cm high, usually perennials, often creeping and smooth; flowering from June to August

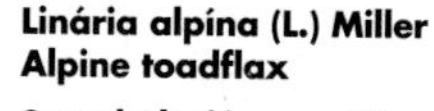

Linária alpína (L.) Miller
Alpine toadflax

Scrophulariáceae – Figwort family

Leaves: Narrow lanceolate, usually blunt-ended, rather fleshy, bluish, sessile, and in whorls of 3–5 but often alternate at top of stems

Inflorescence: Short, many-flowered, terminal racemes

Flowers: Long-stemmed; 5 sepals, lanceolate, greenish and generally with whitish or translucent edges; corolla 8–15 mm long without the long spur, usually violet, with orange-yellow palate; upper lip usually with white margin; tip of lower lip bent down; 4 stamens, enclosed in corolla; superior ovary; fertilised by insects (bumble-bees)

Fruits: Capsules opening by 4–10 apical valves; dispersal by wind

Habitat: In alpine zone (but sometimes washed down to lower ones) on well-drained, base-rich, usually calcareous, mobile scree and gravel poor in fine soil

641 **Ivy-leaved toadflax – C. murális** Plants 10–30 cm long, with thread-like trailing or drooping stems; flowering till October

Cymbalária murális G.M.Sch.
Linária cymbalária (L.) Miller
Ivy-leaved toadflax

Scrophulariáceae – Figwort family

Leaves: Round or heart-shaped, 5- or 7-lobed, sometimes incised nearly to middle, long-stalked, green above, often reddish beneath and alternate

Inflorescence: Solitary flowers in leaf axils

Flowers: 5 sepals, narrow oval and with long points; corolla without spur 3–7 mm long, pale or mid-violet, with pale or dark yellow palate; spur straight and half as long as corolla; upper lip notched and generally flat; lower lip 3-lobed; 4 stamens. enclosed in corolla; superior ovary; fertilised by insects or self-fertile

Fruits: Capsule opening by 2 lateral pores, each with 3 valves

Habitat: In hill (rarely mountain) zone on walls, waste ground, and rocks, on well-drained, often rather damp, usually calcareous stony soil fairly rich in nutrients but poor in fine soil; a typical exploiter of rock and wall crevices; grows in light or half-shade

642 **Narrow-leaved valerian – C. angustifólium** Plants 20–60 cm high, perennials with branching rhizomes; flowering from June to August

Centránthus angustifólium (Miller) DC
Narrow-leaved valerian

Valerianáceae – Valerian family

Leaves: Narrow lanceolate, with narrow base, sessile, usually entire, blunt-tipped or pointed, dark green above, blue-green beneath, and opposite

Inflorescence: Dense, terminal, panicled cymes

Flowers: Calyx of 5 inrolled points, changing to feathery bristles in fruit; corolla tube 5–10 mm long, pink, and with 5 spread lobes and a long spur; 1 stamen; 3 carpels on an inferior ovary, only 1 fertile, the others atrophied

Fruits: Nutlets 4–6 mm long, with pappus bristles 8–10 mm long

Habitat: In hill and mountain zones in limestone scree and wall and rock crevices, on dry soils in sunny places; a western Mediterranean plant, with a range extending south to Algeria; found in the Jura and in southern Savoy

643 **Peony – P. officinális** Plants 50–100 cm high, perennials with woody rhizomes; flowering in May and June

Paeónia officinális L.
Peony

Paeoniáceae – Peony family

Leaves: Roundish in outline, doubly ternate, on stalks, and up to 30 cm in diameter; leaflets narrow oval to lanceolate, entire and pointed

Inflorescence: Large, terminal, solitary flowers

Flowers: 5 sepals, wide oval to lanceolate, often taking different shapes, greenish or reddish and shorter than petals; petals oval (broadest towards tips), up to 5 cm long and red; numerous stamens, united at base into a nectar ring; several (2–3) free-standing superior carpels with many ovules and sessile styles; fertilised by insects

Fruits: Follicles 2–5 cm long, covered in felted hairs

Habitat: In hill and mountain zones in sparse bushes, oak and beech woodlands, and on slopes which are dry in summer, on dry lime-free soils; cultivated varieties planted as ornamentals; range extends north to the Danube

644 **Mountain anemone – P. montána** Perennial plants 10–30 cm high, up to 40 cm high when fruit is ripe; flowering from March to May

Pulsatílla montána (Hoppe) Rchb.
Mountain anemone

Ranunculáceae – Buttercup family

Leaves: Basal leaves broad oval or oval, with leaves proper 36 cm long, on long stalks and multiply pinnate, last leaflets generally lanceolate; stem leaves sessile, fused at the base and consisting of 10–25 narrow leaflets 2–3 cm long

Inflorescence: Solitary terminal flowers on a hairy flower stem

Flowers: 6 petals, brownish-red to dark blue-violet, closed at beginning of flowering but later spread out, hairy, and up to 3 cm long; numerous stamens, some being staminoid and secreting nectar; numerous carpels; fertilised by insects

Fruits: Nutlets

Habitat: In hill and mountain zones in dry grassland on very dry porous soil in sunny places without good ground cover; a plant of central and southern European mountains, found in the Pyrenees, the Alps, the Carpathians, and mountains of the Balkan Peninsula; a gregarious plant

645 **Haller's anemone – P. hálleri** Plants 10–25 cm high, perennials with shaggy stems; flowering from May to July

Pulsatílla hálleri (All.) Willd.
Anémone hálleri All.
Haller's anemone

Ranunculáceae – Buttercup family

Leaves: Basal leaves broad oval or oval, with leaves proper 2–5 cm long, with silky hairs, stalked, pinnate, with pinnately cut leaflets and linear-lanceolate segments, and developing after flowering; stem leaves sessile, with numerous segments

Inflorescence: Solitary, terminal flowers on stems thickly covered in hairs

Flowers: 6 petals, brownish-red to pale violet, erect, closed at start of flowering but later more or less spread out, oval or narrow oval, hairy and pointed or blunt at tips; numerous stamens, some staminoid and secreting nectar; numerous carpels; fertilised by insects

Fruits: Nutlets

Habitat: In subalpine and alpine zones in scree, rock crevices, and meadows lacking complete ground cover, on dry, calcareous, flat soils containing humus; found in the Valais

646 **Alpine squill – S. bifólia**
Plants 10–25 cm high, perennials with bulbs; flowering in March and April

Scilla bifólia L.
Alpine squill

Liliáceae – Lily family

Leaves: Generally 2, linear lanceolate, without stalks, with parallel veins, 5–10 cm long, grooved and with blunt tips

Inflorescence: Raceme of 2–8 flowers; leaves appear at same time as flowers in spring

Flowers: 6 petals, narrow oval, generally blunt- or round-ended, 5–15 mm long, blue, red, or rarely white and spread flat; 6 stamens; violet anthers; superior ovary; fertilised by insects

Fruits: Spherical capsules with 3 compartments; dispersal by ants

Habitat: In hill and mountain zones in oak and beech woods, parks, gardens, and meadows, on well-drained, calcareous, base-rich, loose loam or clay containing nutrients and humus with deep topsoil; likes half-shade; grows on waste ground; a southern European plant, with a range reaching north to central Germany, the Netherlands, and southern Poland, east to Asia Minor, and south to Sardinia, Sicily, and Greece

647 **Spanish hyacinth – H. hispánica**
Plants 10–30 cm high, perennials with bulbs; flowering in March and April

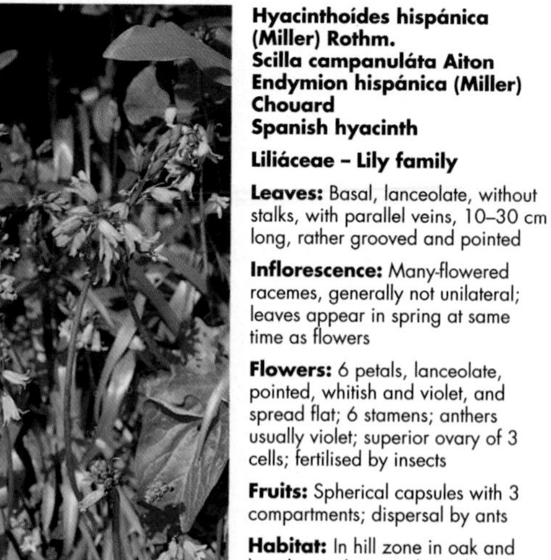

Hyacinthoídes hispánica (Miller) Rothm.
Scilla campanuláta Aiton
Endymion hispánica (Miller) Chouard
Spanish hyacinth

Liliáceae – Lily family

Leaves: Basal, lanceolate, without stalks, with parallel veins, 10–30 cm long, rather grooved and pointed

Inflorescence: Many-flowered racemes, generally not unilateral; leaves appear in spring at same time as flowers

Flowers: 6 petals, lanceolate, pointed, whitish and violet, and spread flat; 6 stamens; anthers usually violet; superior ovary of 3 cells; fertilised by insects

Fruits: Spherical capsules with 3 compartments; dispersal by ants

Habitat: In hill zone in oak and beech woods, in parks and gardens as an ornamental plant, on well-drained, base-rich, often lime-poor, loose loam or clay containing nutrients and humus; grows in half-shade; a native of western Mediterranean area

648 **Tassel hyacinth – M. comósum**
Plants 30–80 cm high, perennials with bulbs; flowering in April and May

Muscári comósum (L.) Miller
Tassel hyacinth

Liliáceae – Lily family

Leaves: Lanceolate, 30–80 cm long, up to 25 mm wide, gradually narrowing towards tips and with finely toothed edges

Inflorescence: Many-flowered raceme up to 10 cm long, with small sterile flowers on long stems

Flowers: Fertile flowers with 6 fused petals 4–8 mm long and violet, with small white teeth at the end; 6 stamens; superior ovary; single pistil has only a small stigma; fertilised by insects

Fruits: Capsules with 3 compartments, each with 2 seeds; capsules not pitted at the tips and with 3 wings

Habitat: In hill and mountain zones in rough pasture, meadows, thickets, by roadsides, and in weed beds, on moderately dry, base-rich, calcareous, loamy or sandy soils; a Mediterranean plant with a range extending north to sunny places in northern Germany and south to north-west Africa; in central Europe only found in very warm places

649 **Dayflower – C. commúnis**
Plants 20–60 cm high, recumbent or upright, rooting at nodes; flowering till October

Commelína commúnis L.
Dayflower

Commelináceae – Spiderwort family

Leaves: Narrow oval or broad lanceolate, on stalks, generally heart-shaped at bases of leaves proper, drawing to gradual points at tips, with parallel veins and 5–10 cm long

Inflorescence: Solitary flowers surrounded by folded bracts or spathes

Flowers: 3 outer petals not fused, narrow oval and green; 3 inner petals not fused, dark or pale blue , round, with short whitish stems and fringed edges; of 6 stamens only those pointing forwards are fertile; superior ovary

Fruits: Capsules with 2 seeds in each of 2 compartments

Habitat: In hill zone on debris, in woody ravines and vineyards, on moist soils containing nutrients in warm locations; an oriental (Japanese and Chinese) plant used as a pot or garden plant and found as an escape, especially in the southern Ticino

650 **Siberian flag – I. sibírica**
Plants 40–100 cm high, perennials with with thin rhizomes; flowering in June and July

Iris sibírica L.
Siberian iris

Iridáceae – Iris family

Leaves: 30–80 cm long, usually basal, ensiform, coming gradually to points and the same colour on both sides

Inflorescence: Solitary flowers or several enclosed in transparent or translucent bracts or spathes

Flowers: 6 petals fused at base into a narrow tube; free part of outer petals erect or bent back, 3–5 cm long and without erect hairs; free part of inner petals narrow oval and longer than pistils; 3 stamens, not fused together; inferior ovary of 3 cells

Fruits: Capsules in 3 compartments with numerous seeds

Habitat: In hill (rarely mountain) zone in ditches, marshes, and flooded waste ground, on intermittently wet ground drying out on the surface in summer and consisting of moderately base-rich clay soil or silt containing nutrients and humus

651 **German iris – I. germánica**
Plants 25–80 cm high, perennials with rhizomes; flowering from May to July

Iris germánica L.
German iris

Iridáceae – Iris family

Leaves: 20–70 cm long, basal and alternate on stem, usually broad ensiform, pointed, and widest in the middle or towards tip

Inflorescence: Solitary flowers or several enclosed in a bract or spathe which is dry and transparent in its upper half

Flowers: 6 petals fused at base into a narrow tube; free part of outer petals erect or bent back, 2–5 cm long, deep violet, with dark veins reaching to edges and with yellow beard; free part of inner petals oval and pale blue-violet; 3 stamens, not fused together; inferior ovary of 3 cells; diverging style branches

Fruits: Capsules in 3 compartments

Habitat: In hill zone naturalised and found wild on rocks, walls, sunny hillsides, and in vineyards, on warm, generally calcareous, base-rich, stony soil

652 **Alpine milkwort – P. alpéstris**
Plants 5–15 cm high, with recumbent or upright stems; flowering in June and July

Polygala alpéstris Rchb.
Alpine milkwort

Polygaláceae – Milkwort family

Leaves: Narrow oval or lanceolate, entire, 5–15 mm long, sessile or short-stalked, narrowing in a wedge towards base, and blunt-ended or pointed; no rosette; lower leaves alternate, upper leaves slightly alternate or opposite and generally longer than lower ones

Inflorescence: Many-flowered racemes

Flowers: 3 of 5 sepals small, 1–2 mm long, blue-green and typical; 2 inner ones petal-like (=wings); wings narrow oval, 4–6 mm long, with scarcely visible lateral, slightly branching veins; 3 fused petals, with lateral ones bent back, with two free upper tips and a lower tip often divided into segments; 8 stamens; filaments fused into a tube at base; superior ovary

Fruits: Capsules 3–4 mm long

Habitat: In subalpine zone in meadows, scrub, and sparse pine woods on moderately well-watered to dry soil

653 **Brooklime – V. beccabúnga**
Plants 30–50 cm high, recumbent or upright perennials; flowering from May to August

Verónica beccabúnga L.
Brooklime

Scrophulariáceae – Figwort family

Leaves: Oval to round, sessile or short-stalked, 1–5 cm long, fleshy, blunt-ended, entire or slightly crenate or toothed, and opposite

Inflorescence: Opposite racemes with few flowers

Flowers: 4 sepals with more or less pointed tips; 4 petals, oval and blue; 2 stamens; superior ovary of 2 cells; fertilised by insects or self-fertile

Fruits: Almost spherical capsules opening by 4 flaps; many seeds

Habitat: From hill to subalpine zones by slowly flowing streams, ponds, marshes, and in wet places in meadows, on flat, usually flooded, often sandy, muddy ground containing more or less nutrients and humus; plant grows in light or half-shade; used in earlier times as a salad and medicinal plant; range extends over all Europe, northern Asia, and northern Africa

654 **Alpine speedwell – V. alpína**
Plants 3–15 cm high, perennials with thin creeping rhizomes; flowering in July and August

Verónica alpína L.
Alpine speedwell

Scrophulariáceae – Figwort family

Leaves: Elliptical, oval, or round, short-stalked, 10–25 mm long, blunt-ended or pointed, entire or with a few short teeth, smooth or with scattered hairs; opposite

Inflorescence: Short umbel-like terminal racemes of up to 20 flowers

Flowers: Hairy, on stems; generally 5 sepals, oval, brownish-black or bluish-black and hairy; 4 petals, oval, 4–7 mm long and blue; 2 stamens; superior ovary of 2 cells; fertilised by insects or self-fertile

Fruits: Almost spherical capsules with many seeds

Habitat: In alpine zone in snow valleys, scree, and grassland, on ground soaked by snow covering it for long periods, on moderately base-rich, generally lime-poor, moderately acid or neutral, stony loam or clay containing nurients and humus; spreads by means of creeping roots; range extends to Iceland and Greenland

655 **Leafless speedwell – V. áphylla**
Plants 2–6 cm high, perennials with thin, spreading rhizomes; flowering in July and August

Verónica áphylla L.
Leafless speedwell

Scrophulariáceae – Figwort family

Leaves: Basal, spathulate or broad oval, with short broad stalks, 10–20 mm long, round-ended and finely toothed; stem leaves rare

Flowers: Racemes of up to 5 flowers on long stem arising from top leaf axil

Flowers: Usually 4 sepals, narrow oval, deep blue or blue-black and hairy; 4 petals, broad oval or round, 5–8 mm long, lilac or violet, with distinct deep violet veins; 2 stamens; superior ovary of 2 cells

Fruits: Many-seeded capsules with disc-shaped seeds

Habitat: In alpine (rarely subalpine) zone in stony rough pasture, meadows, rock crevices, snow valleys, and on scree, on moist to well-drained, calcareous, base-rich, loose, stony loam containing humus; spreads by creeping roots; a central and southern European mountain plant with a range extending to the mountains of the Balkans

656 **Rock speedwell – V. frúticans**
Plants 5–15 cm high, perennials, woody at procumbent base, upright stems; flowering till July

Verónica frúticans Jacq.
Verónica saxátilis Scop.
Rock speedwell

Scrophulariáceae – Figwort family

Leaves: Oblong or oval, generally without stalks, 15–25 mm long, round-ended, narrowing towards base, entire or slightly crenate, usually smooth; opposite

Inflorescence: Loose terminal racemes with up to 10 flowers on stems

Flowers: 4 sepals, oval, greenish and rather hairy; 4 petals, broad oval or round, 4–7 mm long and dark blue; 2 stamens; superior ovary of 2 cells; fertilised by insects or self-fertile

Fruits: Many-seeded capsules up to 9 mm long

Habitat: In subalpine and alpine (rarely mountain) zones on alpine rocks, cliffs, meadows, and stony rough pasture, on moderately well-watered to dry, stony, often lime-poor ground poor in humus and fine soil; an early coloniser of freshly turned soil; an arctic and alpine plant with a range extending to Greenland and Iceland

657 **Shrubby speedwell – V. fruticulósa**
Plants10–20 cm high, perennials, woody at base, upright or ascending; flowering till July

Verónica fruticulósa L.
Shrubby speedwell

Scrophulariáceae – Figwort family

Leaves: Oblong, with very short stalks or sessile, 15–25 mm long, generally round at tips, entire or slightly notched, more or less smooth; opposite

Inflorescence: Fairly dense terminal racemes with up to 10 flowers on stems

Flowers: 4 sepals, narrow oval, dark green, and covered like flower stems in glandular hairs; 4 petals, broad oval or round, 4–7 mm long, and pale pink (with darker veins!); 2 stamens; superior ovary of 2 cells; fertilised by insects or self-fertile

Fruits: Many-seeded capsules up to 7 mm long

Habitat: In subalpine (rarely mountain) zone in rock crevices and scree on moderately well-drained to dry, loose, lime-rich, rocky soil; a central and southern European mountain plant, with a range extending to the mountains of northern Spain, the Pyrenees, Corsica, and Croatia; not very common in the Alps and the Jura

658 **Creeping speedwell – V. filifórmis**
Plants 10–40 cm long, perennials with very thin stems, covered in glandular hairs; flowering April to August

Verónica filifórmis Sm.
Creeping speedwell

Scrophulariáceae – Figwort family

Leaves: Round, short-stalked, tending to heart-shape at base, blunt-ended or round, toothed, light green and rather hairy; small bulbils in leaf axils help plants to spread

Inflorescence: Long-stemmed solitary flowers in leaf axils

Flowers: 4 sepals, narrow oval and with glandular hairs at base; 4 petals, round or oval, 3–5 mm long, pale lilac or pale blue, with dark veins and with a rather paler tip on bottom petal; 2 stamens; superior ovary of 2 cells; fertilised by insects

Fruits: Many-seeded capsules, 3–7 mm long, up to 6 mm wide and covered in sparse hairs

Habitat: In hill and mountain zones on well-drained, generally lime-poor, sandy or pure loam containing nutrients and humus; originated in northern Anatolia; has been naturalised in Europe since 1930, and has now spread over much of world; a weed in grassland

659 **Large speedwell – V. teúcrium**
Plants 10–40 cm high, ascending or upright perennials; flowering from May to July

Verónica teúcrium L.
Large speedwell

Scrophulariáceae – Figwort family

Leaves: Oblong to narrow oval, 3–7 cm long, sessile, rounded or heart-shaped at base, usually blunt-ended, coarsely toothed; opposite

Inflorescence: Flowers in long-stemmed racemes rising from leaf axils

Flowers: Usually 5 sepals, green and often hairy; 4 petals, round to rhomboid, 3–6 mm long and dark blue with rather darker veins; 2 stamens; superior ovary of 2 cells

Fruits: Capsules with numerous disc-shaped seeds

Habitat: In hill and mountain (rarely subalpine) zones in sparse woodlands, by roadsides and wood margins, on dry grassland and on the edges of sunny thickets, on moderately dry, usually calcareous, neutral, loose loam or loess containing humus and with an average depth of topsoil; a European and Asian plant with a range extending north to the Netherlands and central Russia and south to central Spain and Tuscany

660 **Germander speedwell – V. chamaédrys**
Plants 10–30 cm high, perennials with thin rhizomes; flowering from April to August

Verónica chamaédrys L.
Germander speedwell

Scrophulariáceae – Figwort family

Leaves: Oval, short-stalked, or sessile, up to 35 mm long, blunt-ended, coarsely toothed with teeth pointing towards tips; hairy or smooth; opposite

Inflorescence: Long-stemmed, many-flowered racemes arising from leaf axils

Flowers: 4 sepals, narrow oval, 2–6 mm long, with scattered hairs and green; 4 petals, oval or round, 4–6 mm long, usually azure, with darker veins and often whitish edges; 2 stamens; superior ovary of 2 cells; fertilised by insects

Fruits: Many-seeded triangular capsules opening by 2 flaps; dispersal by animals

Habitat: From hill to subalpine zones in meadows, sparse woodland, thickets, by camp sites and along roadsides and wood margins, on damp to moderately dry, more or less base-rich, generally neutral loam containing more or less nutrients and humus; spreads by rooting at nodes of recumbent stem and by creeping roots

661 **Spiked speedwell – V. spicáta**
Plants 10–30 cm high, perennials with short rhizomes; flowering from July to September

Verónica spicáta L.
Spiked speedwell

Scrophulariáceae – Figwort family

Leaves: Lanceolate to narrow oval, short-stalked or sessile, 3–8 cm long, narrowing towards base, blunt- or round-ended, entire or with blunt teeth; smooth or somewhat hairy; opposite

Inflorescence: Long, dense, many-flowered racemes at end of stem

Flowers: 4 sepals, green and with glandular hairs; 4 petals, lanceolate (especially bottom 3), with a long corolla tube and blue; 2 stamens; superior ovary of 2 cells; fertilised by insects

Fruits: Capsules with scattered hairs, containing numerous flat oval seeds

Habitat: From hill to mountain zones in dry grassland, on dry rocks, at margins of sparse thickets and on dry gravel, on dry, infertile, often lime-poor, more or less neutral, stony or sandy ground; plant grows in light or half-shade; range extends north to England

662 **Alpine flax – L. perénne**
Plants 10–30 cm high, perennials with ascending or upright stems; flowering till July

Linum perénne L. ssp. alpínum (Jacq.) Ockendon
L. alpínum Jacq.
Alpine flax

Linaceae – Flax family

Leaves: Linear lanceolate, 10–25 mm long, broadest towards base, drawing to long points, entire, smooth, sessile, blue-green, and alternate

Inflorescence: Flowers in branching cymes

Flowers: 5 sepals, narrow oval, pointed, entire, 4–7 mm long and 3-veined; 5 petals, narrow oval (broadest at tips), often slightly fringed, 10–20 mm long, and pale blue (rarely white); 5 stamens, fused in lower part and with 5 nectar glands at base; superior ovary with 5 cells and 5 styles; fertilised by insects

Fruits: Capsules up to 8 mm long, exploding to throw out seeds when ripe

Habitat: In subalpine (rarely mountain or alpine) zone in unmanured meadows, sunny grassland, and dry woodlands, on dry, neutral, lime-rich, loose, sandy or stony ground containing humus, in sunny places

663 **Wild fennel – N. arvénsis**
Plants 10–40 cm high, annuals, erect, generally branching, smooth; flowering from June to Sept.

Nigélla arvénsis L.
Wild fennel

Ranunculáceae – Buttercup family

Leaves: Stem leaves only, doubly or trebly pinnate and blue-green; last leaflets up to 1 mm wide, with short points

Inflorescence: Solitary terminal flowers

Flowers: No bracts; 5 petal-like sepals, light blue, with green veins, broad oval or round, narrowing to long claws and often with short points at tips; nectaries about a quarter the length of petals, with a thread-like upper lip and a rather larger lower lip with 2 lobes and scattered hairs; numerous stamens, elongated into points; carpels united only below middle; fertilised by insects (bees)

Fruits: Cylindrical follicles 10–15 mm long, with many seeds

Habitat: In hill (rarely mountain) zone in cornfields and fallow land, on moderately dry, lime-rich, generally stony, loose, warm loam containing nutrients and more or less humus; roots up to 60 cm deep

664 **Wood scabious – K. silvática**
Plants 20–100 cm high, perennials with rhizomes; flowering from June to September

Knaútia silvática (L.) Duby
Knaútia dipsacifólia Kreutzer
Wood scabious

Dipsacáceae – Teasel family

Leaves: Lowest broad lanceolate to elliptical, usually with blunt teeth, on stalks, rough-haired, with fine hairs on margins and pointed; middle and top leaves narrow lanceolate to oval, with blunt teeth or coarsely notched, sessile, hairy or smooth and pointed

Inflorescence: Flowers in flat capitula; outer bracts in several rows, lanceolate, often with hairs 1–2 mm long and single glandular hairs

Flowers: Calyx with fine hairs at base; calyx bristles 2–3 mm long, ending in fine hairs; 4 petals, fused into a tube, with 4 points and generally violet (rarely pink or white); 4 stamens fused into a tube; 1-celled inferior ovary

Fruits: Nutlets 5–7 mm long, hairy and with an appendage at the base to attract ants

Habitat: In mountain and subalpine (rarely hill) zones in scrub, mountain meadows, and along shady wood margins, on damp to well-drained soils

665 **Sheep's-bit – J. montána**
Plants 10–60 cm high, biennials branching into several stems, without runners; flowering till August

Jasióne montána L.
Sheep's-bit

Campanuláceae – Bellflower family

Leaves: Lanceolate to narrow oval, wavy-edged, with blunt teeth, sessile, usually with round tips

Inflorescence: Terminal capitula 10–25 mm across with numerous short-stemmed flowers; bracts lanceolate, entire or coarsely toothed, and often somewhat hairy

Flowers: 5 sepals, lanceolate; 5 petals, initially fused, later forming band-like points; 5 stamens, not widening at base; inferior ovary of 2 cells; fertilised by insects

Fruits: Capsule opening by valves; dispersal by wind

Habitat: In hill and mountain zones in grassy places, rough pasture, on heaths, cliffs, and banks, and by roadsides, on dry, lime-poor, stony soil poor in fine soil and humus and warm in summer; an early coloniser of newly turned soil which can root up to 1 m deep; found especially in the Black Forest, the Vosges, and in valleys of central and southern Alps

666 **Shining scabious – S. lúcida**
Plants 10–30 cm high, perennials with sterile leaf rosettes; flowering in July and August

Scabiósa lúcida Vill.
Shining scabious

Dipsacáceae – Teasel family

Leaves: Basal and lower stem leaves narrow oval, long-stalked, toothed (with teeth pointing towards tips) blunt-ended or pointed, somewhat hairy on edges and veins, but otherwise smooth; middle and upper stem leaves similar but pinnately cut

Inflorescence: Terminal flat capitula up to 4 cm across; flower stem generally longer than rest of plant; bracts slightly smaller or rather larger than outer florets and narrow lanceolate

Flowers: Outer calyx 1–2 mm long, transparent and indistinctly toothed; calyx bristles 4–8 mm long, with wedge-shaped vein on inner sides, flattened and generally dark in colour; 5 petals, fused into a tube, with 5 points, reddish or bluish-violet; 4 stamens; inferior ovary of 2 cells

Fruits: Cylindrical nutlets with 8 furrows; dispersal by wind

Habitat: Common in subalpine and alpine zones in mountain meadows, grassy slopes, and rubbish dumps, on well-drained soils more or less rich in nutrients

667 **Field scabious – K. arvénsis**
Plants 30–100 cm high, perennials with branching rhizomes; flowering from May to September

Knaútia arvénsis (L.) Coulter em. Duby
Field scabious

Dipsacáceae – Teasel family

Leaves: Lower leaves lanceolate, stalked, entire, toothed or pinnately cut, pointed and somewhat hairy; middle and upper leaves lanceolate to oval, sessile and usually pinnately cut; leaflets lanceolate and pointed

Inflorescence: Flowers in flat capitula (marginal larger than central); outer bracts lanceolate and hairy

Flowers: Calyx with fine hairs at base; calyx bristles 2–3 mm long , ending in hairs up to 1 mm long; 4 petals, fused into a tube, with 4 points and red- to blue-violet; 4 fused stamens in corolla tube; inferior ovary with one cell; fertilised by insects

Fruits: Nutlets 4–6 mm long, with hairs up to 1.5 mm long and with an appendage at base to attract ants to eat it

Habitat: In hill and mountain zones in dry grassy fields, dry pasture, banks, by roadsides and wood margins, on well-drained to moderately dry, base-rich, slightly acid soils containing humus and nutrients

668 **French scabious – S. gramúntia**
Plants 10–50 cm high, perennials with branching stems; flowering from June to August

Scabiósa gramúntia L.
Scabiósa triándra L.
French scabious

Dipsacáceae – Teasel family

Leaves: Basal and lower stem leaves narrow oval, stalked, hairy on edges, veins, and underside of stalk and doubly pinnate; middle and upper stem leaves doubly or trebly pinnate, with lanceolate leaflets

Inflorescence: Flat capitula 15–30 mm across, marginal florets being larger than inner ones; bracts half as long as or rather longer than outer florets

Flowers: Outer calyx 1–2 mm long, transparent and indistinctly toothed; usually 5 calyx bristles, light to mid-brown and 1–3 mm long; corolla with 5 points; 4 stamens; inferior ovary of 2 cells; fertilised by insects

Fruits: Nutlets 2–3 mm long; with 8 furrows; slightly hairy

Habitat: In hill and mountain zones on grassland, dry hillsides, rough pasture, cliffs, and sparse pine woods, on dry, loose soils; found especially in the Ticino, Graubuenden, and the Valais, but not often

669 **Grass-leaved scabious – S. graminifólia**
Plants 20–45 cm high, perennials with woody rhizomes and sterile shoots; flowering until August

Scabiósa graminifólia L.
Grass-leaved scabious

Dipsacáceae – Teasel family

Leaves: Grass-like, entire, dark green, with close hairs on both sides; opposite

Inflorescence: Flat capitula 3–5 cm across, marginal florets larger than inner ones; bracts narrow lanceolate and half as long as outer florets

Flowers: Outer calyx 2–4 mm long, transparent and indistinctly toothed; calyx bristles 3–5 mm long and yellowish; 5 petals, fused into a tube and violet; 4 stamens; inferior ovary of 2 cells

Fruits: Cylindrical nutlets or achenes with 8 visible grooves

Habitat: In hill and mountain zones on rocky slopes, scree and ridges on dry, stony soils in warmer places; a central and southern European mountain plant, with a range extending south to the Pyrenees, the Apennines, and the Balkan mountains; rare, but found in Switzerland in the southern Ticino

670 **Greater periwinkle – V. major**
Plants 20–50 cm high, perennials with spreading rhizomes; flowering in April and May

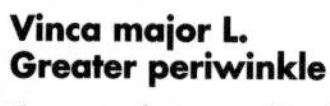

Vinca major L.
Greater periwinkle

Apocynáceae – Periwinkle family

Leaves: Narrow oval or heart-shaped, 3–10 cm long, rounded or heart-shaped at base, pointed at tips, hairy on edges (especially in earliest stages), short-stalked, decussate, leathery, and evergreen

Inflorescence: Solitary flowers on stems arising from leaf axils

Flowers: 5 sepals, fused in lower part, each with a calyx bristle up to 15 mm long; 5 petals, fused into a tube at base, 10–20 mm long and violet-blue; calyx lobes oval (broadest towards tips) and spread flat; 5 stamens inserted on the corolla; ovary of 2 carpels, free below but with a common style; fertilised by insects

Fruits: Cylindrical follicles; seeds without plumes

Habitat: In hill zone in thickets, hedges, and wood margins, on well-drained, more or less base-rich, clay or loam containing humus and more or less nutrients; a naturalised garden plant, found in the wild mainly in the Valais, the southern Ticino, and the southern foothills of the Alps

671 **Lesser periwinkle – V. minor**
Plants 5–20 cm high, perennials with spreading rhizomes; flowering in April and May

Vinca minor L.
Lesser periwinkle

Apocynáceae – Periwinkle family

Leaves: Lanceolate to narrow oval, 2–5cm long, narrowing gradually at base, pointed at tip, smooth, with edges slightly bent downwards, sessile or short-stalked, dark green and glossy above and matt green beneath

Inflorescence: Solitary flowers on stems arising from leaf axils

Flowers: 5 sepals, fused as far as the middle, each with a point 4–5 mm long; usually 5 petals, pale or reddish-violet; calyx lobes oval (broadest towards tips) and spread flat; 5 stamens inserted on the corolla; ovary of 2 carpels, free below but with a common style; fertilised by insects

Fruits: Cylindrical follicles; seeds without plumes, dispersed by wind; plant also spreads by runners

Habitat: In hill and mountain zones in deciduous woods, copses, and hedgerows on well-drained, base-rich, calcareous clay or loam containing humus and nutrients; plant grows in half-shade

672 **Garden forget-me-not – M. sylvática**
Plants 15–45 cm high, perennials with stems thickly covered in hairs; flowering from May to July

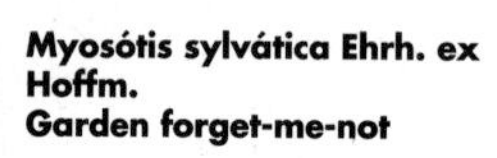

Myosótis sylvática Ehrh. ex Hoffm.
Garden forget-me-not

Boragináceae – Borage family

Leaves: Lower ones oval or narrow oval, 4–10 cm long and on stalks; upper leaves narrow oval or lanceolate, 4–8 cm long, sessile, pointed, hairy, and alternate

Inflorescence: Branching panicles

Flowers: Calyx in 5 fused parts, dark brown, with erect or backward-pointing hairs and 5 pointed segments; 5 petals, fused into a tube with flat lobes and blue or pale violet; lobes oval or shaped like an inverted egg with greatest width towards tips, often somewhat incised; 5 stamens enclosed in corolla tube; superior ovary of 2 cells; fertilised by insects

Fruits: Nutlets with 4 points and sections up to 1.7 mm long

Habitat: In mountain and subalpine zones in moist grassland, scrub, by roadsides, wood margins, and cow-sheds, and in forest clearings, on damp, base-rich, loose soils containing humus and nutrients

673 **Forget-me-not – M. scorpioídes**
Plants 10–70 cm high, perennials with creeping rhizomes; flowering from May to July

Myosótis scorpioídes L. em. Hill
Myosótis palústris (L.) Hill
Forget-me-not

Boragináceae – Borage family

Leaves: Lanceolate or narrow oval, 3–10 cm long, sessile, often rounded at base, blunt-ended or pointed and more or less thickly covered with erect hairs

Inflorescence: Branching panicles, always without leaves

Flowers: Calyx in 5 fused parts up to two-thirds of length, 2–4 mm long, dark brown, with close hairs or smooth and with triangular segments; 5 petals, fused into tube with flat lobes, and blue; lobes oval or round; 5 stamens enclosed in corolla tube; superior ovary of 2 cells; fertilised by insects

Fruits: Nutlets with 4 sections up to 1.7 mm long

Habitat: From hill to subalpine zones on river banks, in ditches, in wet grassland, moors, and scrub, on wet or moist sandy or pure loam or clay containing nutrients and humus; common throughout the northern hemisphere up to 68 degrees

674 **Rehstein's forget-me-not – M. rehsteineri**
Plants 2–8 cm high, perennials with creeping rhizomes; flowering in April and May

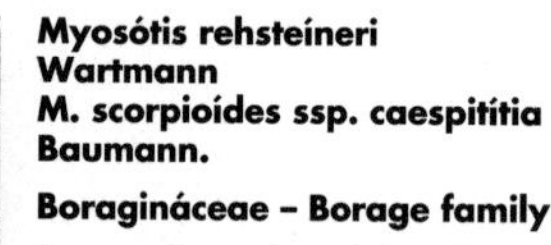

Myosótis rehsteineri Wartmann
M. scorpioídes ssp. caespititia Baumann.

Boragináceae – Borage family

Leaves: Lanceolate, 2–6 cm long, blunt-ended or pointed, hairy, with hairs pointing towards tips on tops and undersides of upper stem leaves, and towards base on undersides of lower leaves

Inflorescence: Short panicles or racemes with few flowers and no leaves

Flowers: 5 fused sepals, reddish-green or green, each with a pointed tip; corolla 6–12 mm in diameter, bluish-violet or pale reddish-violet, and in 5 parts; lobes oval (broadest towards tips) and spread flat; 5 yellow scales enclosing throat; 5 stamens enclosed in corolla tube; superior ovary of 2 cells; fertilised by insects

Fruits: Nutlets with 4 sections up to 2 mm long

Habitat: In hill zone on gravelly, flat, seasonally flooded banks of lakes in lower Alps

675 **Marsh felwort – S. perénnis**
Plants 15–35 cm high, perennials with square stems; flowering in July and August

Swértia perennis L.
Marsh felwort

Gentianáceae – Gentian family

Leaves: Narrow oval or lanceolate, on stalks in lower part of plant, sessile in upper part, blunt-ended or pointed, entire, blue-green, and opposite

Inflorescence: Flowers in loose racemes and/or panicles

Flowers: 5 sepals, free almost to base, narrow triangular and greenish or brownish; 5 petals, narrow oval or narrow rectangular, coming to short points, deep violet (also with white stripes), often with darker dots, 10–15 mm long and usually somewhat greenish at base; 5 stamens; anthers usually deep violet; superior ovary of 2 cells; fertilised by insects

Fruits: Egg-shaped capsules 10–15 mm long; dispersal by wind or animals

Habitat: In mountain and subalpine zones in moorland and marshes on wet, more or less base-rich, generally calcareous, moderately acid, and often peaty soils containing more or less nutrients; plants grow in light or half-shade

676 **Borage – B. officinális**
Plants 20–70 cm high, annuals with tap-roots and fleshy stems; flowering till August

Borágo officinális L.
Borage

Boragináceae – Borage family

Leaves: Narrow oval or oval, 3–10 cm long, narrowing to winged stalks (often decurrent in topmost leaves), coming to short points at tips, rough-haired, with irregular fine teeth

Inflorescence: Many-flowered cymes

Flowers: Flower stem 2–4 cm long and covered in erect hairs; 5 sepals, lanceolate, 10–15 mm long, dark or blackish-green, swollen during flowering period and clearly covered in erect hairs; 5 petals, narrow oval, pointed, fused at base into short tube, with lobes spread flat and blue or pale violet; white scales protruding from corolla tube; superior ovary of 2 cells; fertilised by insects; attracts bees

Fruits: 4 warty nutlets; dispersal by ants

Habitat: In hill zone (less often in mountain zone, mainly in gardens), common in gardens, vineyards, and waste ground, on well-drained soils containing nutrients in places enjoying mild winters; used as a medical plant from early times

677 **Venus's looking-glass – L. spéculum-véneris**
Plants 10–30 cm high, annuals with branching stems, usually smooth; flowering in June and July

Legoúsia spéculum-véneris (L.) Chaix
Venus's looking-glass

Campanuláceae – Bellflower family

Leaves: Lanceolate, narrow oval or oval, narrowing to stalks in lower part of plant, sessile in upper part, with wavy or bluntly toothed edges and usually with rounded tips

Inflorescence: Short-stemmed flowers in sparse panicles or racemes

Flowers: 5 sepals, narrow lanceolate, pointed, green, often rather longer than petals; these are roundish, spread flat, 20–25 mm wide, deep violet and with short points; 5 unfused stamens, widening somewhat towards base; inferior ovary of 3 cells; pistil with 3 thread-like stigmata; fertilised by insects or self-fertile

Fruits: Capsules 10–15 mm long; dispersal by wind

Habitat: In hill zone in arable fields and vineyards on moderately well-drained to dry, base-rich, usually calcareous, loam or clay containing nutrients; range extends to northern Germany

678 **Alpine eryngo – E. alpínum**
Plants 30–70 cm high, perennials with branching stems; flowering from July to September

Eryngium alpínum L.
Alpine eryngo

Apiáceae (Umbellíferae) – Carrot family

Leaves: Basal leaves long-stalked, triangular to oval, deeply heart-shaped at base, often irregularly incised at tips, with teeth having short prickles; stem leaves sessile or short-stalked, deeply incised or cut, with spiny teeth

Inflorescence: Blue pinnately cut bracts, segments with long prickles, with protruding, dense, cylindrical flower head

Flowers: 5 calyx teeth with prickles 2–4 mm long; 5 petals, smaller than sepals; inferior ovary; fertilised by insects

Fruits: Schizocarps of 2 cells, each containing 1 seed, with indistinct, scaly, longitudinal ridges with wrinkled surface in between

Habitat: In subalpine zone in scrub and rough grassland, on moist to well-drained, calcareous clay containing nutrients; used as an ornamental garden plant; a central and southern European mountain plant found rarely and in isolated locations

679 **Amethyst eryngo – E. amethystinum**
Plants 50–100 cm high, perennials with much-branching stems; flowering from July to September

Eryngium amethystínum L.
Amethyst eryngo

Apiáceae (Umbellíferae) – Carrot family

Leaves: Basal leaves on stalks, triangular or pentagonal, up to 25 cm long, usually divided into 3 lobes and without teeth at base; lobes in segments with spiny points; middle and top leaves lack amplexicaul ears

Inflorescence: Bracts narrow lanceolate, pointed, much longer than flower head and greenish or bluish; flower head spherical, up to 20 mm across and bluish, especially in upper part of plant

Flowers: 5 calyx teeth with prickles 2–4 mm long; 5 petals, about half as long as sepals; inferior ovary; fertilised by insects

Fruits: Schizocarps of 2 cells, each containing 1 seed with indistinct, longitudinal ridges covered in white scales

Habitat: In hill zone in rough grassland, on roadsides, in waste places and vineyards, on dry, sandy, or stony ground in hot places; a south-east European plant with range extending north to the Po valley; cultivated as a garden plant

680 **Liverwort – H. nóbilis**
Plants 5–15 cm high, perennials, usually with almost upright rhizomes; flowering from March to May

Hepática nóbilis Schreber
Hepática trilóba Chaix
Liverwort

Ranunculáceae – Buttercup family

Leaves: Basal leaves heart-shaped, 3-lobed, mid- or dark green above, reddish-brown or violet beneath, with a hairy stalk; the 3 stem leaves lie close to the petals, resembling a calyx, and are oval, entire, up to 1 cm long, and sessile

Inflorescence: Solitary flowers on several stems

Flowers: 5–10 petals, oval (broadest at tips) and blue, pink, or white; numerous stamens; numerous carpels with 1 ovule each; fertilised by insects

Fruits: Hairy nutlets; dispersal by ants

Habitat: In hill and mountain zones in mixed deciduous woods with good ground cover, beech or coniferous woods, on well-drained to fairly dry, base-rich, usually calcareous, neutral, loose loam containing nutrients; an indicator of loam; a European plant with range extending north to Scandinavia (63 degrees); not found in British Isles

681 **Sweet violet – V. odoráta**
Plants 4–10 cm high, perennials with thin runners above ground, rooting at ends; flowering in March and April

Víola odoráta L.
Sweet violet

Violáceae – Violet family

Leaves: All basal, round or kidney-shaped, oval or broad oval, usually heart-shaped at base, blunt-ended or coming to short points, covered in scattered hairs or smooth, and coarsely toothed; stipules smaller than the stalk, broad lanceolate or oval, with fringed margins

Inflorescence: Long-stemmed solitary flowers rising from leaf axils

Flowers: 5 unfused sepals, lanceolate, pointed, with appendages at base, 4–7 mm long (including appendages) and pale or mid-green; 5 petals, free-standing and deep violet; bottom petals usually has straight spur 10–18 mm long; 5 stamens, with short filaments thickening at base; superior ovary of 3 cells; single pistil; fertilised by insects

Fruits: Hairy capsules with 3 valves; dispersal by capsules falling to the ground and spilling out seeds

Habitat: In hill and mountain zones along wood margins, in thickets and orchards, on damp or well-drained loamy soils containing nutrients and humus

682 **Long-spurred pansy – V. calcaráta**
Plants 4–15 cm high, perennials with stems creeping underground; flowering from June to August

Víola calcaráta L.
Long-spurred pansy

Violáceae – Violet family

Leaves: Lanceolate or oval, with blunt-ended or rounded tips, on stalks, with notched edges and usually basal; stipules half as long as leaves, pinnately cut, toothed, or entire, often with up to 2 points on each side

Inflorescence: Solitary flowers on long smooth stems

Flowers: 5 sepals, lanceolate or narrow oval, pointed, smooth and up to 15 mm long including appendages; 5 petals, not fused, up to 4 cm long and violet; bottom petal up to 3 times as long as sepals and with spur 20–35 mm long; 5 stamens; superior ovary; fertilised by insects

Fruits: Capsules with 3 valves

Habitat: In subalpine and alpine zones on meadows, rough grassland, scree, and snow gullies, on moist, base-rich, neutral, stony soil containing nutrients, humus, and topsoil and covered by snow for long periods; binds soil with its runners; plant grows in light or half-shade

683 **Tall violet – V. elátior**
Plants 20–50 cm high, perennials with erect stems and rhizomes; flowering in May and June

Víola elátior Fr.
Tall violet

Violáceae – Violet family

Leaves: All stem leaves, narrow oval or oval, on stalks, flat at base of leaves proper or narrowing towards stalks, pointed at tips, notched, pale to mid-green, often with short hairs on veins

Inflorescence: Solitary flowers arising from leaf axils

Flowers: 5 sepals with long points, often hairy on margins and 6–12 mm long including appendages; 5 petals, not fused, pale blue or whitish, often with short hairs on margins; bottom petal with straight spur up to 25 mm long; 5 stamens; superior ovary; fertilised by insects

Fruits: Capsules with 3 valves

Habitat: In hill zone in marshy and reedy meadows, bogs, riverbanks, wood margins, and roadsides, on intermittently wet or moist, generally calcareous, base-rich clay soil containing little humus; plant grows in light or half-shade, on alluvial soil; range extends to northern Italy; a rare plant which seems to be becoming extinct

684 **Trumpet gentian – G. acaúlis**
Plants 5–10 cm high, perennials with thin rhizomes; flowering from May to August

Gentiána acaúlis L. s.str.
Gentiána kochiána Perr. and Song
Trumpet gentian

Gentianáceae – Gentian family

Leaves: Rosette leaves elliptical, oval or inverted oval broadest at tips, narrowing to short wide stalks, blunt-ended or pointed at tips, entire, glossy, and 4–10 cm long; stem leaves lanceolate, pointed, much shorter than rosette leaves

Inflorescence: Solitary terminal flowers on stems

Flowers: 5 fused sepals, often divided into 5 sections to half length, with long pointed teeth; 5 sepals fused into a narrow bell, 3–7 cm long, with spreading tips, usually dark blue with olive green stripes along the corolla tube; 5 stamens; superior ovary of 2 cells

Fruits: Many-seeded capsules opening along divisions

Habitat: In subalpine and alpine zones on meadows and rough grassland, on damp or well-drained, lime-poor, rather acid, loamy soil often containing peaty humus

685 **Willow gentian – G. asclepiádea**
Plants 30–80 cm high, perennials with rhizomes, without rosettes; flowering from August to October

Gentiána asclepiádea L.
Willow gentian

Gentianáceae – Gentian family

Leaves: Lanceolate or narrow oval, flat or rounded at base, 3–8 cm long, usually with 5 distinct veins

Inflorescence: Flowers almost without stems in axils of top leaves

Flowers: 5 sepals fused into a tube, reddish-brown or dark green, with short sharp points; 5 petals fused into a narrow bell, with spreading tips, reddish-violet or black dots, and pale longitudinal stripes inside corolla tube; 5 stamens; superior ovary of 2 cells; fertilised by insects

Fruits: Many-seeded capsules opening along divisions

Habitat: From hill to subalpine zones on moist wooded slopes, in thickets, wet meadows, and mixed mountain forests, on moist to well-drained calcareous soils

686 **Marsh gentian – G. pneumonánthe**
Plants 15–70 cm high, perennials with rhizomes, without rosettes; flowering from July to September

Gentiána pneumonánthe L.
Marsh gentian

Gentianáceae – Gentian family

Leaves: Lanceolate, sessile, usually blunt-tipped, 2–5 cm long, entire, with edges often slightly rolled down, usually single-veined and opposite

Inflorescence: Solitary flowers on stems in leaf axils, or several grouped at end of stem

Flowers: 5 fused sepals, generally pale green, with long sharp points; 5 petals fused into a narrow bell, with tips spread out and blue; 5 green-spotted stripes inside corolla tube; 5 stamens; anthers fused into a tube; superior ovary of 2 cells; fertilised by insects (bumble-bees)

Fruits: Many-seeded capsules opening along divisions

Habitat: In hill (rarely mountain) zone on wet heathland, reedy meadows, and rough grassland on intermittently flooded, base-rich, generally lime-free or calcareous, neutral, or rather acid clay or peaty soil rich in humus; plant grows in light; used as a medicinal plant in earlier times

687 **Field gentian – G. campéstris**
Plants 5–20 cm high, usually biennials with tap-roots and branching stems; flowering till October

Gentiána campéstris L.
Gentianélla campéstris (L.) Boerner
Field gentian

Gentianáceae – Gentian family

Leaves: Basal leaves narrow oval or spathulate, usually broadest towards tips, sessile, usually blunt-ended, entire, in rosettes, often dying off before flowering; many opposite leaves on stem

Inflorescence: Flowers on stems in leaf axils and forming a terminal raceme

Flowers: Calyx divided into 4 almost to base and pale green; outer 2 calyx teeth lanceolate and pointed, inner 2 narrow lanceolate; 4 or 5 fused petals, violet, each with a lobe, usually spreading and 3–6 mm long; throat of corolla tube bearded; 4 or 5 stamens with free anthers; superior ovary; fertilised by insects or self-fertile

Fruits: Many-seeded capsules on short stems

Habitat: In mountain and subalpine zones on rough grassland, waysides, and dunes, on well-drained to dry, moderately base-rich soil fairly rich in nutrients

688 **Spring gentian – G. vérna**
Plants 2–8 cm high, perennials with thin rhizomes, forming mats; flowering in April and May

Gentiána vérna L.
Spring gentian

Gentianáceae – Gentian family

Leaves: Basal leaves in rosette and narrow oval or oval, sessile on a widened base, round- or blunt-ended or pointed, entire, 10–30 mm long, matt and soft to the touch; stem leaves generally lanceolate, pointed, and shorter

Inflorescence: Terminal solitary flowers on rectangular stems

Flowers: 5 fused sepals, with winged ridges, pale or mid-green, with a lanceolate points; 5 dark azure blue petals fused into a corolla tube 15–30 mm long, with rounded lobes; 5 stamens; superior ovary of 2 cells; fertilised by insects

Fruits: Many-seeded capsules divided into compartments

Habitat: From mountain to alpine zones on meadows, heaths, and wet moorland, in calcareous rough pasture and subalpine stony grassland, on well-drained to moderately dry, usually lime-rich, neutral, often stony clay or loam containing humus; a central and southern European mountain plant

689 **Short-leaved gentian – G. brachyphylla**
Plants 2-10 cm high, perennials with thin rhizomes, forming mats; flowering from June to August

Gentiána brachyphylla Vill.
Short-leaved gentian

Gentianáceae – Gentian family

Leaves: Basal leaves in rosette, rhomboidal, broad oval or round, sessile on a widened base, usually pointed, entire, up to 10 mm long and soft to the touch; stem leaves rather smaller

Inflorescence: Terminal solitary flowers on rectangular stems

Flowers: 5 fused sepals, without or with only small winged ridges, pale or mid-green, with lanceolate or narrow lanceolate tips; 5 dark blue petals fused into corolla tube 15–22 mm long; lobes usually narrow oval and greenish or whitish on outer side; 5 stamens with free anthers; superior ovary of 2 cells

Fruits: Many-seeded capsules divided into compartments

Habitat: In alpine zone on grassland, rocks, cliffs, and scree, on well-drained, base-rich, lime-poor, neutral or slightly acid, finely divided, stony soil; an alpine and Pyrenean plant extending as far east as Styria; not very common

690 **Fairies' thimble s- C. cochleariifólia**
Plants 5–15 cm high, perennials with ascending stems; flowering from June till August

Campánula cochleariifólia Lam.
Campánula púsilla Haenke
Fairies' thimbles

Campanuláceae – Bellflower family

Leaves: Stem leaves in lower part lanceolate, usually pointed, coarsely toothed, on stalks, and close together; middle and upper stem leaves narrow oval to lanceolate, pointed, with distinct stalks, and widely spaced

Inflorescence: Nodding flowers, solitary or in sparse racemes

Flowers: 5 sepals, narrow lanceolate and drawing to fine points, 5 petals fused into a bell, 10–20 mm long and pale or bluish-lilac; lobes broad triangular with short points; 5 stamens, white, and at base enclosing pistil in a ring; inferior ovary; fertilised by insects

Fruits: Capsules opening by 3 lateral pores

Habitat: From mountain to alpine zones on scree, rock crevices, and in the gravel of mountain streams, on moist, usually calcareous scree or stony soil containing more or less topsoil; grows in light or half-shade; a central and southern European mountain plant with a range south-east to Albania

691 **Meadow bellflower – C. pátula**
Plants 20–50 cm high, biennials with much-branching stems; flowering from June until August

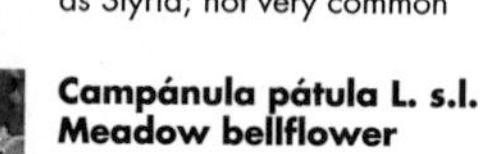

Campánula pátula L. s.l.
Meadow bellflower

Campanuláceae – Bellflower family

Leaves: In lower part of plant narrow oval or lanceolate, 10–15 mm long, entire or slightly toothed, and in upper part lanceolate

Inflorescence: Flowers on stems in loose much-branching cymes; 2 bracteoles above middle of lateral flower stems

Flowers: 5 sepals, narrow triangular, with long points, fused only in lower part; 5 petals fused into a cone halfway up length, 15–25 mm long, smooth and pale or mid-bluish violet; 5 stamens, at base enclosing pistil in a ring; inferior ovary; fertilised by insects

Fruits: Smooth upright capsules opening by 3 lateral pores near tip

Habitat: In hill and mountain (rarely subalpine) zones in copses, hedgerows, meadows, and by roadsides, on damp or well-drained, generally lime-poor, often sandy soil containing nutrients and humus

692 **Diamond-leaved bellflower – C. rhomboidális**
Plants 20–60 cm high, perennials with scattered hairs on stem; flowering from July to August

Campánula rhomboidális L.
Diamond-leaved bellflower

Campanuláceae – Bellflower family

Leaves: Basal leaves no longer present during flowering period; stem leaves more or less widely spaced according to location, oval or broad lanceolate, sessile or short-stalked, pointed at tips, toothed, hairy on both sides

Inflorescence: Nodding flowers in generally unilateral raceme

Flowers: 5 sepals, fused in lower part, with long pointed tips; shallow notches between tips; 5 petals, fused into a wide bell, up to 2 cm long and blue-violet; 5 stamens; inferior ovary; fertilised by insects

Fruits: Nodding capsules, smooth, opening by 3 pores near base

Habitat: In mountain and subalpine zones in meadows, between rocks, on damp to well-drained, usually calcareous and loamy soil rich in nutrients; fairly common in the Alps and Jura; the plant shown here is growing in a relatively dry location

693 **Bearded bellflower – C. barbáta**
Plants10–35 cm high, perennials with rough-haired stems; flowering in July and August

Campánula barbáta L.
Bearded bellflower

Campanuláceae – Bellflower family

Leaves: Lanceolate to narrow oval, sessile, narrowing gradually towards base, usually blunt-ended, entire, or finely toothed, mid to dark green above and usually grey-green beneath

Inflorescence: Short-stemmed flowers in sparse unilateral racemes

Flowers: Initially erect, later nodding; 5 sepals, fused in lower part, rough-haired, with narrow triangular points; 5 petals, pale blue or blue and 15–30 mm long; corolla tube bell-shaped, widening a little at base; the lobes, often bent backwards, are bearded inside; 5 stamens; inferior ovary; fertilised by insects

Fruits: Hairy capsules, curved downwards and opening by 3 pores near base

Habitat: From mountain to alpine zones in meadows, sparse woodland, and rough pasture, on well-drained, base-rich, rather acid and infertile clay or loam

694 **Nettle-leaved bellflower – C. trachélium** Plants 30–90 cm high, perennials with acute-angled stems covered in stiff hairs; flowering till September

Campánula trachélium L.
Nettle-leaved bellflower

Campanuláceae – Bellflower family

Leaves: Basal and lower stem leaves on long stems; leaves proper heart-shaped, pointed, doubly toothed, with scattered stiff hairs; top stem leaves heart-shaped or oval and short-stalked or sessile

Inflorescence: Flowers in leafy panicle with short branches of 1–4 flowers on short stems

Flowers: 5 fused sepals, wide lanceolate, with long points and long, erect hairs; 5 sepals usually fused into a narrow cone, 25–40 mm long and pale blue or pale violet; corolla lobes with long hairs on margins; 5 stamens; inferior ovary

Fruits: Nodding capsules covered with stiff hairs and opening by 3 pores near base

Habitat: In hill and mountain zones in oak and beech woods with good ground cover, thickets, forest clearings, and wood margins, on damp to well-drained, base-rich, neutral, loose, often stony loam containing nutrients and humus; plant grows in shade or half-shade

695 **Diamond-leaved bellflower – C. rhomboidális** Plants 20–40 cm high, perennials with scattered stem hairs; flowering June to August

Campánula rhomboidális L.
Diamond-leaved bellflower

Campanuláceae – Bellflower family

Leaves: Basal leaves die before flowering starts; density of stem leaves varies in different parts; oval to broad lanceolate, sessile or short-stalked, pointed at tips, toothed, and hairy on both sides

Inflorescence: Obliquely slanting or nodding flowers in usually unilateral raceme

Flowers: 5 sepals, fused at base, each with long pointed tip, with shallow notches between tips; 5 petals, fused into a wide bell, up to 2 cm long and blue-violet; 5 stamens; inferior ovary; fertilised by insects

Fruits: Nodding capsules opening by pores near base

Habitat: In mountain and subalpine zones in meadows and among rocks, on damp to well-drained, usually calcareous, loamy soil rich in nutrients; the plant shown here, growing much more luxuriantly than the one in Fig. 692, is in a fairly damp place

696 **Clustered bellflower – C. glomeráta** Plants 15–50 cm high, perennials with smooth or slightly hairy stems; flowering till September

Campánula glomeráta L.
Clustered bellflower

Campanuláceae – Bellflower family

Leaves: Lanceolate to oval, rounded or heart-shaped at base, finely toothed, dark or blue-green, 3–10 cm long, sessile in top part of stem, with stalks winged in middle, long and unwinged towards base

Inflorescence: Terminal flower head, often with solitary flowers or few flowers on branches in leaf axils below terminal head

Flowers: 5 fused sepals, hairy on margins, each with a narrow triangular point; pointed notches between tips; 5 petals fused into bell or cone, 15–30 mm long, smooth or rather hairy and blue-violet; 5 stamens; inferior ovary; pistil shorter than corolla; fertilised by insects

Fruits: Erect hairy capsules opening by 3 pores towards base

Habitat: In hill and mountain zones in dry grassland, thickets, meadows, and by roads and paths, on moderately well-drained, base-rich, calcareous soils containing more or less nutrients

697 **Bristly bellflower – C. cervicária** Plants 20–80 cm high, perennials with stems covered in stiff hairs; flowering from June to August

Campánula cervicária L.
Bristly bellflower

Campanuláceae – Bellflower family

Leaves: Lower leaves narrow lanceolate, narrowing gradually to winged stems, usually blunt-ended, with stiff hairs; upper leaves usually rather narrower, sessile or partially amplexicaul

Inflorescence: Terminal flower head, often with solitary flowers or few flowers on branches in leaf axils below terminal head

Flowers: 5 sepals, whitish, partly fused, hairy, each with blunt triangular tip with margins rolled back; pointed notches between tips; 5 petals fused into bell or cone, 10–20 mm long, with hairy veins and pale blue-violet; 5 stamens; inferior ovary; pistil longer than corolla; fertilised by insects

Fruits: Erect hairy capsules opening by 3 pores

Habitat: In hill and mountain zones on the margins of sunny thickets, sparse oak and pine woods, and shady meadows, on intermittently moist or well-watered base-rich soils

698 **Anchusa – Anchusa officinális** Plants 20–90 cm high, perennials covered in erect hairs; flowering from May to September

Anchúsa officinális L.
Anchusa

Boragináceae – Borage family

Leaves: Long or linear lanceolate, often somewhat heart-shaped at base, blunt-ended or pointed, 5–15 cm long and entire or irregularly toothed

Inflorescence: Numerous many-flowered lateral cymes in upper part of stem

Flowers: 5 fused sepals, with 5 points and distinct white hairs; 5 petals, fused into a corolla tube 6–10 mm long, with rounded lobes; blue or violet; 5 stamens, enclosed in corolla; superior ovary of 2 cells; fertilised by insects or self-fertile

Fruits: 4 nutlets with oblique ridges; dispersal by ants

Habitat: In hill and mountain (rarely subalpine) zones in dry grassland, fields, roadsides, on waste ground and banks, on moderately dry, generally lime-poor, sand or gravel which contains nutrients, may or may not contain humus, and is warm in summer

699 **Blue stoneweed – B. púrpuro-caeruleum** Plants 20–50 cm high, perennials with rhizomes; flowering in May and June

Buglossoídes púrpuro-caeruleum (L.) I.M. Johnston
Blue stoneweed

Boragináceae – Borage family

Leaves: Lanceolate, with long points, 5–10 cm long, narrowing at base, dark green above with distinct main veins, but with only middle vein visible beneath

Inflorescence: Terminal panicles surrounded by bracts

Flowers: 5 sepals fused in calyx divided almost to base, greenish and with narrow strap-like tips; 5 petals, up to 15 cm in diameter, fused into a tube in lower part and initially red, later brilliant blue-violet; petal lobes narrow oval or narrow triangular, usually rounded at tips; 5 hairy stripes inside corolla tube; 5 stamens, not protruding from corolla tube; superior ovary of 2 cells; fertilised by insects

Fruits: 4 smooth, glossy, whitish nutlets each 4–5 mm long

Habitat: In hill (rarely mountain) zone in oak woods, on hot slopes, and in oak copses on moderately dry soil

700 **Viper's bugloss – E. vulgáre**
Plants 30–140 cm high, perennials thickly covered in bristly hairs; flowering from May to September

Echium vulgáre L.
Viper's bugloss

Boragináceae – Borage family

Leaves: Lanceolate, pointed, hairy, in lower part of stem narrowing to broad-winged stalks, and in upper part rounded at base, sessile, and partially amplexicaul

Inflorescence: Many-flowered cyindrical cymes, elongating after flowering into racemes up to 50 cm long

Flowers: 4 (or 5) sepals, fused only in lower part, pale green, with long hairs; covered thickly in hairs at the tips between long bristles; 5 fused petals, blue or violet, up to 20 mm long and with 2 more or less distinct lips; corolla tube without throat scales; 5 stamens distinctly protruding beyond corolla tube; superior ovary of 2 cells

Fruits: 4 nutlets with toothed ridges

Habitat: In hill and mountain (rarely subalpine) zones on banks, in gravel pits, by roadsides, near railway embankments, docks, and quarries, on moderately dry, stony soils which are warm in summer

701 **Vervain – V. officinális**
Plants 30–70 cm high, perennials with rectangular stems; flowering from June to September

Verbéna officinális L.
Vervain

Verbenáceae – Vervain family

Leaves: In upper part of stem linear or oval, sessile and irregularly toothed; in middle part oval, ternately or pinnately cut to middle and on stalks; in lower part coarsely toothed and on stalks; all leaves rough to touch and opposite

Inflorescence: Terminal or axillary many-flowered narrow spikes

Flowers: 5 sepals fused into calyx tube, covered in glandular hairs and 2–3 mm long; 5 fused petals, 3–5 mm long and pale lilac; 4 stamens; superior ovary of 2 cells; fertilised by insects

Fruits: Schizocarp of 4 cylindrical or egg-shaped nutlets; dispersal by ants

Habitat: In hill and mountain zones by roadsides, in gardens, weed beds, rubbish dumps, banks, on walls, and by fences, on well-drained to fairly dry, base-rich, usually calcareous, stony or pure clay or loam with nutrients; grows in half-shade; used as a medicinal plant

702 **Jacob's ladder – P. caerúleum**
Plants 30–100 cm high, perennials with horizontal rhizomes; flowering from June to August

Polemónium caerúleum L.
Jacob's ladder

Polemoniáceae – Phlox family

Leaves: Oval, unequally pinnate, on stalks, in lower part up to 15 cm long and alternate; leaflets lanceolate, 5–14 on each side and 1–2 cm long

Inflorescence: Racemes 10–20 cm long

Flowers: 5 sepals, fused in lower part only, green, and covered in glandular hairs; 5 fused petals, with oval lobes, 15–20 mm long and usually brilliant blue; 5 stamens, all fused to top edge of short corolla tube; superior ovary of 3 cells; fertilised by insects or self-fertile

Fruits: Many-seeded capsules in 3 compartments

Habitat: In hill and mountain zones by roadsides, on walls, in scrub, thickets, and hedgerows, on moist to well-drained, base-rich, usually calcareous, stony or pure clay or loam containing nutrients and humus; plants grow in light or half-shade; also grown as an ornamental garden plant; range extends over north and central Europe and Siberia

703 **Alpine snowbell – S. alpína**
Plants 5–15 cm high, perennials with short rhizomes; flowering from May to July

Soldanélla alpína L.
Alpine snowbell

Primuláceae – Primrose family

Leaves: All basal, round or kidney-shaped, with long hairy stems, round-ended, usually entire with flat edges and leathery to the touch; leaves proper up to 35 mm across

Inflorescence: 2–3 flowers on leafless stem

Flowers: 5 fused sepals, reddish–brown with narrow lanceolate tips; 5 petals, fused into a tube in lower part, funnel-shaped, 7–15 mm long, blue-violet with fringed lobes; 5 stamens, fused in lowest third of length; filaments shorter than anthers; superior ovary; fertilised by insects

Fruits: Capsules 10–15 mm long, opening by 10 teeth

Habitat: In subalpine and alpine zones in snow valleys, hollows, wet meadows, sparse woodland and wet scrub, on wet, cool soils more or less rich in bases and nutrients

704 **Round-headed rampion – P. orbiculáre**
Plants 10–35 cm high, perennials with swollen roots; flowering from May to July

Phyteúma orbiculáre L.
Round-headed rampion

Campanuláceae – Bellflower family

Leaves: Basal leaves lanceolate to oval, on stalks, heart-shaped or rounded at base of leaf proper, blunt-ended or pointed, finely toothed, smooth or slightly hairy; stem leaves rather narrower and sessile, especially towards top

Inflorescence: Compound flower head of 20–30 florets; bracts entire or slightly toothed and half as long as the flower head

Flowers: 5 sepals, narrow lanceolate and 2–3 mm long; 5 petals, initially fused at base and tip, 10–15 mm long and dark blue; 5 free stamens; inferior ovary of 3 cells; fertilised by insects

Fruits: Capsules opening by lateral pores

Habitat: In mountain and subalpine zones in rough grassland, sparse woodland, and moorland, on well-drained, infertile, generally lime-rich, neutral, stony or pure clay or loam containing humus; range extends north to parts of southern England

705 **Betony-leaved rampion – P. betonicifólium**
Plants 25–70 cm high, perennials with turnip-like roots; flowering from June to August

Phyteúma betonicifólium Vill.
Betony-leaved rampion

Campanuláceae – Bellflower family

Leaves: Basal leaves oval, on stalks, heart-shaped or rounded at base of leaf proper, pointed, blunt, or rounded at tips, finely toothed and smooth or hairy; lower stem leaves rather narrower; top stem leaves narrow lanceolate and sessile

Inflorescence: Short or elongated spikes; bracts narrow, lanceolate

Flowers: 5 sepals up to 3 mm long and smooth or less often rather hairy; 5 petals, blue-violet; 5 free stamens; inferior ovary of 3 cells; fertilised by insects

Fruits: Capsules opening by lateral pores

Habitat: In subalpine and alpine zones in meadows, thickets, sparse woodland, and rough pasture, on well-drained, lime-poor, rather acid clay or loam containing humus and topsoil; a good indicator of infertility and acidity; an alpine plant, with range extending to Carinthia and Styria

706 **Perennial cornflower – C. montána** Plants10–60 cm high, perennials with felted hairs on stems; flowering from May to August

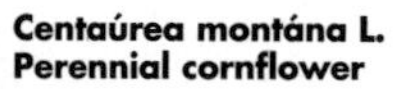

Centaúrea montána L.
Perennial cornflower

Asteráceae (Compósitae) – Sunflower family

Leaves: Linear to narrow oval, entire, sessile and (especially in middle and upper parts of stems) decurrent, pointed at tips, with felted hairs

Inflorescence: Solitary terminal composite flowers 15–20 mm across; bracts green, up to 4 mm long, overlapping like tiles, with dark brown or black margins; middle ones also have 5–9 black hairs projecting from margins on each side

Flowers: Only tubular florets present, those at edge neuter, inner ones hermaphrodite; pappus bristles replace sepals; tubular corolla of 5 fused petals, those at edge enlarged, and blue, violet, or purple; 5 stamens; anthers fused into a tube; inferior ovary of 2 cells

Fruits: Achenes

Habitat: In mountain and subalpine zones in fertile meadows, sparse woodland, scrub, and thickets, on moist to well-drained, base-rich soil containing nutrients

707 **Cornflower – C. cyánus** Plants 20–60 cm high, annuals or biennials with white felted hairs; flowering from June to October

Centaúrea cyánus L.
Cornflower

Asteráceae (Compósitae) – Sunflower family

Leaves: Linear lanceolate, on stalks in lower parts of stems, with isolated small teeth or pinnately cut; in upper parts of stems not divided, sessile, and not decurrent

Inflorescence: Solitary terminal composite flowers; bracts green, with light brown, violet, or whitish margins, and comb-like appendages

Flowers: Only tubular florets present; pappus bristles replace sepals; tubular corolla of 5 fused petals, those at edge enlarged, and blue; 5 stamens; inferior ovary of 2 cells; freely visited by bees

Fruits: Achenes; dispersal by wind or ants

Habitat: In hill and mountain zones in corn fields, on waste ground, and by roadsides, on well-drained to dryish, often lime-poor, alkaline or neutral clay or loam containing nutrients but little humus; prefers warmer locations; a cornfield weed, formerly common but now becoming rare

708 **Italian aster – A. améllus** Plants 20–50 cm high, perennials branching in bottom third of stems; flowering in August and Sept.

Aster améllus L.
Italian aster

Asteráceae (Compósitae) – Sunflower family

Leaves: Lanceolate to narrow oval, hairy beneath, entire in upper parts of plants, and sessile on narrow bases, in lower parts coarsely toothed, with teeth pointing forwards, narrowing to winged stems

Inflorescence: Composite flowers 2–3 cm across in loose panicles; bracts generally spathulate, close, green, often with brownish tips and hairy margins

Flowers: Ray-florets lanceolate, forming a single row; bluish-pink; tubular florets hermaphrodite and in 5 fused parts; pappus yellowish and up to 5 mm long; 5 stamens; inferior ovary of 2 cells; fertilised by insects

Fruits: Hairy achenes 2–3 mm long

Habitat: In hill and mountain zones in dry grassland, sparse oak and pine woods, at margins of thickets and by roadsides, on moderately dry, usually base-rich, loamy soil containing nutrients and humus and warm in summer

709 **Mountain aster – A. alpínus** Plants 5–20 cm high, perennials with thin rhizomes; flowering from June to August

Aster alpínus L.
Mountain aster

Asteráceae (Compósitae) – Sunflower family

Leaves: Lanceolate or narrow oval, broadest towards tips, entire, narrowing to stalks in lower and middle parts but fluffy and sessile in upper part

Inflorescence: Solitary terminal composite flowers 30–45 mm across; bracts hairy at margins and blunt or pointed

Flowers: Ray-florets lanceolate, in a single row, bluish-violet; tubular florets hermaphrodite and in 5 fused parts; yellowish pappus up to 6 mm long; 5 stamens; inferior ovary of 2 cells; fertilised by insects

Fruits: Hairy achenes 2–3 mm long; dispersal by wind

Habitat: In subalpine and alpine zones (rarely penetrating to valleys) in meadows, sunny scree, and on rock ridges, on well-drained, generally lime- and base-rich, flat, stony clay or loam with humus; this arctic/alpine plant is very variable, especially in its leaf shapes and degree of hairiness

710 **Chicory – C. intybus** Plants 30–120 cm high, sparsely branching perennials with spreading or ascending stems; flowering till September

Cichórium intybus L.
Chicory

Asteráceae (Compósitae) – Sunflower family

Leaves: Basal leaves form rosette, narrow oval, usually pinnate, with scattered rough hairs beneath, especially on veins, and narrowing to narrow winged stems; terminal leaflet very large, lateral ones triangular and pointing backwards; stem leaves coarsely and irregularly toothed, pointed and partially amplexicaul

Inflorescence: Composite flowers, or sessile clusters of 2–3 in axils of upper leaves; 2 rows of bracts; receptacle lacks scales

Flowers: All ray-florets; pappus in 2 rows; corolla in 5 fused parts, lanceolate, with 5 faint teeth at end, 15–25 mm long and pale blue; 5 stamens; inferior ovary of 2 cells

Fruits: Light brown achenes 2–3 mm long

Habitat: In hill and mountain zones on waste ground, in meadows, dry grassland, fields, and on roadsides, on well-drained or moderately dry soil rich in nutrients

711 **Blue sow-thistle – C. alpína** Plants 50–30 cm high, perennials with hollow stems and rhizomes; flowering till August

Cicérbita alpína (L) Wallr.
Blue sow-thistle

Asteráceae (Compósitae) – Sunflower family

Leaves: Oval, with long points, irregularly pinnate almost to middle vein, dark green above, blue-green beneath, with large triangular end leaflet and several lateral ones perpendicular to main vein; leaflets notched; only lowest leaves have stalks

Inflorescence: Numerous composite flowers in simple or compound raceme; bracts narrow triangular, deep red or dark green, with erect brown glandular hairs; receptacle lacks scales

Flowers: All ray-florets; single-row pappus; corolla in 5 fused parts, lanceolate, with 5 distinct teeth at end, 15–20 mm long and blue-violet; 5 stamens; inferior ovary of 2 cells; fertilised by insects

Fruits: Achenes 4–5 mm long

Habitat: In subalpine (rarely mountain) zone in scrub, woody ravines, thickets, and mixed deciduous woodland, on moist, base-rich, loose clay or loam containing nutrients and humus

712 **Ground globe daisy – G. nudicaúlis**
Plants 10–25 cm high, perennials with woody branching rhizomes; flowering from June to August

Globulária nudicaúlis L.
Ground globe daisy

Globulariáceae – Globe daisy family

Leaves: Basal, in rosettes, narrow oval, gradually narrowing to stalks, usually blunt or round at ends, entire, and 5–15 cm long

Inflorescence: Terminal compound flower heads 15–25 mm across; bracts small, lanceolate, and pointed

Flowers: 5 sepals, fused into a tube, smooth or with scattered hairs and with lanceolate tips; 5 petals, fused into a tube, blue, 4–8 mm long, with lanceolate lobes; upper lip with 2 strap-shaped lobes and lower lip with 3; 4 stamens (2 bottom ones longer than top) fused to the corolla tube and protruding from it; superior ovary of 2 cells; fertilised by insects or self-fertile

Fruits: Nutlets up to 2 mm long

Habitat: In subalpine (rarely mountain or alpine) zone in dry grassland and calcareous rough pasture, on moderately well-drained to dry, lime-rich, neutral, stony clay or loam containing humus

713 **Alpine calamint – A. alpínus**
Plants 10–20 cm high, perennials with thin woody rhizomes; flowering from July to September

Acínos alpínus (L.) Moench
Saturéja alpínus (L.) Scheele
Alpine calamint

Lamiáceae (Labiátae) – Mint family

Leaves: Elliptical or oval, short-stalked, blunt- or round-ended, usually entire, 10–20 mm long, hairy, and not rolled down at edges

Inflorescence: Short-stemmed flowers, usually in threes, in half-whorls in axils of top leaves

Flowers: 5 sepals, fused into a tube, with 13 veins, 4–7 mm long, with erect hairs, slightly swollen towards base, reddish-brown and with upper and lower lips with 2 and 3 lobes respectively; lobes of lower lip distinctly longer than those of upper one; 5 petals fused into 2 lips, 10–20 mm long, and purple-violet; 4 stamens; superior ovary of 2 cells; fertilised by insects (bees)

Fruits: Schizocarps of 4 nutlets

Habitat: In subalpine zone in dry grassland, sparse pine woods, sunny rough pasture, on dunes, rocks, and excavations, on damp (or dry in summer) base-rich soils with humus

714 **Common sage – S. officinális**
Plants 40–70 cm high, perennials with woody lower stems; flowering from May to July

Sálvia officinális L.
Common sage

Lamiáceae (Labiátae) – Mint family

Leaves: Linear to oval, base of leaves proper often round, blunt, or pointed, entire or finely toothed, with wrinkled surface, thickly covered in hair beneath, sessile towards top, on stalks towards base

Inflorescence: Several whorls with 4–8 flowers at end of stem; bracts very short

Flowers: 5 sepals, fused into tube, veined, reddish-brown, 8–12 mm long, with upper lip having 3 points; 5 petals, violet, formed into 2 lips, 15–25 mm long and with almost straight upper lip; 2 stamens; superior ovary of 2 cells; fertilised by insects

Fruits: Schizocarps dividing into 4 nutlets 2–3 mm long

Habitat: In hill zone on rocky plains, in mats on dry ground, and in gardens (where it is grown as a medicinal plant and as a herb); a southern European plant that occurs as an escape from cultivation but is naturalised only in very warm locations; S. tomentósa, also found in gardens, has leaves heart-shaped at their bases

715 **Meadow clary – S. praténsis**
Plants 30–50 cm high, perennials with thick tap-roots; flowering from May to August

Sálvia praténsis L.
Meadow clary

Lamiáceae (Labiátae) – Mint family

Leaves: Basal leaves oval, base of leaves proper rather heart-shaped, pointed or blunt at tips, with irregular coarse teeth, on stalks, both sides hairy or smooth, and in rosettes; stem leaves sessile

Inflorescence: Several whorls with 4–8 flowers at end of stem; bracts heart-shaped, with scattered hairs and greenish or reddish

Flowers: Hermaphrodite or female, usually on separate plants; 5 sepals, fused into tube, veined, dark reddish-brown, 8–12 mm long, with erect hairs and 3 points on upper lip; 5 petals fused into tube; hermaphrodite flowers blue-violet, in 2 lips, 15–25 mm long, with upper lip compressed to form a hood; female flowers smaller (10 mm); 2 stamens (if present); superior ovary of 2 cells; fertilised by insects (bumble-bees)

Fruits: Schizocarps dividing into 4 nutlets 2–3 mm long

Habitat: In hill and mountain zones on banks, in thickets, in mats on dry ground, warm meadows, and roadsides on moderately well-drained to dry, usually calcareous, loose loam containing humus and a moderate amount of nutrients and warm in summer; very deep-rooted

716 **Whorled sage – S. verticilláta**
Plants 30–50 cm high, branching perennials with long rhizomes; flowering from June to September

Sálvia verticilláta L.
Whorled sage

Lamiáceae (Labiátae) – Mint family

Leaves: Stem leaves usually heart-shaped, stalked, pointed at tips, 4–12 cm long, irregularly toothed, with rather wrinkled top surface, rather hairy and with 2 erect stipules

Inflorescence: Axillary whorls with up to 24 flowers, forming an interrupted terminal spike; bracts lanceolate

Flowers: On short stems; 5 sepals fused into a narrow bell-shaped tube with 2 lips; upper lip has short middle lobe and 2 narrow, pointed, lateral lobes; 5 fused petals, generally violet, with 2 lips, 10–15 mm long and with an almost straight upper lip narrowing like a stem above the middle; 2 stamens; superior ovary of 2 cells; pistil clearly emerging above upper lip; smaller female flowers occur

Fruits: Schizocarps

Habitat: In hill and mountain zones in weed beds, in mats on dryish ground, in thickets and tips, on moderately dry soil with nutrients

717 **Dragon mouth – H. pyrenaícum**
Plants 10–30 cm high, perennials with woody rhizomes, sweet-scented; flowering till August

Hormínum pyrenaícum L.
Dragonmouth

Lamiáceae (Labiátae) – Mint family

Leaves: Basal leaves oval, on stalks, blunt- or round-ended, with blunt teeth, smooth, with wrinkled surfaces, mid or dark green above and whitish-green beneath; stem leaves narrow triangular to oval, short and sessile

Inflorescence: Short-stemmed flowers, in half-whorls of 2–6 in axils of top leaf pairs

Flowers: 5 sepals fused into a tube, with 13 nerves, with 2 lips, 6–10 mm long, each with a pointed tooth; calyx tube smooth inside; 5 fused petals, violet and with 2 lips; ring of hairs inside corolla tube; upper lip clearly truncated; 4 stamens lie close to upper lip; superior ovary of 2 cells; fertilised by insects

Fruits: Schizocarps with shiny egg-shaped nutlets

Habitat: In mountain and subalpine zones in rough, stony, or calcareous grassland, on well-drained, lime-rich, loose and usually stony loam or clay containing humus; plant prefers light

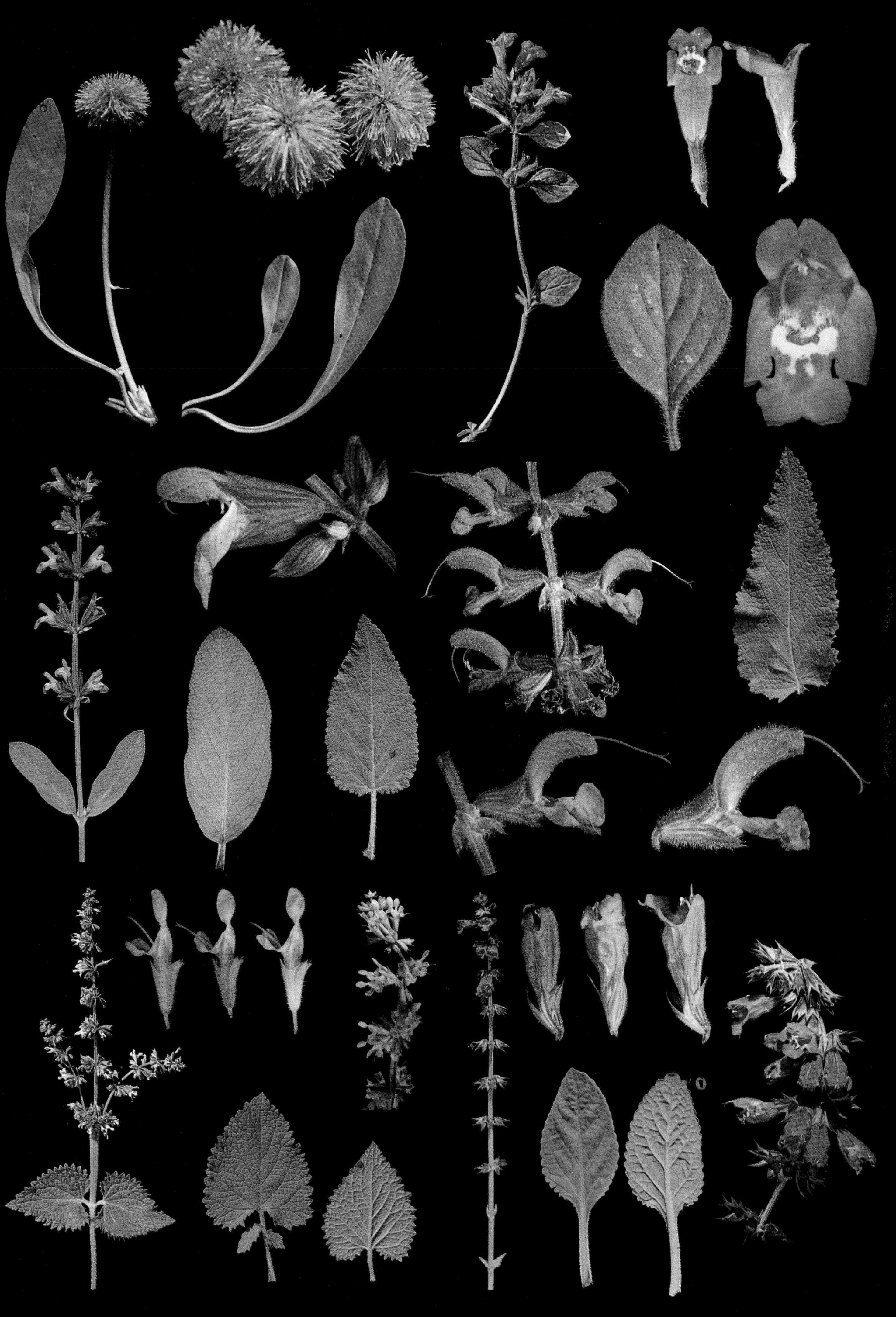

718 **Ground ivy – G. hederácea**
Plants 5–20 cm high, perennials with wide spreading creeping stems; flowering in April and May

Glechóma hederácea L.
Ground ivy

Lamiáceae (Labiátae) – Mint family

Leaves: Kidney- or heart-shaped, on stalks, rounded at tips, coarsely and bluntly toothed, 2–4 cm wide, smooth to very hairy; opposite

Inflorescence: Short-stemmed flowers in groups of 2–3 in leaf axils

Flowers: 5 sepals fused into narrow bell, with 15 veins, covered in short hairs, 3–7 mm long, greenish and deep purple-violet and more or less 2-lipped; 5 fused petals, with lobes of corolla tube pointing forwards, blue-violet and 10–20 mm long; upper lip flat, straight, notched at tip, with short hairs; lower lip in 3 lobes, with middle one enlarged; 4 stamens reaching under upper lip; superior ovary; fertilised by insects

Fruits: Smooth egg-shaped three-cornered nutlets

Habitat: In hill and mountain zones in meadows, deciduous woodland, on roadsides and hedgerows, on wet to well-drained, base-rich, loose clay soil containing nutrients and humus; spreads by means of rooting stems; an indicator of fertile soil

719 **Common bugle – A. réptans**
Plants 10–25 cm high, perennials with overground runners; flowering from April to July

Ajúga réptans L.
Common bugle

Lamiáceae (Labiátae) – Mint family

Leaves: Narrow oval to oval, sessile or short-stalked, entire or with blunt teeth, 3–8 cm long, smooth or with scattered hairs and smaller towards top of stem; leaves within inflorescence entire

Inflorescence: In groups of 2–6 in axils of stem leaves; at tip of stem flowers form a spike

Flowers: 5 sepals, fused into a bell, with many nerves, each with a sharp point, 3–5 mm long, with scattered hairs; 5 fused petals, blue, 10–15 mm long and 2-lipped; upper lip short, straight and divided into 2 lobes; lower lip divided into 3 lobes, middle one largest; 4 stamens protruding beyond corolla; superior ovary; fertilised by insects or self-fertile

Fruits: Egg-shaped nutlets up to 2.5 mm long; plant reproduces by means of rooting runners and dispersal of seeds by ants

Habitat: From hill to subalpine zones in meadows and woodland containing many species of plants, on well-drained, neutral or moderately acid loam containing nutrients and humus; an indicator of fertility; plant grows in light or half-shade

720 **Alpine larkspur – D. elátum**
Plants 50–150 cm high, perennials with knotty rhizomes; flowering from July to August

Delphínium elátum L.
Alpine larkspur

Ranunculáceae – Buttercup family

Leaves: Lower and middle stem leaves have long stalks; leaves proper polygonal, palmately cut into 3–7 leaflets with divisions reaching almost to middle, and dark or blue-green; leaflets irregularly and deeply toothed, usually with long points; top stem leaves generally lanceolate and sessile

Inflorescence: Dense terminal racemes

Flowers: 5 blue petals, 4 lowest oval and top one with backward-pointing spur; 4 nectaries, dark brown, unfused and with white hairs on inner side, the 2 top ones spurred; numerous stamens, with filaments widening towards base; 3–10 superior carpels with numerous ovules; fertilised by insects

Fruits: Follicles up to 1 cm long

Habitat: In subalpine (rarely mountain) zone in scrub, hillsides, and hollows on moist, calcareous, stony soil rich in nutrients and with high humidity; a plant of the Euro-Siberian plain, with a range extending to Mongolia; like all delphiniums, this plant is poisonous

721 **Monkshood – A. napéllus**
Plants 40–140 cm high, perennials with tuberous roots; flowering from July to August

Aconítum napéllus L s.l.
Aconítum compáctum (Rchb.) Gayer
Monkshood

Ranunculáceae – Buttercup family

Leaves: Roundish, on stalks and palmate; lobes without stalks, deeply divided, wedge-shaped at base, with long, very narrow tips

Inflorescence: Very dense, unbranching, terminal racemes

Flowers: Zygomorphic, on stems; 5 petals, more or less hairy and violet; 4 lateral petals oval or round, top petal forming a helmet, with bottom edge straight or rather bent; nectaries almost horizontal on forward-curving tips of erect stems; numerous stamens; 3–5 carpels, with numerous ovules; fertilised by insects (bumble-bees)

Fruits: Follicles

Habitat: In all zones in scrub and grassland and on stream banks on wet to well-drained, base-rich, neutral loam or clay containing nutrients and humus; plant grows in light or half-shade; also grown as an ornamental plant; this plant is poisonous

722 **Lucerne – M. sativa**
Plants 30–80 cm high, with underground runners; flowering till August. Also cultivated

Medicágo satíva L.
Lucerne, alfalfa

Fabáceae (Papilionáceae) – Pea family

Leaves: Divided into 3 leaflets and on stalks; leaflets narrow oval or oval, blunt- or round-ended, 10–30 mm long, on stalks (the 2 lateral leaflets with shorter stems) and with scattered hairs; bractds at base of leaf stalks often have few short teeth

Inflorescence: Short-stemmed flowers in short racemes on stalks

Flowers: 5 sepals, each with pointed tip; 5 petals, lilac to violet, 6–10 mm long, and consisting of standard, 2 short wings and a straight keel; 10 stamens, with topmost filaments free; superior ovary of 1 cell

Fruits: Pods opening by 2 valves, 3–6 mm long, in a spiral of 1.5–3 turns

Habitat: In hill and mountain zones in dry meadows, banks, thickets and on roadsides, on fairly dry, calcareous, base-rich loam or loess containing nutrients and with deep topsoil

723 **Alpine milk-vetch – A. alpínus**
Plants 5–15 cm high, perennials with woody rhizomes and ascending stems; flowering till August

Astrágalus alpínus L.
Alpine milk-vetch

Fabáceae (Papilionáceae) – Pea family

Leaves: Unequally pinnate, on stalks, with 8-12 pairs of leaflets, which are narrow oval, short-stalked, blunt-ended, and hairy or smooth

Inflorescence: Oval or round many-flowered racemes on stalks arising from leaf axils

Flowers: 5 fused sepals, greenish, often with reddish overlay, with dark green notched or pointed tips; 5 petals, bluish-white and with a blue-violet standard, 2 entire wings and a keel; 10 stamens, topmost free; superior ovary of one cell; fertilised by wind

Fruits: Pods 10–15 mm long, thick, sessile, with dark hairs, not constricted between seeds

Habitat: In subalpine and alpine zones in meadows and stony grassland, on well-drained or moderately dry, lime-rich, neutral, stony clay soil containing humus; an arctic/alpine plant, also native to North America and Asia

724 **Bush vetch – V. sépium**
Plants 20–50 cm high, perennials with upright or trailing stems; flowering till July

Vícia sépium L.
Bush vetch

Fabáceae (Papilionáceae) – Pea family

Leaves: With 4–7 pairs of leaflets, narrow oval or oval, 6–30 mm long, usually rounded at both ends, often on short stalks, at tip also slighly notched or with short narrow points, and with scattered hairs beneath; end leaflet as branching tendrils

Inflorescence: Short racemes of 3–6 flowers

Flowers: 5 fused sepals, often reddish and with teeth of unequal length; 5 petals, dull violet or pink, and formed of standard, 2 wings and keel (shorter than wings); 10 stamens, with topmost free; superior ovary of 1 cell; fertilised by insects

Fruits: Flat pods 20–35 cm long, containing 3–6 seeds

Habitat: In hill and mountain zones in grassy places, hedges, and thickets with good ground cover, on moist to well-drained, base-rich, loose loam or clay containing nutrients and humus; an indicator of fertile soil; grows in light or half-shade

725 **Tufted vetch – V. crácca**
Plants 20–110 cm high, perennials, usually with ascending stems; flowering from June to August

Vícia crácca L. s.l.
Tufted vetch

Fabáceae (Papilionáceae) – Pea family

Leaves: Divided, with 6–11 pairs of leaflets, narrow oval or linear, 10–25 mm long, on short stalks, generally rounded at base, rounded at tips or with short narrow points, with scattered hairs or smooth; end leaflet as branching tendrils

Inflorescence: Racemes on stalks, with 15–40 short-stemmed flowers

Flowers: 5 fused sepals, with teeth 1–2.5 mm long of unequal length, hairy or smooth and greenish or reddish; 5 petals, blue-violet, with standard, 2 wings and keel; 10 stamens, with topmost free; superior ovary of 1 cell; fertilised by insects

Fruits: Flat pods 15–25 mm long, containing 2–8 seeds and without spiny points

Habitat: From hill to subalpine zones in meadows, fields, woodland, on wood margins and river banks, on well-drained to fairly dry, more or less base-rich, neutral, stony or sandy soil with more or less nutrients

726 **White false hellebore – V. álbum**
Plants 50-150 cm high, perennials with thick rhizomes, smooth in lower parts; flowering in July

Verátrum álbum L. ssp. lobeliánum (Bernh.) Rchb
White false hellebore

Liliáceae – Lily family

Leaves: Oval to broad oval, sessile, partially amplexicaul, pointed at tips, up to 20 cm long, with parallel veins, often slightly hairy beneath; upper stem leaves narrow oval to lanceolate

Inflorescence: Dense many-flowered racemes up to 50 cm in length

Flowers: 6 petals, oval, 10–15 mm long, pointed, yellowish-green, with dark green veins; 6 stamens; superior ovary of 3 cells; 3 styles; single-sexed flowers also present; bracts much longer than flower stems; fertilised by insects

Fruits: 3-celled many-seeded capsules, opening inwards

Habitat: From hill to subalpine zones on alpine meadows, by cattle sheds, in scrub and moorland, on wet to well-drained, calcareous loam or clay containing nutrients and humus and with deep topsoil; a poisonous weed; range extends nothwards to 71 degrees but not found in the British Isles, Ireland, or Iceland

727 **Bur reed – S. eréctum**
Plants 30–140 cm high, perennials with creeping rhizomes; flowering from June to August

Spargánium eréctum L. s.l.
Spargánium ramósum Hudson
Bur reed

Sparganiáceae – Bur reed family

Leaves: Grass-like, rigid, erect, triangular in section in lower parts and coming to more or less blunt tips; upper stem leaves do not widen into sheaths

Inflorescence: Male and female flowers in different spherical capitula; male capitula borne above the female on the branches of the inflorescence

Flowers: Male flowers generally with 3 petals and 3 or more stamens; female flowers in axils of bracts, with 3–6 scale-like petals and a 1-celled ovary with a single ovule

Fruits: Dark brown nutlets, varying widely in shape in different varieties

Habitat: In hill and mountain zones on mud or in shallow water in ponds, ditches and slow-moving rivers and on ungrazed marshland, on a muddy or sandy bottom

728 **Blue spurge – E. myrsinítes**
Plants 30–120 cm high, perennials with rhizomes; flowering from April to July

Euphórbia myrsinítes L.
Blue spurge (with greenish flowers)

Euphorbiáceae – Spurge family

Leaves: Narrow oval or oval, sessile, 2–3 cm long, coming to short or blunt points, entire, blue-green, and either alternate or opposite; no bracts; leaves exude latex when cut

Inflorescence: Solitary flowers in leaf axils in upper part of stem, and in a spike at end of stem; each long-stemmed solitary flower (or cyathium) is surrounded by round bracts; glands of cyathium without appendages

Flowers: Bracts enclosing a female flower, generally hanging on a long stem, and several male flowers of one stamen only; ovary on stem and of 3 cells, with 3 styles fused at the base and each in 2 parts; compare Fig. 328 for a form with yellow flowers

Fruits: Capsules separating into 3 parts, leaving a persistent centre

Habitat: Found mainly in the Mediterranean region (and in Balearic Isles) and strongly represented in the Orient; in recent years this smooth, grey-blue or greyish-white herb has often been planted in gardens on rockeries

729 **Seguier's spurge – E. seguieriána**
Plants 20–50 cm high, perennials with long woody rhizomes; flowering from May to July

Euphórbia seguieriána Necker
Euphórbia gerardiána Jacq.
Seguier's spurge

Euphorbiáceae – Spurge family

Leaves: Linear or linear lanceolate, usually sessile, with fine points at tips, entire, 10–30 mm long, blue-green and alternate

Inflorescence: Terminal umbel with up to 15 primary stems; each flower surrounded by bracts formed into a cup which has oval, yellow, semicircular glands on the outer side

Flowers: Bracts enclosing a female flower on a long stem consisting of an ovary with 3 cells, and several male flowers consisting of 1 stamen only

Fruits: Capsules separating by 2–3 valves into compartments 2–4 mm long

Habitat: From hill to subalpine zones in dry grassland, sunny, dry, flat ground, on dunes and banks, on dry, base-rich, calcareous, neutral soil containing humus

730 **Cypress spurge – E. cyparíssias**
Plants 10–50 cm high, perennials with long underground runners; flowering from April to July

Euphórbia cyparíssias L.
Cyress spurge

Euphorbiáceae – Spurge family

Leaves: Narrow lanceolate, with parallel veins in middle parts, 15–30 mm long, usually entire, sessile, smooth, and yellowish or dark green

Inflorescence: Terminal umbels with 10–20 rays, often with axillary flowering branches below; bracts of individual flowers oval or semicircular; yellow cyathium with sickle-shaped yellow glands

Flowers: Cyathium includes a single female flower on a stem, with 3-celled ovary, and several male flowers of 1 stamen each; no perianth

Fruits: Schizocarp, breaking into 3 hemispherical nutlets with cylindrical warts

Habitat: From hill to subalpine zones in rough pasture, grassland, waste ground, along roadsides, and in thickets, on moderately dry, calcareous, base-rich (but also on lime-free ground!) loose loam or loess containing humus; a poisonous weed

731 **Marsh spurge – E. palústris**
Plants 50–140 cm high, perennials with thick rhizomes; flowering in May and June

Euphórbia palústris L.
Marsh spurge

Euphorbiáceae – Spurge family

Leaves: Lanceolate to narrow oval, usually entire, smooth, sessile, rounded at tips, 4–8 cm long, dark green above, blue-green beneath; alternate

Inflorescence: Terminal umbels; bracts not fused together; cyathium with yellow, oval glands

Flowers: Inside cyathium are a single female flower on a stem, with a 3-celled ovary, and several male flowers of 1 stamen each; no perianth

Fruits: Schizocarp 4–6 mm long, with cylindrical warts, breaking into 3 nutlets

Habitat: In hill zone in marshes, ditches, by streams and ponds, in weed beds and water meadows, on wet, generally calcareous, turfy soil with humus and more or less nutrients; a plant growing on alluvial soil; it prefers light and indicates clay; a plant of the Euro-Siberian plain with range extending north to the Netherlands and southern Scandinavia

732 **Seguier's spurge – E. seguieriána**
Plants 20–50 cm high, perennials with long woody rhizomes; flowering from May to July

Euphórbia seguieriána Necker
Euphórbia gerardiána Jacq.
Seguier's spurge

Euphorbiáceae – Spurge family

Leaves: Linear or linear-lanceolate, generally sessile, with fine point at tip, entire, 10-30 mm long, blue-green with lighter main vein and alternate

Inflorescence: Terminal umbel with up to 15 primary stems; each flower surrounded by bracts formed into a cup which has oval, yellow, semicircular glands on outer side; with lateral axillary flower heads (compare Fig. 729)

Flowers: Bracts enclosing a female flower on a long stem consisting of an ovary with 3 cells, and several male flowers of 1 stamen only

Fruits: Capsules separating by 2–3 valves into compartments 2–4 mm long

Habitat: From hill to subalpine zones in dry grassland, sunny, dry, flat ground, on dunes and banks, on dry, base-rich, calcareous, neutral soil containing humus

733 **Cuckoo-pint – A. maculátum**
Plants 20–50 cm high, perennials with tuberous rhizomes; flowering in April and May

Arum maculátum L.
Cuckoo-pint, lords and ladies

Aráceae – Aroid family

Leaves: Hastate, 10–20 cm long, pointed or rounded at tips, entire, with long-stalks

Inflorescence: Terminal spadix surrounded by a yellow-green, rolled, conical bract or spathe; spadix 5–10 cm long and usually purple in the upper part

Flowers: All unisexual; lower ones female with a single-celled ovary; upper flowers male with 1 or more stamens; sterile flowers present above and below the male flowers; fertilised by flies

Fruits: Poisonous red berries up to 6mm across

Habitat: Common in hill (rarely mountain) zone in beech and mixed forests with good ground cover and in hedgerows, on water-bearing or well-drained calcareous, loose, neutral or moderately acid loam or clay which contains nutrients and humus and has deep topsoil; prefers shade; an indicator of fertile ground; attracts flies and midges

734 **Dog's mercury – M. perénnis**
Plants 10–30 cm high, perennials without lateral branches, with branching roots; flowering till July

Mercuriális perénnis L.
Dog's mercury

Euphorbiáceae – Spurge family

Leaves: Lanceolate or long oval, on stalks, with glands at base of leaves proper, often rounded, pointed at tips, with irregular blunt teeth, 5–12 cm long; opposite

Inflorescence: Plants are dioecious; male flowers in many-flowered clusters; female flowers solitary or in groups of 2–3 on long stalks

Flowers: Male flowers with up to 20 stamens; anthers spherical; female flowers with 3 yellow calyx points, 3 greenish staminodes, and 2 carpels; fertilised by insects or wind

Fruits: Schizocarp of 2 egg-shaped bristly nutlets; dispersal by ants

Habitat: In hill and mountain (rarely subalpine) zones in beech and mixed deciduous woods and scrub, on moist, base-rich, neutral to moderately acid, loose, well aerated, often stony or sandy loam containing nutrients and humus; spreads by runners

735 **Herb Paris – P. quadrifólia**
Plants 15–30 cm high, perennials with underground creeping rhizomes; flowering in April and May

Paris quadrifólia L.
Herb Paris

Liliáceae (Trilliáceae) – Lily family

Leaves: Generally 4 in a terminal whorl, oval or round, sessile, coming to short points and 5–10 cm long

Inflorescence: Solitary flowers at end of stem 3–6 cm long

Flowers: Sepals usually 4, green, lanceolate, 20–35 mm long, with long points; petals usually 4, rather shorter than sepals, very narrow and coming to long points; 8 stamens; filaments narrowing to awn-like points; superior ovary

Fruits: Round, dark blue berries opening to show many compartments

Habitat: From hill to subalpine zones in damp woods of most kinds, on well-drained, more or less base-rich, loose clay or loam containing humus and more or less nutrients; an indicator of water seepage; binds exhausted soil; roots up to 50 cm deep; range of this plant of Europe and Asia extends north to Iceland and Scandinavia to 71 degrees, south to northern Spain and southern Italy

736 **Good King Henry – C. bónus-henrícus**
Plants 30–80 cm high, perennials with glandular simple hairs; flowering till August

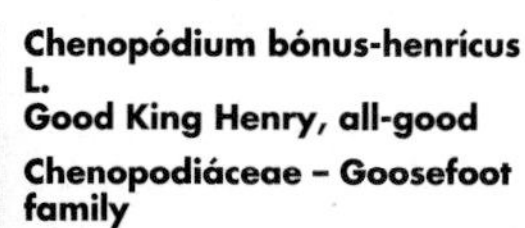

Chenopódium bónus-henrícus L.
Good King Henry, all-good

Chenopodiáceae – Goosefoot family

Leaves: Triangular or hastate, blunt or pointed at tips, entire or with slightly wavy edges, on long stalks; alternate

Inflorescence: Mainly terminal tapering flower head of numerous small cymes, leafless except at base

Flowers: Hermaphrodite and female flowers present; 3–5 petals, greenish, with irregularly incised and toothed margins; stamens usually fewer than 5 and protruding beyond petals; superior ovary of 2 cells

Fruits: Flat nutlets with dark red shiny seeds

Habitat: From hill to subalpine zones in rich pastures, farmyards, roadsides, etc. (usually near buildings), on well-drained, extremely nitrogen-rich, sandy or pure clay or loam containing nutrients and humus; also used as a vegetable (=wild spinach) and as pig fodder; range extends north to central Scandinavia and England (local and rather rare in Scotland and Ireland)

737 **Quinoa – C. quinóa**
Plants 30–90 cm high, annuals with numerous flower heads in upper parts; flowering till July

Chenopódium quinóa Willd.
Quinoa

Chenopodiáceae – Goosefoot family

Leaves: Oval, entire, deeply toothed or irregularly lobed, on stalks, narrowing in a broad wedge towards bases of leaves proper, round, blunt-ended or coming to short points at tips, with leaves proper up to 7 cm long, blue-green above and grey-green beneath

Inflorescence: Terminal flower head with upright lateral branches, all covered with numerous small cymes

Flowers: Hermaphrodite and female flowers present; 3–5 petals, greenish or yellowish, rather translucent and entire or with toothed margins; stamens usually fewer than 5 and protruding beyond petals; superior ovary of 2 cells; style with 2 long, slightly papillose stigmata

Fruits: Nutlets

Habitat: Introduced as a leaf vegetable from the High Andes of Colombia, Chile, and the Argentine; these countries are still the main places where this species is grown, especially the white-fruited, cultivated form

738 **Fat hen – C. álbum**
Plants 30–120 cm high, perennials with erect branches; flowering from July to September

Chenopódium álbum L.
Fat hen

Chenopodiáceae – Goosefoot family

Leaves: Lanceolate, oval or rhomboidal, with base narrowing in wedge shape towards stalk, blunt or pointed, entire or (especially in lower part of plant) irregularly toothed, same colour on both sides or grey-green beneath (because of thickly set glandular hairs)

Inflorescence: Terminal racemes of numerous small cymes

Flowers: 5 petals, greenish, with raised ridges running down back; usually 5 stamens; superior ovary

Fruits: Flat nutlets, with brownish-red or black shiny seeds

Habitat: From hill to subalpine zone; an invader of forest clearings, arable land, gardens, waysides, waste land, and river banks, on well-drained to dry, sandy or loamy soil with nutrients and humus (also grows on fresh soil); an early coloniser of newly turned soil with its deep roots; used as an edible plant in Neolithic times and has accompanied cultivation ever since; now spread worldwide

739 **Sticky goosefoot – C. bótrys** Plants 20–50 cm high, annuals, sticky to touch, thickly covered in glandular hairs; flowering July /Aug.

Chenopódium bótrys L.
Sticky goosefoot

Chenopodiáceae – Goosefoot family

Leaves: Lanceolate or oval, deeply and irregularly notched, short-stalked, 3–7 cm long and usually grey-green beneath; lobes entire or irregularly and coarsely toothed; hardly any leaves in flower head

Inflorescence: Terminal and lateral racemes of numerous small cymes

Flowers: 5 petals, greenish-yellow and entire; 5 stamens; superior ovary

Fruits: Flattish nutlets, with dark brown, smooth, shiny seeds

Habitat: In hill and mountain zones in vineyards, waste places, and rubbish dumps on moderately dry or dry, loamy, sandy or stony soil containing more or less nutrients but no humus, in warm places; an early coloniser of newly turned soil; found in the whole Mediterranean area, but has spread north to central Alpine valleys, and to America and Australia

740 **Lady's mantle – A. vulgáris**
Plants 10-40 cm high, perennials, never with silky hairs; flowering from May to July

Alchemílla vulgáris L. s.l.
Lady's mantle

Rosáceae – Rose family

Leaves: Basal leaves with 7–13 lobes, on stalks, dark or blue-green above, often rather lighter beneath, smooth or hairy in various ways and up to 15 cm across; lobes rounded at end, with numerous teeth extending into the intervals between lobes; stem leaves short-stalked or sessile, often deeply heart-shaped at base

Inflorescence: Compound terminal cyme of dense or lax terminal cymes

Flowers: 4 outer petals, yellowish-green, narrow oval, pointed and 1–2 mm long; 4 inner sepals, green or yellowish, pointed and 1–2 mm long; calyx and sepals smooth; no petals; 4 stamens; single carpel with 1 pistil, 1 stigma, and 1 ovary; nectar secreted by ring on receptacle

Fruits: Nutlets enclosed by a smooth calyx

Habitat: Found in all zones in damp grassland, by springs, on moist scree, on moist to well-drained soils containing nutrients and humus

741 **Yellow-green lady's mantle – A. xanthochlóra** Plants 20–60 cm high, perennials with erect hairy stems; flowering from May to July

Alchemílla xanthochlóra Rothmaler s.l.
Alchemílla vulgáris L. s.l.
Yellow-green lady's mantle

Rosáceae – Rose family

Leaves: Basal leaves with 7–11 lobes, on stalks, green or blue-green above, pale or yellow-green beneath, hairy on main veins and 4–13 cm across; lobes rounded at ends, with 7–12 teeth extending into intervals between lobes

Inflorescence: Compound terminal cymes of dense or lax terminal cymes

Flowers: 4 (sometimes 3) outer petals, yellowish-green, usually pointed and 1–2 mm long; 4 (sometimes 3) inner sepals, green or yellowish, usually pointed and 1–2 mm long; no petals; 4 stamens; single carpel with 1 pistil, 1 stigma, and 1 ovary; nectar secreted by ring on receptacle

Fruits: Nutlets enclosed by smooth calyx

Habitat: From hill to subalpine zones in damp grassland, by springs, streams and ditches, on moist to well-drained soils with nutrients and humus; a European plant with range extending to north of England

742 **Alpine lady's mantle – A. alpína**
Plants 10–30 cm high, perennials with ascending shoots; flowering in June and july

Alchemílla alpína L. s.l.
Alpine lady's mantle

Rosáceae – Rose family

Leaves: Basal leaves palmately divided almost or quite to base into 5–7 lobes, on stalks, mid to dark green above, thickly covered in close silvery hairs beneath; lobes narrow oval, sessile, blunt or round at ends, with 2–6 sharp teeth on either side; stalk hairy; stem leaves with 3 segments simple in upper parts

Inflorescence: Short-stemmed flowers in rather dense clusters forming terminal cymes

Flowers: 4 inner sepals, usually yellowish, pointed, and 1–2 mm long; 4 outer petals, smaller, narrower, and pale yellow; sepals covered in erect hairs on outer side; no petals; 4 stamens; single carpel with 1 pistil, 1 stigma, and 1 ovary; nectar secreted by ring on receptacle

Fruits: Nutlets enclosed by soft, smooth calyx

Habitat: In alpine zone on grassland, rock crevices, scree, and mountain tops, on well-drained to dry, base-poor, acid loam containing nutrients and humus; range extends north to Greenland and Iceland

743 **Ribwort plantain – P. lanceoláta**
Plants 10–40 cm high, perennials with spherical or cylindrical spikes; flowering from April to Sept.

Plantágo lanceoláta L.
Ribwort plantain

Plantagináceae – Plantain family

Leaves: Lanceolate or narrow oval, with pointed tips, entire or weakly toothed, with 3–7 veins, erect and forming basal rosettes

Inflorescence: Many-flowered spike at end of long stem (longer than leaves); the stem has about 5 deep grooves beneath the spike

Flowers: Bracts oval, pointed, without awns, hairy at tips, smooth-edged and rather longer than the 4 sepals; these are of unequal length; 4 fused petals, 2–4 mm long, smooth and with brownish tips; 4 stamens, fused into corolla tube, yellowish, with long filaments; superior ovary of 2 cells; fertilised by wind

Fruits: Egg-shaped capsules, containing 2 seeds, with top opening like a lid; dispersal by sticky seeds adhering to animals' fur

Habitat: Common in all zones except the alpine one, in grassland, arable fields, and by roadsides, on moist to fairly dry soils

744 **Great plantain – P. major**
Plants 15–30 cm high, perennials with erect or oblique leaves; flowering till October

Plantágo major L. ssp. major
Great plantain

Plantagináceae – Plantain family

Leaves: Broad oval, on stalks, round or heart-shaped at base of leaves proper, entire or weakly toothed, usually rounded at tips, with 5–9 veins; forming basal rosettes

Inflorescence: Many-flowered, narrow, cylindrical, terminal spikes up to 10 cm long

Flowers: Bracts oval, rather pointed, without awns, smooth-edged and shorter than the 4 sepals; these are free almost to base; 4 fused petals, 2–4 mm long, smooth , with yellowish tips; 4 stamens, initially pale lilac, later yellowish-brown; superior ovary of 2 cells; fertilised by wind

Fruits: Egg-shaped capsules, 3–4 mm long, containing 8 seeds, with tops opening like lids

Habitat: From hill to subalpine zones on trampled ground in farmyards, by roads, in cultivated ground, and on riverbanks, rarely in grassy places, on well-drained thick loam or clay containing nutrients; an early coloniser of trampled ground, rooting up to 80 cm deep; endures salt in soil; used in earlier times as a medicinal plant

745 **Hoary plantain – P. media**
Plants 20–35 cm high, perennials, usually with recumbent leaves; flowering till July

Plantágo média L.
Hoary plantain

Plantagináceae – Plantain family

Leaves: Narrow to broad oval, short-stalked or sessile, blunt or pointed ends, weakly toothed, with 7–9 veins, covered sparsely or thickly in hairs on both sides, and forming basal rosettes

Inflorescence: Cylindrical terminal spikes 3–8 cm long; stem of spike has close hairs

Flowers: Bracts oval, without awns, smooth and only slightly shorter than the 4 sepals; these are free almost to base; 4 fused petals, 3–4 mm long, smooth, with white tips; 4 lilac stamens; superior ovary of 2 cells; fertilised by insects or self-fertile

Fruits: Egg-shaped capsules containing 3–8 seeds; dispersal by wind

Habitat: In hill and mountain (rarely subalpine) zones in grassland and on roadsides, on well-drained or intermittently well-drained sandy or pure loam containing nutrients and usually with deep topsoil; less often found in Mediterranean region

746 **Alpine plantain – P. alpína**
Plants 5–15 cm high, perennials with ascending or erect leaves; flowering till July

Plantágo alpína L.
Alpine plantain

Plantagináceae – Plantain family

Leaves: Linear or linear lanceolate, narrowing to stem-like bases, sessile, pointed at tips, usually entire, with 3 veins, covered in scattered hairs and forming basal rosettes

Inflorescence: Cylindrical terminal spikes 15–30 mm long

Flowers: Bracts oval, slightly pointed, without awns, smooth or hairy only on edges, as long as or longer than the 4 sepals; these are free almost to base; 4 fused petals, 2–3 mm long, smooth, with whitish tips; corolla tube somewhat hairy on lower part of exterior; 4 stamens, yellow; superior ovary of 2 cells; fertilised by wind

Fruits: Capsules containing up to 10 seeds; dispersal by sticky seeds adhering to animals' fur

Habitat: In subalpine and alpine zones on all types of grassland, on at least moderately well-drained, lime-poor, rather acid and stony loam containing humus and a moderate amount of nutrients; roots up to 1 m deep; a good fodder plant

747 **Serpentine plantain – P. serpentína**
Plants 10–30 cm high, perennials with ascending or erect leaves; flowering till August

Plantágo serpentína All.
Serpentine plantain

Plantagináceae – Plantain family

Leaves: Linear or linear lanceolate, sessile, fleshy, coming to grooved points at tips, entire, with 3 veins, smooth, and forming basal rosettes

Inflorescence: Cylindrical terminal spikes 2–10 mm long

Flowers: Bracts rather pointed, without awns, smooth-edged, as long as or longer than the 4 sepals; these are free almost to base, hairy and with a fairly well-marked ridge; 4 fused petals, 2–3 mm long, smooth. with whitish or brownish tips; 4 stamens; superior ovary of 2 cells

Fruits: Many-seeded, rather pointed capsules

Habitat: In subalpine (rarely hill and mountain) zones in grassland, scree, and by roadsides on moderately well-drained to dry, weakly acid or basic, and stony or gravelly soil in warm places; a central and southern European mountain plant, with a range extending east to the Tyrol and the Dolomites

748 **Stinging nettle – U. dioíca**
Plants 30–150 cm high, perennials, usually branching; flowering from June to September

Urtica dioíca L.
Stinging nettle

Urticáceae – Stinging nettle family

Leaves: Oval, on stalks, heart-shaped or round at base, with pointed tooth at tip, coarsely toothed, 5–10 cm long, covered in stinging hairs and many shorter hairs; with wavy surface and opposite

Inflorescence: Numerous many-flowered racemes rising from leaf axils

Flowers: Dioecious plants (male and female flowers on different plants); male flowers with 4 greenish petals and 4 stamens protruding from them; female flowers have 2 smaller outer and 2 larger inner petals with a superior ovary and a feathery stigma

Fruits: Lentil-shaped nutlets

Habitat: In all zones except the alpine one in hedgerows, woods, grassy places, fens, and near buildings, especially where ground is covered with litter or rubble, on moist to well-drained, particularly nitrogenous, neutral clay or loam containing nutrients and humus and with deep topsoil; the young shoots can be eaten like spinach

749 **Salad burnet – S. minor**
Plants 30–50 cm high, perennials with hairy stems towards base; flowering till August

Sanguisórba minor Scop.
Salad burnet

Rosáceae – Rose family

Leaves: Basal leaves in rosette and unequally pinnate, with 5–15 pairs of leaflets, oval or roundish, sessile or on stalks, dark green above and paler green beneath, with up to 8 coarse pointed teeth on either side; stem leaves have oval or narrow oval leaflets

Inflorescence: Dense flower heads, usually spherical

Flowers: Lower flowers male, middle hermaphrodite, upper female; sepals oval, green, often with reddish-brown margins and 2–4 mm long; no petals; in male flowers 10–30 stamens, in hermaphrodite 1–4; in female flowers stigma on pistil of ovary is constructed of feathery filaments; fertilised by wind

Fruits: Nutlets enclosed by receptacle formed from calyx; dispersal by wind

Habitat: From hill to subalpine zones on calcareous grassland, in thickets, ditches, and by roadsides on moderately dry soils

750 **White bryony – B. álba**
Perennial plants with turnip-like roots and rough climbing stems; flowering in June and July

Bryónia álba L.
White bryony

Cucurbitáceae – Gourd family

Leaves: Pentagonal, lobed, on short stalks, generally heart-shaped at base, with pointed tips, rough-haired on both sides, and with sharp teeth

Inflorescence: Plants are monoecious with female flowers in umbels and male flowers in racemes

Flowers: Male flowers with 5 green fused sepals with pointed tips; 5 petals, oval, pointed, divided almost to middle, generally spread flat and greenish-yellow; 5 stamens. Female flowers with rather smaller petals and inferior ovary; smooth styles; fertilised by insects

Fruits: Black berries

Habitat: In hill zone in hedgerows, thickets, by fences and roadsides, on well-drained, loose loam containing nutrients and warm in summer; poisonous; formerly a medicinal plant; an eastern European plant, with range extending north to warm places in Scandinavia; found in central Alpine valleys; also found as an escape from gardens

751 **White bryony – B. dioíca**
Perennial plants with thickened roots, up to 4 m long, flowering from June to September

Bryónia dioíca Jacq.
White bryony, red bryony

Cucurbitáceae – Gourd family

Leaves: Pentagonal in outline, with triangular or oval lobes, 4–10 cm wide, with hairy stem, deeply heart-shaped at base, rough-haired on both sides

Inflorescence: Plants dioecious, with female flowers in umbels and male flowers in racemes or panicles on different plants

Flowers: Male flowers with 5 green fused sepals with pointed tips; 5 petals, oval, pointed, divided almost to middle, generally spread flat and greenish-yellow; 5 stamens. Female flowers with rather smaller petals and inferior ovary; hairy styles; fertilised by insects

Fruits: Berries, green when unripe, red when ripe; spread by birds

Habitat: In hill (rarely mountain) zone in sparse woodland and occasionally orchards, in hedgerows and thickets, on well-drained, loose loam with nutrients and humus; range extends north to England

752 **Stinking hellebore – H. foétidus**
Plants 20–60 cm high, perennials with basal leaves staying green through winter; flowering till April

Hellebórus foétidus L.
Stinking hellebore

Ranunculáceae – Buttercup family

Leaves: Basal leaves kidney-shaped or round, divided almost to base into 3–9 leaflets, long-stemmed and dark green to blackish-green; segments lanceolate, coming to long points, usually toothed and short-stalked or sessile; stem leaves have fewer leaflets higher up stem; top stem leaves oval, entire, and up to 5 cm long

Inflorescence: Numerous hanging flowers in terminal, branched cymes

Flowers: 5 petals, bent to form a bell, overlapping at edges, green, often with reddish tips and 1–2 cm across; 5–15 nectaries, green, forming a cone and shorter than petals; numerous stamens; 3–8 many-seeded carpels fused at base; fertilised by insects

Fruits: Follicles with long beak-like points

Habitat: In hill and mountain zones in beech and oak woods with good ground cover, at wood margins and in blackthorn thickets, on well-drained to moderately dry, base-rich soils containing nutrients

753 **Green hellebore – H. víridis**
Plants 20–50 cm high, variable perennials, often flowering in February (in flower February to April)

Hellebórus víridis L.
Green hellebore

Ranunculáceae – Buttercup family

Leaves: Usually only 2 basal leaves, not staying through winter, divided almost to base in 7–11 leaflets, on stalks, dark green and usually no longer present during flowering; segments toothed and partially divided into 2–3 lobes; stem leaves also divided, but often not to base

Inflorescence: Nodding, solitary, terminal flowers

Flowers: 5 petals, more or less spread flat, broad oval or round, rounded at tips, green and overlapping; several nectaries, green; numerous stamens; 3–8 many-seeded carpels fused at base; fertilised by insects

Fruits: Follicles with beak-like points; without these, up to 25 mm long

Habitat: In hill zone in sparse woodland (occasionally orchards) and scrub, on well-drained to dry, base-rich, usually calcareous, stony, loose loam containing nutrients in warmer locations; used as a medicinal plant in earlier times

754 **Starry hare's ear – B. stellátum**
Plants10–35 cm high, perennials with rhizomes covered in leaf sheaths; flowering till August

Bupleúrum stellátum L.
Starry hare's ear

Apiáceae (Umbelliferae) – Carrot family

Leaves: Basal leaves linear lanceolate, narrowing toward base, with blunt ends or short points, 5–30 cm long and with clearly protruding middle veins beneath; 2 veins run along edges; between these main veins a distinct network of veins; middle and upper stem leaves partially amplexicaul

Inflorescence: Primary and secondary umbels at ends of main and lateral stems; 2–4 bracts on primary umbel, usually of unequal size and smaller than stem leaves; secondary umbels usually connate, forming a slightly concave sheath

Flowers: 5 yellowish petals; 5 stamens; inferior ovary of 2 cells; fertilised by insects

Fruits: Schizocarp; both nutlets have 5 distinct ridges, slightly winged

Habitat: In subalpine and alpine zones in meadows, on scree, and in rough grassland, on dry, acid, lime-poor soil

755 **Mugwort – A. vulgáris**
Plants 20–120 cm high, perennials with an unpleasant smell; flowering from July to September

Artemísia vulgáris L.
Mugwort

Asteráceae (Compósitae) – Sunflower family

Leaves: Oval, 4–12 cm long, with or without stalks, often eared, singly or doubly pinnate, dark green above and distinctly covered with cottony hairs beneath; last leaflets lanceolate, with rolled down margins

Inflorescence: Dense panicles of numerous composite flowers, 2–4 mm long and sessile or short-stemmed; bracts arranged like tiles, hairy, with white, dry, translucent margins

Flowers: All florets tubular, yellowish or brownish-red, female in outer parts of flower head and hermaphrodite in inner parts; fertilised by wind

Fruits: Achenes up to 2 mm long, usually with longitudinal grooves; dispersal by wind or by sticky seeds attaching themselves to animnals' fur

Habitat: In hill and mountain zones on waste places, river banks, roadsides, weed beds with many species and in thickets, on moist to well-drained soils with nutrients and more or less humus; cultivated in ancient times; formerly used as a medicinal plant and an aromatic flavouring

756 **Pineapple weed – M. discoídes**
Plants 5–35 cm high, annuals with hollow conical receptacles; flowering from May to October

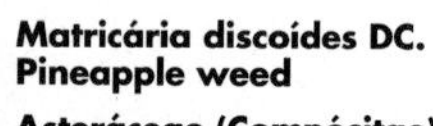

Matricária discoídes DC.
Pineapple weed

Asteráceae (Compósitae) – Sunflower family

Leaves: Narrow oval, doubly or trebly pinnate, 2–6 cm long, broad in middle (especially in upper part), sessile and strongly aromatic

Inflorescence: Panicles of numerous composite flowers 5–10mm across; bracts pale green, smooth and arranged like tiles

Flowers: Only tubular florets present; petals fused together in tubes, greenish, with 4 teeth; inferior ovary of 2 cells; fertilised by insects or self-fertile

Fruits: Smooth achenes 1–2 mm long, usually with dark hairy stripes and toothed margin

Habitat: From hill to subalpine zones by roadsides, on waste land, in gravel pits, on trampled ground, by railway yards, and close to human habitation, on damp to well-drained, solid or loose loam or clay containing nutrients and more or less humus; has been naturalised since about 1850; originally an oriental plant

757 **Field eryngo – E. campéstre**
Plants 20–60 cm high, perennials with much-branching stems; flowering from July to September

Eryngium campéstre L.
Field eryngo

Apiáceae (Umbellíferae) – Carrot family

Leaves: Basal leaves triangular or pentagonal, on long stalks, 10–20 cm long and multiply pinnate; teeth on leaflets have spines; middle and upper stem leaves sessile , with spiny tips, longer than flower heads

Inflorescence: Dense whitish-green flower heads, usually spherical and up to 15 mm across, at end of branching stems; bracts whitish-green, with spines on margins, and longer than flower heads

Flowers: Long spiny bracts, up to 10 mm long, protruding beyond flowers; 5 stamens, with distinct teeth; 5 petals, about half as long as calyx teeth; 5 stamens; inferior ovary of 2 cells; fertilised by insects

Fruits: Schizocarps, with nutlets covered in white scales; dispersal by wind or by sticky seeds attaching themselves to animals' fur

Habitat: In hill zone in rough pasture and by roadsides, on dry, often sandy soils

758 **Monk's rhubarb – R. alpínus**
Plants 30–100 cm high, perennials branching from middle of stems; flowering in July and August

Rumex alpínus L.
Monk's rhubarb

Polygonáceae – Buckwheat family

Leaves: Basal leaves oval or round, 30–50 cm long, on red stalks, heart-shaped at base of leaves proper and with wavy edges; stem leaves smaller, narrow oval, and with halves of leaves proper often ending at different points on stalk

Inflorescence: Numerous dense panicles at ends of main and lateral stems; lateral branches themselves branching and erect

Flowers: Hermaphrodite; 6 petals; outer 3 petals small and thin; inner 3 petals entire or irregularly toothed and reddish-brown; 5 stamens; superior ovary of 3 cells; 3 stigmata, bent back and reddish–brown

Fruits: Nutlets 2–3 mm long, glossy and yellowish brown

Habitat: In sub-alpine zone near buildings and besides streams and roads, on moist to well-drained loam with nutrients and humus; a good indicator of nitrogen; often used as pig fodder; naturalised in North America

759 **Sorrel – R. acetósa**
Plants 20–100 cm high, perennials with fringed or toothed leaf sheaths; flowering from May to August

Rumex acetósa L.
Sorrel

Polygonáceae – Buckwheat family

Leaves: Of many shapes; lower leaves arrow-shaped, lobes directed down, up to 25 cm long, usually with blunt ends and on stalks; upper stem leaves sessile

Inflorescence: Dense panicles; lateral branches point obliquely upwards

Flowers: Unisexual, rarely hermaphrodite; 6 petals; inner 3 petals 3–5 mm long, oval or round, entire, with backward-pointing tubercle at base; outer 3 petals oval, whitish-yellow, usually with reddish margins and bent back to lie close to flower stem; in male flowers 6 stamens; in female flowers superior ovary

Fruits: Dark brown nutlets 2–2.5 mm long

Habitat: From hill to subalpine zone in grassland, weed beds, and by streams and roads, on moist to well-drained, loose loam or clay with nutrients, humus, and deep topsoil; an indicator of nitrogen

760 **Snow dock – R. nivális**
Plants 10–25 cm high, upright or ascending; flowering in July and August

Rumex nivális Hegetschw.
Snow dock

Polygonáceae – Buckwheat family

Leaves: Basal leaves in spring oval in outline, in summer arrow-shaped; basal leaves on stalks, with large round ears at base, round at tips, usually entire and 15–30 mm long; stem carries only 1–2 leaves

Inflorescence: Panicle without branches

Flowers: Unisexual and hermaphrodite: plants may be dioecious; 6 petals; inner 3 petals roundish, heart-shaped at flower stem, 3–5 mm wide, entire, with backward-pointing tubercle and red; outer 3 petals bent back close to flower stem; in male and hermaphrodite flowers 6 stamens; in female and hermaphrodite flowers superior ovary;

Fruits: Yellowish or reddish glossy nutlets 2–3 mm long

Habitat: In alpine zone on ground subject to long periods of snow cover, on calcareous soil irrigated by melted snow and containing nutrients, humus, and fine scree; an eastern Alpine plant; range extends to Albanian mountains

761 **Rubble dock – R. scutátus**
Plants 15–50 cm high, upright perennials with underground runners; flowering till July

Rumex scutátus L.
Rubble dock

Polygonáceae – Buckwheat family

Leaves: Usually arrow-shaped, 3–5 cm long, on stalks, often with 2 diverging pointed lobes at base and narrowing thereafter, spread flat, entire, and green or blue-green

Inflorescence: Loose panicles with upright flower heads

Flowers: Usually unisexual, less often hermaphrodite; 6 petals; outer 3 petals close to inner ones during flowering; inner petals round, heart-shaped at flower stem, 4–6 mm long, entire, without tubercle, translucent, and reddish

Fruits: Grey nutlets 3–4 mm long

Habitat: From hill to subalpine zones on old walls, grassland, and scree, on well-drained to moderately dry, base-rich, lime-free or calcareous, loose, coarse, or fine scree; often an escape from gardens; sometimes used in salads and sauces

762 **Mountain sorrel – O. digyna**
Plants 5–25 cm high, perennials, usually with several stems; flowering in July and August

Oxyria digyna (L.) Hill
Mountain sorrel

Polygonáceae – Buckwheat family

Leaves: Basal leaves oval or round, on stalks, heart- or kidney-shaped at base, 15–40 mm wide, with wavy edges, dark grey-green above and pale grey-green beneath; usually no stem leaves

Inflorescence: Panicles of small cymes

Flowers: All hermaphrodite; 4 green petals, 2 inner much larger than 2 outer and pressed to winged edges of fruit; 4 stamens, rarely 6; superior ovary

Fruits: Lentil-shaped nutlets with wide winged margins

Habitat: In alpine and subalpine zones on mobile scree and moraines, and in damp rocky places on mountains, especially beside streams, on well-drained, lime-poor, open soil; an early coloniser of newly turned soil; an arctic/alpine plant, especially common in the central and southern Alps but also found as far as Spitzbergen, Ireland, Iceland, and Scandinavia

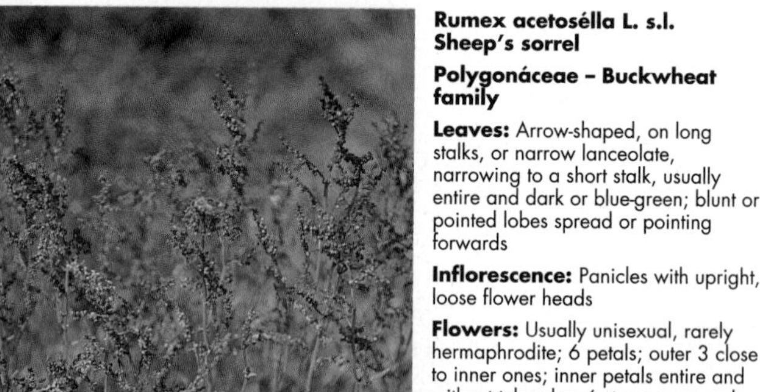

763 **Sheep's sorrel – R. acetosélla**
Plants 10–30 cm high, perennials, erect or ascending; flowering from May to August

Rumex acetosélla L. s.l.
Sheep's sorrel

Polygonáceae – Buckwheat family

Leaves: Arrow-shaped, on long stalks, or narrow lanceolate, narrowing to a short stalk, usually entire and dark or blue-green; blunt or pointed lobes spread or pointing forwards

Inflorescence: Panicles with upright, loose flower heads

Flowers: Usually unisexual, rarely hermaphrodite; 6 petals; outer 3 close to inner ones; inner petals entire and without tubercles; 6 stamens on male flowers; superior ovary on female flowers

Fruits: Brown-red nutlets 1–2 mm long

Habitat: From hill to subalpine zones on grassland and cultivated land, in forest clearings, on heaths and roadsides and banks, on moderately well-drained to dry, base-poor, acid, sandy or loamy soil poor in nutrients; an indicator of acidity and infertility; poor fodder plant; generally distributed throughout Europe and most temperate parts of the world, including Greenland, Japan, and the USA

764 **Bird's nest orchid – N. nídus-ávis** Plants 15–5 cm high, perennials with creeping rhizomes, fleshy lateral roots (bird's nest); flowering till July

Neóttia nídus-ávis (L.) R. Br.
Bird's nest orchid

Orchidáceae – Orchid family

Leaves: No leaves proper, but stem clothed at base with numerous brownish, thin, dry, sheathing scales

Inflorescence: Raceme 5–15 cm long, loose in lower part and dense in upper part

Flowers: Bracts narrow lanceolate, usually reaching middle of ovary; 5 petals, pale brown, bent together, oval, generally blunt and 4–6 mm long; labellum twice as long as petals, darker brown, with 2 small lateral teeth near base, with 2 lobes extending almost halfway and diverging, straight or outwardly curved; single stamen and stigmata borne on special structure (the column); inferior ovary of 3 cells; fertilised by insects or self-fertile

Fruits: Capsules with numerous small seeds

Habitat: In hill and mountain (rarely subalpine) zones in beech, oak, pine, and mixed deciduous woodland on moist to dry, generally calcareous soil containing nutrients and humus

765 **Mare's tail – H. vulgáris**
Plants 20–80 cm high, perennials with spreading, creeping rhizomes; flowering from May to August

Hippúris vulgáris L.
Mare's tail
Hippuridáceae – Mare's tail family

Leaves: Linear, sessile, generally 10–30 mm long, pointed or blunt at tips and in whorls of 8–12; limp in water, but rigid above water surface

Inflorescence: Solitary flowers in axils of leaves above water

Flowers: Sessile and hermaphrodite; calyx has 2–4 notches in margin; no petals; single stamen with long filament and anthers pointing inwards; inferior ovary with pistil, single cell and single ovule; fertilised by wind

Fruits: Egg-shaped single-seeded achenes, 1–2 mm long and dark brown; dispersal by wind, water, or birds; plant also has over-wintering buds

Habitat: From hill to subalpine zones in standing or slowly flowing water which may or may not contain nutrients and is calcareous and clear, over a muddy bottom containing humus; plant's range is worldwide

4. English species, genus and family names

5. Bibliography

Binz, A. and Becherer, A. *Schul- und Excursionsflora für die Schweiz,* 18th ed. Basle: Schwabe, 1986.

Bonnier, G. *La Grande Flore en couleur.* Neuchâtel: Delachaux and Niestlé, 1990.

Czihak, G. *Biologie: ein Lehrbuch für Studenten der Biologie.* Berlin: Springer, 1976.

Eichler, A.W. *Blütendiagramme.* Leipzig: Wilhelm Engelmann, 1878. Reprinted, Eppenheim: Otto Koeltz, 1954.

Hegi, G. *Illustrierte Flora Mitteleuropas.* 7 vols. Munich: J.F.Lehmann, 1908–31. 2nd ed., 1957–

Hess, H.E. and Landolt, E. *Flora der Schweiz.* Basel and Stuttgart: Birkhäuser, 1967.

Lippert, W. *Fotoatlas der Alpenblumen.* Munich: Gräfe und Unzer, 1981.

Oberdorfer, E. *Pflanzensoziologische Exkursionsflora.* Stuttgart: Eugen Ulmer, 1983.

Strasburger, E. *Lehrbuch der Botanik.* 30th ed. Stuttgart: Gustav Fischer, 1971.

Zander, R. *Handwörterbuch der Pflanzennamen.* Stuttgart: Eugen Ulmer, 1980.

The following books in English were used by the translator:

Clapham, A.R. et al. *Flora of the British Isles.* 3rd ed. Cambridge: Cambridge University Press, 1987.

Huxley, A. *Mountain Flowers.* London: Blandford, 1967.

Polunin, O. *Flowers of Europe.* Oxford: Oxford University Press, 1969.

Taylor, P.G. *British Ferns and Mosses.* London: Eyre and Spottiswoode, 1960.

Tutin, T.G. et al., eds. *Flora Europaea.* 6 vols. Cambridge: Cambridge University Press, 1964–80.

Photographs taken by the author are among the illustrations in this book

Notes